Hot Topics in Crystal Engineering

Hot Topics in Crystal Engineering

Edited by

Kari Rissanen
Department of Chemistry,
University of Jyväskylä,
Jyväskylä, Finland

Elsevier
Radarweg 29, PO Box 211, 1000 AE Amsterdam, Netherlands
The Boulevard, Langford Lane, Kidlington, Oxford OX5 1GB, United Kingdom
50 Hampshire Street, 5th Floor, Cambridge, MA 02139, United States

Library of Congress Cataloging-in-Publication Data
A catalog record for this book is available from the Library of Congress

British Library Cataloguing-in-Publication Data
A catalogue record for this book is available from the British Library

ISBN: 978-0-12-818192-8

For information on all Elsevier publications visit our website at
https://www.elsevier.com/books-and-journals

Publisher: Oliver Walter
Acquisitions Editor: Sean Simms
Editorial Project Manager: Lindsay Lawrence
Production Project Manager: Niranjan Bhaskaran
Cover Designer: Matthew Limbert

Typeset by TNQ Technologies

Contents

List of contributors

Christer B. Aakeröy
Department of Chemistry, Kansas State University, Manhattan, KS, United States

Geetha Bolla
Department of Chemistry, National University of Singapore, Singapore

Neil R. Champness
School of Chemistry, University of Birmingham, Birmingham, United Kingdom

Marijana Đaković
Department of Chemistry, Faculty of Science, University of Zagreb, Zagreb, Croatia

Antonio Frontera
Universitat de les Illes Balears, Palma de Mallorca (Baleares), Spain

Rosa M. Gomila
Universitat de les Illes Balears, Palma de Mallorca (Baleares), Spain

Tiddo J. Mooibroek
van 't Hoff Institute for Molecular Sciences, Amsterdam, the Netherlands

Ashwini K. Nangia
School of Chemistry, University of Hyderabad, Hyderabad, Telangana, India; CSIR-National Chemical Laboratory, Pune, Maharashtra, India

Daniel O'Nolan
Department of Chemical Sciences, Bernal Institute, University of Limerick, Limerick, Ireland

C. Malla Reddy
Department of Chemical Science, Indian Institute of Science Education and Research (IISER), Mohanpur, Kolkata, India

Bipul Sarma
Department of Chemical Sciences, Tezpur University, Tezpur, Assam, India

Kashyap Kumar Sarmah
Department of Chemistry, Gauhati University, Guwahati, Assam, India; Department of Chemistry, Behali Degree College, Borgang, Biswanath, Assam, India

Ranjit Thakuria
Department of Chemistry, Gauhati University, Guwahati, Assam, India

Michael J. Zaworotko
Department of Chemical Sciences, Bernal Institute, University of Limerick, Limerick, Ireland

Preface

The *COVID-19* year 2020 will be remembered to be effecting seriously everyday life, culture, and science activities, especially in experimental sciences, due to lockdowns of many public and research facilities. This book was also greatly affected and also delayed due to the outbreak of the corona virus. The planning of this book started already in 2018, and some chapters were delivered in 2019, yet majority of the present six chapters were delivered during the heat of the corona outbreak in spring 2020. I very much appreciate the contributors of this book for their dedication to deliver their chapters within quite narrow submission schedule, despite the havoc caused to their personal and scientific life. The volume was originally planned to contain 10 chapters, but due to the situation in spring 2020, some of the planned contributors were not able to deliver their chapters at all, which, of course, was very unfortunate.

Crystal engineering, as an elemental part of X-ray crystallography, especially single-crystal X-ray crystallography (SCXRD), has become one of the most active research areas in solid-state science due to the speed and accuracy offered by the very powerful tools for detailed structural analysis of single crystalline materials. Much of this is made possible by the recent technical advances in the instrumentation (more powerful radiation sources and more sensitive detectors) and in the computing power and sophisticated crystallographic programs, including automated data collection and structure solution protocols, which have eased up especially the routine and but also the nonroutine structural analysis, allowing a nearly "black-box" structure solution and refinement. While this is certainly a general advancement of structural science itself, e.g., by enabling research that can report as many as 25 X-ray structures in a single publication, it also opens a door to factory-like production of X-ray structures without a deeper insight into the interactions that lay embedded in every single crystal. On the other hand, and at the same time, also the deeper understanding about the phenomena leading to crystals (crystal growth) and knowledge of the nature of the intermolecular interactions have greatly advanced over the past 10 years.

This book is not intended to cover all aspects of *crystal engineering*, a mission impossible even within a 1000 page book, and it endeavors to highlight those areas of *crystal engineering*, which have surfaced during the past decade. The research presented in the chapters of this book has opened new paths and expanded the field of *crystal engineering* considerably. This book opens with quite an unorthodox research topic, namely 2-D crystallography, in which Professor Neil Champness (School of Chemistry, University of Nottingham, United Kingdom) discusses how scanning probe microscopies are used to study hydrogen-bonded assemblies at molecular resolution in contribution of "Surface Self-Assembly of Hydrogen-Bonded Frameworks." In the second chapter "Crystal Engineering of Coordination Networks," Professor Mike Zaworotko and Dr. Daniel O'Nolan (Bernal Institute, Department of Chemical Sciences, University of Limerick, Limerick, Ireland) review how the crystal engineering of metal–organic frameworks (MOFs) has evolved from the initial design principles to the better understanding of structure—function relationships to improve the performance of MOFs. Chapter 3 titled "From Mechanical Properties to Mechanochemistry of Molecular Crystals: Role of Nanoindentation and AFM Indentation Techniques in Crystal Engineering" by C. Malla Reddy (Department of Chemistry, Indian Institute of Science Education and Research (IISER) Kolkata, India) and Kashyap Kumar Sarmah and Ranjit Thakuria (Department of Chemistry, Gauhati University, Assam, India) probe comprehensively a very intersting new topic in crystal engineering, namely nanoindentation. They propose that the mechanical effects in plastic, superplastic, superelastic, ferroelastic, elastic, and brittle deformation processes should be considered as part of the mechanochemistry due to the fact that these processes also involve the reorganization of supramolecular interactions that respond to mechanical stimuli. New interactions such as halogen and chalcogen bonding have emerged strongly over the past 10 years, and Antonio Frontera (Universitat de les Illes Balears, Palma de Mallorca, Spain), Tiddo J. Mooibroek, and Rosa M. Gomila (van 't Hoff Institute for Molecular Sciences, Universiteit van Amsterdam, The Netherlands) provide new insights by combining theoretical calculations and Cambridge Structural Database (CSD) data in Chapter 4 titled "A Combined Theoretical and CSD Perspective on σ-Hole Interactions With Tetrels, Pnictogens, Chalcogens, Halogens, and Noble Gases." Chapter 5 titled "Crystal Engineering and Pharmaceutical Crystallization" by Ashwini K. Nangia (School of Chemistry, University of Hyderabad, Hyderabad and CSIR-National Chemical Laboratory, Pune, India), Geetha Bolla (Department of Chemistry, National University of Singapore, Singapore), and Bipul Sarma (Department of Chemical Sciences, Tezpur University, Tezpur, India) discusses deeply into the principles of cocrystal and salt design to the applications in pharmaceuticals and also manufacturing improvements with continuous flow processes. This concise six-chapter book is closed with an innovative treatise by Marijana Djakovic (Department of Chemistry, Faculty of Science, University of Zagreb, Zagreb, Croatia) and Christer Aakeröy (Department of Chemistry, Kansas State University, Manhattan, United States) on how molecular electrostatic potential surfaces (MEPS) of intermolecular interactions can offer robust practical guidelines for hierarchical crystal engineering in the closing chapter titled "From Molecular Electrostatic Potential Surfaces to Practical Avenues for Directed Assembly of Organic and Metal-Containing Crystalline Materials."

On the whole, the contributors to this volume have provided personal accounts on the "hot topics" in contemporary *crystal engineering*. As editor, I have been truly privileged to have the opportunity to work with my very distinguished colleagues, who have been in the forefront of their research areas for many years. Thanks to their meticulous dedication, I am very much convinced that this book will be a valuable and widening reading experience for researchers and students willing to learn more about the "new frontiers" in crystal engineering.

Kari Rissanen
Jyväskylä, February 2021

CHAPTER

Surface self-assembly of hydrogen-bonded frameworks

1

Neil R. Champness
School of Chemistry, University of Birmingham, Birmingham, United Kingdom

1. Introduction

The history of employing hydrogen bonding interactions to create two-dimensional self-assembled frameworks on surfaces has an extensive history [1–3]. Success in the field results from the rich diversity of chemistry using hydrogen bonds, which has been developed across many fields, in particular supramolecular chemistry [4]. Hydrogen bonds have been used to develop many complex supramolecular structures including capsules and cages [5], polymeric and oligomeric constructs [6,7], and interlocked arrangements [8]. In the solid state, a diversity of two- and three-dimensional structures has been constructed using hydrogen bonding interactions [9]. Many of these structures have been inspired by nature, whether from important biological systems, e.g., DNA [10] or proteins [11], or from naturally occurring materials as simple as ice [12]. In this chapter, the application of hydrogen bonding to the formation of self-assembled frameworks on surfaces is discussed. We will illustrate that it is possible to design framework structures using the knowledge of reliable and robust hydrogen bonding interactions but also illustrate the complexity of design that leads to unexpected and exciting directions of research.

There are many and varied approaches to designing molecules that contain hydrogen bonding groups and to design suitable supramolecular synthons for the construction of self-assembled framework structures. Supramolecular synthons have been defined and described as "spatial arrangements of intermolecular interactions" that "play the same focusing role in supramolecular synthesis that conventional synthons do in molecular synthesis" [13], a principle that facilitates a design strategy that can be adapted for making complex structures. The supramolecular synthon approach is particularly useful in the creation of hydrogen-bonded frameworks on surfaces and has been used to create many diverse structures. This chapter will focus on important and illustrative examples of supramolecular synthons that have been employed to create 2D arrays on surfaces and will compare surface self-assembly with more traditional solid-state crystal engineering.

In contrast to solid-state crystal engineering, for which single-crystal X-ray diffraction (SCXRD) is commonly used to establish long-range order, scanning

Hot Topics in Crystal Engineering. https://doi.org/10.1016/B978-0-12-818192-8.00002-0

probe microscopies, often scanning tunneling microscopy (STM), are the techniques that are employed to interrogate self-assembly on surfaces. Whereas SCXRD is an averaging technique that requires long-range order throughout the crystal, scanning probe microscopies allow identification of structure at the level of individual molecules or even submolecular resolution [14]. Thus, in surface self-assembly, it is possible to identify the arrangement of small clusters of molecules and to establish whether a single phase of self-assembled array is observed or whether a more diverse variety of phases is present. Although polymorphism is a prevalent issue in solid-state crystal engineering [15], long-range order of the different phases is required to allow identification by X-ray diffraction techniques, SCXRD or powder X-ray diffraction (PXRD). Recent advances in electron microscopy [16] may represent a new approach to characterizing the crystal structure of small samples with small domain sizes but even so it is likely to be sometime before it is possible to achieve the level of detail that is observed using scanning probe microscopies.

Firstly, the use of the prototypical example of hydrogen bonding synthons, those based on DNA nucleobases, thymine (T), guanine (G), cytosine (C), adenine (A), and also uracil (U) (Fig. 1.1), will be considered. The 2D self-assembly of DNA bases has been studied over a number of years [17], and a variety of hydrogen bonds observed for such systems are many and varied. The use of DNA nucleobases in supramolecular chemistry has been a persistent theme including the seminal work of Seeman [18], which has been expanded by a number of groups to great effect [19]. Indeed, Seeman and Winfree demonstrated the specific preparation of 2D crystals on surfaces using his "sticky-end" approach [20].

Initial studies at the molecular level focused predominantly on the use of the individual DNA nucleobases to create self-assembled structures on surfaces. The assembly of each of the bases has been studied on a variety of different substrates, and the assemblies observed adopt a large number of supramolecular synthons including Watson–Crick [10] and Hoogsteen [21] hydrogen bonding. The self-assembly of guanine on surfaces has received prominent attention [22–25], for example, the adsorption of guanine onto an Au(111) surface [22,25]. Imaging by STM reveals the formation of guanine quartet through Hoogsteen-style hydrogen bonding; the quartets are in turn associated through further N–H···N hydrogen bonds to give

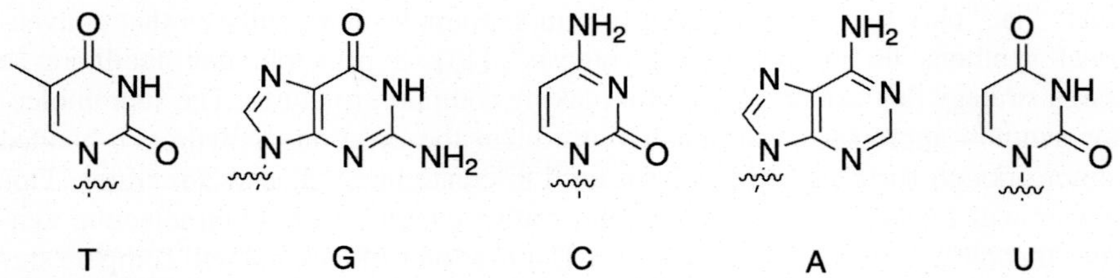

FIGURE 1.1

DNA nucleobases that have been employed in 2D surface supramolecular chemistry, thymine (T), guanine (G), cytosine (C), adenine (A), and uracil (U).

rise to two-dimensional supramolecular structures. This example illustrates the complexity of using DNA bases due to the variety of potential hydrogen bonding motifs that can be adopted, from classic Watson–Crick pair formation, Hoogsteen interactions, and reverse Watson–Crick and reverse Hoogsteen arrangements.

The diverse and complex possibilities of hydrogen bonding that are open to DNA nucleobases can be controlled by functionalization, effectively restricting the plethora of options for hydrogen bonding interactions by blocking certain arrangements. An example of such a strategy is given by the study of DNA nucleobase-functionalized porphyrin molecules (tetra-TP and tetra-AP) in which the porphyrin core is functionalized in each *meso*-position by a phenylthymine, or phenyladenine, moiety such that each nucleobase presents a hydrogen bonding face *exo* to the porphyrin core [26,27]. The molecules self-assemble on a highly oriented pyrolytic graphite (HOPG) substrate to give rise to two-dimensional grid structures. In the case of tetra-TP, the molecules interact through $R^2_2(8)$ intermolecular thymine ... thymine hydrogen bonds (Fig. 1.2A and B) and adopt a chiral arrangement when adsorbed onto the HOPG surface as a result of the asymmetric arrangement of the thymine groups. The almost perfectly square 2D unit cell observed for the tetra-TP structure suggests that all of the thymine groups within an individual tetra-TP molecule adopt the same orientation with respect to the porphyrin and that individual domains contain only molecules of the same handedness [26,27]. Tetra-AP also forms a 2D grid structure [27]; in this case, through alternating orientations of the adenine, appendages are observed, which allows each adenine group to adopt three hydrogen bonds including an $R^2_2(8)$ interaction between the termini of the adenine groups (Fig. 1.2C and D). Coassembly of tetra-TP and tetra-AP was also successfully achieved although STM imaging was only successful when the tetra-TP was metallated with a Zn cation, Zn-tetra-TP (providing sufficient contrast between the two different porphyrin molecules). In this instance, the nucleobases adopt hydrogen bonds such that an ATAT adenine–thymine quartets (Fig. 1.2E and F). Thus, each AT pair adopts a Watson–Crick hydrogen bond, but each pair also interacts with an adjacent pair to form the quartet structure. Overall an alternating chessboard pattern of Zn-tetra-TP and tetra-AP molecules can be constructed across the HOPG surface (Fig. 1.2E and F). It is noteworthy that the anticipated and desired molecular pattern is created although the exact nature of the hydrogen bonding pattern that is adopted is less straightforward to predict. A complex interplay of interactions is at play, not just the hydrogen bonding components, but also molecule–substrate interactions that can drive close-packing, in turn maximizing the energetic contribution of surface adsorption.

A related approach [28], which had previously been proven in the solution phase [29–31], is to use rod-shaped molecules appended by DNA nucleobases, G, C, A, U (Fig. 1.3). Opposing ends of the rods are designed such that complementary hydrogen bonding groups (G:C or A:U) interact to form cyclic structures (Fig. 1.3C and D), and the resulting surface-based constructs on an HOPG substrate indicates the successful employment of the complementary hydrogen bonding moieties when imaged by STM. Interestingly, the A:U system forms an AUAU quartet

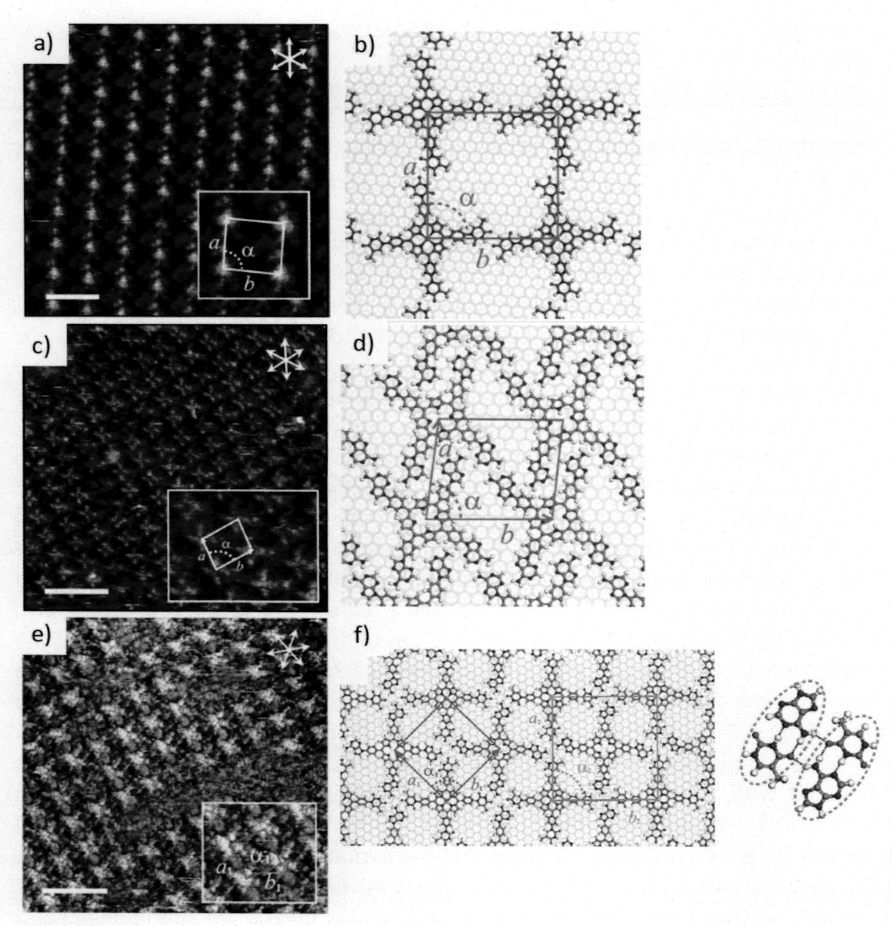

FIGURE 1.2

STM images at the liquid–solid interface between 1,2,4-trichlorobenzene (TCB) (A) or THF/TCB (C, E) and HOPG. (A) Zn-tetra-TP; (C) Tetra-AP; (E) Zn-tetra-TP and tetra-AP network. Insets show a magnified view of the structure with the 2D unit cell marked. Bright features correspond to porphyrin cores. (B), (D), and (F) show molecular models of the networks from molecular mechanics simulations.

Reproduced with permission from M.O. Blunt, Y. Hu, C.W. Toft, A.G. Slater, W. Lewis, N.R. Champness. J. Phys. Chem. C 122 (2018) 26070–26079. Copyright 2018 American Chemical Society.

structure (Fig. 1.3D) similar to the ATAT quartet observed for the Zn-tetra-TP:tetra-AP assembly discussed earlier [27]. The rod molecules are functionalized with alkyl chains to occupy residual space, inhibiting close-packed arrangements, such as those observed in the tetra-AP systems discussed earlier. However, despite the presence of

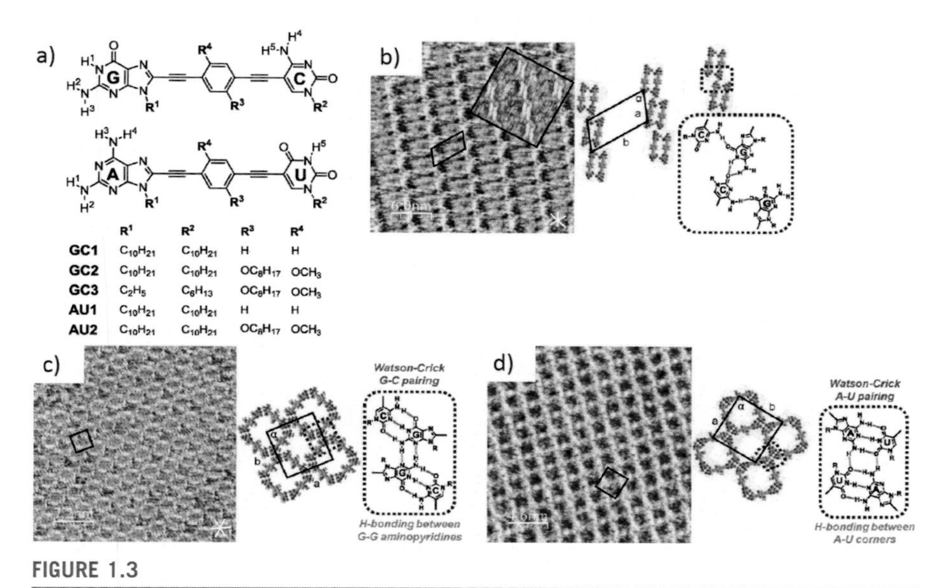

FIGURE 1.3

(A) Mixed guanine–cytosine and adenine–uracil hydrogen bonding rods employed for self-assembly [28]. (B) STM image of GC1 on HOPG showing pairs of monomers and model of observed structure. (C) STM image of GC2 on HOPG and model of observed cyclic structure. (D) STM image of AU2 on HOPG and model of observed cyclic structure [28].

Reproduced from N. Bilbao, I. Destoop, S. De Feyter, D. González-Rodríguez. Angew. Chem. Int. Ed. 55 (2016) 659–663, with permission from Wiley-VCH Verlag GmbH and Co. KGaA.

these alkyl chain appendages, the cyclic GC structures are capable of adsorbing coronene within the array with the alkyl chains assumed to vacate the space due to preferential physisorption of the polyaromatic guest onto the HOPG surface.

As is observed with DNA nucleobases, it is possible to envisage synthetic hydrogen bonding groups that produce heteromolecular supramolecular synthons. An early example of a bimolecular network was prepared from perylene-3,4,9,10-tetracarboxylicdiimide (PTCDI) and melamine [2,32]. PTCDI is an aromatic, perylene, group that is functionalized with two imide groups at opposing ends of the molecule (Fig. 1.4A). These imide groups present both N-H hydrogen bond donors and carbonyl hydrogen bond acceptors, in an analogous arrangement to that observed in thymine and uracil. The hydrogen bonding acceptor–donor–acceptor (ADA) arrangement is complementary to the arrangement of donors and acceptors (DAD) observed in melamine. PTCDI and melamine can be codeposited onto different substrates, e.g., Ag/Si(111) [32] or Au(111) [33–36], and self-assemble to form a honeycomb array with triple hydrogen bonds formed between the PTCDI imide moieties and melamine. Interestingly, in the case of the PTCDI/melamine

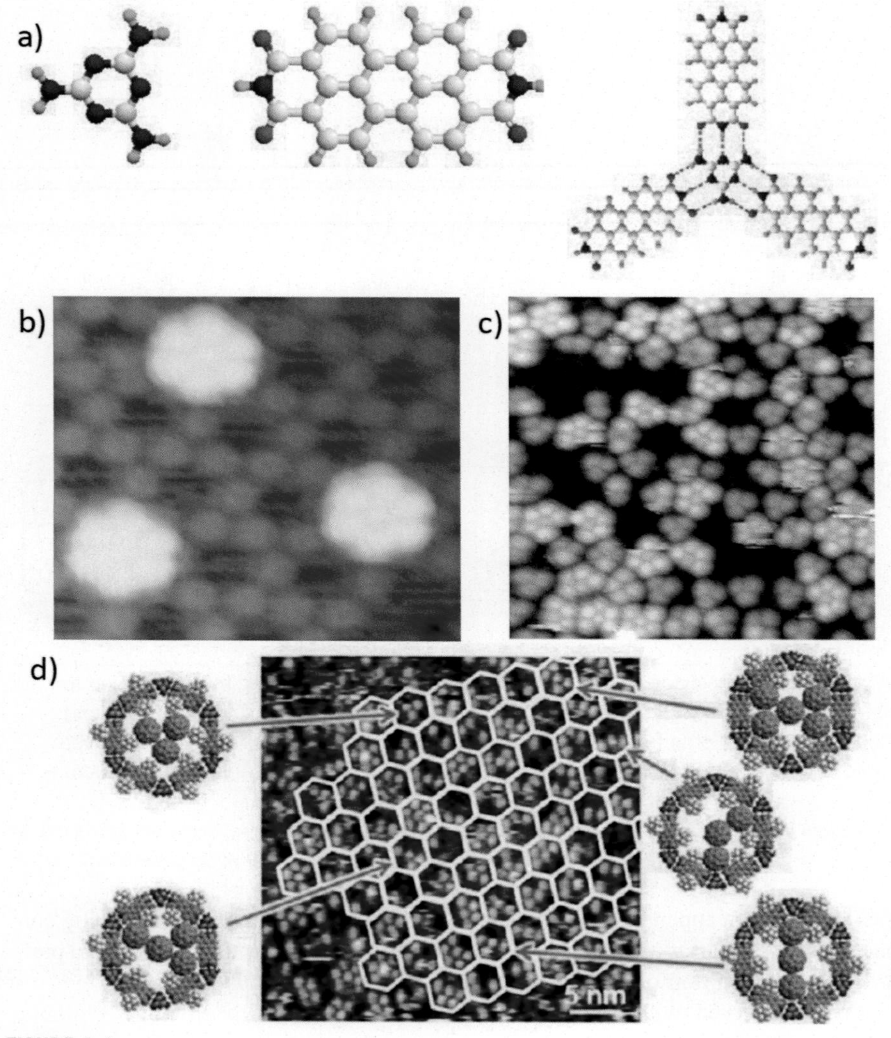

FIGURE 1.4

(A) Melamine and PTCDI and the threefold vertex that is formed through hydrogen bonding and leads to the formation of a honeycomb network. N, blue; O, red; C, gray; H, cyan. STM images of (B) C_{60} heptamers [2] and (C) C_{84} clusters [44] trapped within the PTCDI–melamine array on Ag/Si(111), in (C) clusters with two to seven molecules may be clearly resolved; (D) C_{60} entrapment in a $(SAdam)_2$–PTCDI/melamine network on Au(111); the honeycomb network is indicated as a guide to the eye. Schematic figures illustrate and identify the different arrangements of C_{60} within the pores of the structure that arise due to missing adamantyl units [38].

Images reproduced with permission from (A) J.A. Theobald, N.S. Oxtoby, M.A. Phillips, N.R. Champness, P.H. Beton. Nature 424 (2003) 1029–1031, with permission from the Nature Publishing Group; (B) A.G. Slater, L.M.A. Perdigao, P.H. Beton, N.R. Champness. Acc. Chem. Res. 47 (2014). 3417–3427. Copyright 2014 American Chemical Society; (C) J.A. Theobald, N.S. Oxtoby, N.R. Champness, P.H. Beton, T.J.S. Dennis. Langmuir 21 (2005) 2038–2041. Copyright 2005 American Chemical Society; (D) M.T. Räisänen, A.G. Slater (née Phillips), N.R. Champness, M. Buck. Chem. Sci. 3 (2012) 84–92, with permission from The Royal Society of Chemistry.

structure formed on Au(111), annealing the structure at higher temperatures leads to the observation of a parallelogram phase, which has the same stoichiometric ratio as the honeycomb structure, but is more densely packed [34].

The basic self-assembled PTCDI/melamine array can be decorated by functionalizing the PTCDI building block such that the pore of the structure presents a different chemical functionality or is simply restricted in dimensions [2,37–39]. Indeed, when disubstituted PTCDI building blocks are used, it is common to observe unimolecular honeycomb arrays [37,39,40] in which a trimolecular hydrogen-bonded junction leads to the formation of arrays with pores that are smaller than those observed for the bimolecular PTCDI/melamine arrays.

The porous nature of the PTCDI/melamine arrays offers exciting opportunities for guest encapsulation and ordering, mirroring a widely studied research theme in crystal engineering using metal–organic frameworks (MOFs) [41], covalent–organic frameworks (COFs) [42], and hydrogen-bonded organic frameworks (HOFs) [43]. Indeed, a range of guests have been captured using the surface-based PTCDI/melamine array including fullerenes, e.g. C_{60} [32], C_{84} [44], and Lu@C_{82} [45]. The number of fullerenes captured by the framework depends on both the size of the fullerene and the size of the pore in the framework. Thus, the parent PTCDI/melamine array hosts heptameric C_{60} clusters [32] (Fig. 1.4B), but smaller numbers of C_{84} molecules are trapped and the growth of the clusters can even be observed (Fig. 1.4C) [44]. The parallelogram phase, which is more densely packed, hosts only two C_{60} molecules, due to the constricted shape of the framework pore. Functionalization of the PTCDI network can also be used to restrict pore size and therefore the ability of the array to trap guest molecules. For example, self-assembly of thioadamantyl-functionalized PTCDI, $(SAdam)_2$-PTCDI, and melamine leads to the anticipated honeycomb array [38]. However, upon formation of a $(SAdam)_2$-PTCDI/melamine network, some of the thioadamantyl groups are cleaved from the PTCDI moieties giving rise to a variety of different pore sizes and arrangements, which can be visualized by STM following C_{60} adsorption onto the network. Using this approach, it is possible to identify the different orientations of the molecular C_{60} clusters, ranging from dimers to heptamers, within each pore that is determined by the degree of thioadamantyl cleavage (Fig. 1.4D). The PTCDI/melamine family of self-assembled arrays [2] has been employed to trap a variety of guest molecules including thiols [35,36,46], $Mn_{12}O_{12}(O_2CCH_3)_{16}(H_2O)_4$ clusters [47], thiol-functionalized porphyrins [48], complex functionalized polyoxometalates (POMs) [49], and polyaromatic molecules [50–52].

As hydrogen bonds can be relatively weak interactions, there is an enhanced possibility of structural diversity and polymorphism in 2D self-assembled arrays. In a series of elegant studies, Kandel and coworkers [53–55] have studied a range of compounds based on the indole skeleton (Fig. 1.5). The compounds have been systematically varied to present a variety of hydrogen bond donors and acceptors, including strong hydrogen bond donors (NH, OH) and weak hydrogen bond donors (CH) [56]. Such an approach has been adopted in the field of crystal engineering and

indole-2-COOH   indole-3-COOH   1H-indole-2,3-dione (isatin)   3-methyl-2-oxindole (3MTO)

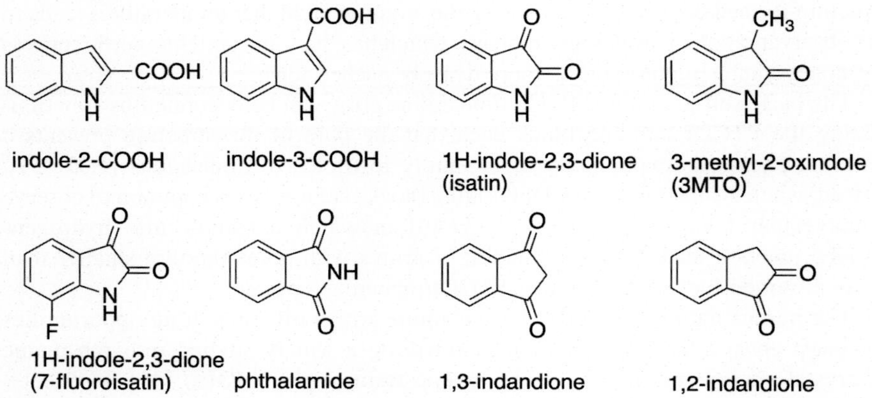

1H-indole-2,3-dione (7-fluoroisatin)   phthalamide   1,3-indandione   1,2-indandione

FIGURE 1.5

The range indole and related compounds studied by Kandel and coworkers [53–55]. 2D self-assembly leads to a variety of structures, which are influenced by both strong and weak hydrogen bonding interactions.

provides a mechanism for understanding the interplay between different weak interactions [56,57]. The target molecules are studied by deposition onto a surface Au(111) surface and the self-assembly of the molecules studied by STM.

Interestingly the most common motif observed for this family of compounds is a cyclic pentamer structure. This basic pentameric structure is observed for indole-2-COOH [53], indole-3-COOH [53], isatin [54,55], and 3M2O [54] and is stabilized by a mixture of N–H···O and C–H···O hydrogen bonds between adjacent molecules (Fig. 1.6). Although this rather unusual pentameric arrangement is the favored form, an observation supported by discrete Fourier transform (DFT) calculations, STM imaging confirms that other phases, or polymorphs, are observed. Thus, whereas the energetically favored form of indole-2-COOH contains pentamers, this compound also forms hexamers and catemer chains under the conditions of the experiment. In contrast, indole-3-COOH monolayers are generally disordered.

Similar to indole-2-COOH, isatin forms pentamers, and calculations show that the pentamer structure is more stable than the corresponding dimer by 12 kJ/mol [54]. Comparison of the behavior of isatin with that exhibited by 3M2O reveals that the latter forms an array, which contains a mixture of catemer chains and pentamers. The pentamers of 3M2O exhibit a distinct structure from those observed for isatin (Fig. 1.7). In contrast to isatin, 7-fluoroisatin does not form pentamers as a result of the presence of the fluoro-substituent blocking the adoption of favorable C–H···O hydrogen bonds observed in the pentamers of isatin. Thus, 7-fluoroisatin is observed to form both close-packed, ordered, domains and hexamer clusters (Fig. 1.7). A further study [55] investigates two pairs of isomers, isatin and phthalimide and 1,3-indandione and 1,2-indandione (Fig. 1.5). Although isatin forms pentamers, phthalimide, an isomeric arrangement, self-assembles into close-packed arrays and

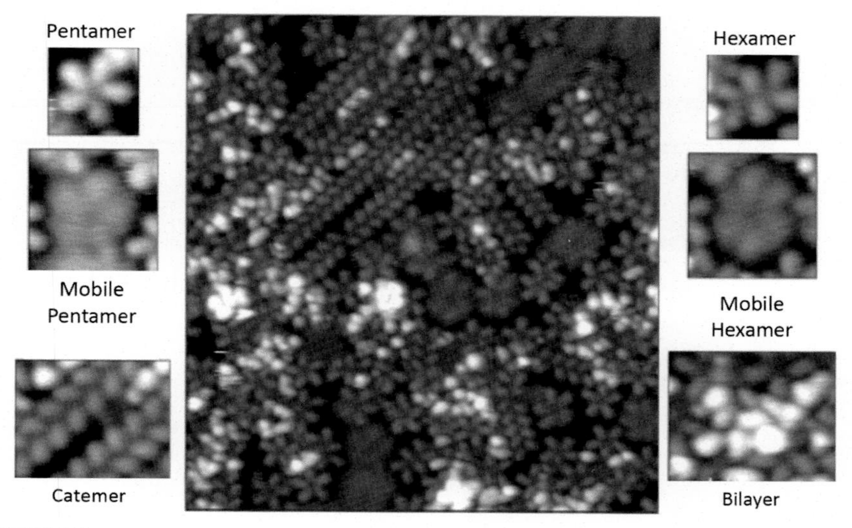

FIGURE 1.6

STM images of indole-2-COOH [53], central image over a range of 250 Å × 275 Å. Magnified images of pentamers, hexamers, catemer chains, and bilayers of aggregated molecules are also shown. In some instances, images of pentamers or hexamers can be blurred due to motion on the surface.

Reproduced with permission from N.A. Wasio, R.C. Quardokus, R.D. Brown, R.P. Forrest, C.S. Lent, S.A. Corcelli, J.A. Christie, K.W. Henderson, S.A. Kandel. J. Phys. Chem. C 119 (2015) 21011–21017. Copyright 2015 American Chemical Society.

self-assembled networks that contain tetramer subunits (Fig. 1.7). Interestingly 1,3-indandione also forms close-packed arrays and self-assembled tetramer networks despite removal of the N—H hydrogen bond donor, in comparison with phthalimide. The self-assembly of 1,2-indandione does not lead to a structure that resembles those displayed by the other related compounds, rather close-packed areas and disordered regions without distinguishable order are observed.

These series of studies display the complexity that is often observed in families of compounds in crystal engineering and demonstrate the power of using an iterative approach to appreciating the influence of chemical structure on 2D self-assembly processes. It is also acknowledged that the coexistence of different self-assembled structures is at least in part a result of the nonequilibrium conditions, which are present during both the deposition and assembly processes when forming 2D self-assembled arrays.

The formation of these highly unusual pentamers has also been observed for the self-assembly of ferrocenecarboxylic acid [Fc(COOH)] on an Au(111) substrate [58]. Fc(COOH) forms pentagonal arrangements through intermolecular carboxylic acid—carboxylic acid O—H···O hydrogen-bonding interactions, stabilized by additional C—H···O hydrogen bonds (Fig. 1.8). As with indole-2-COOH [53] and isatin

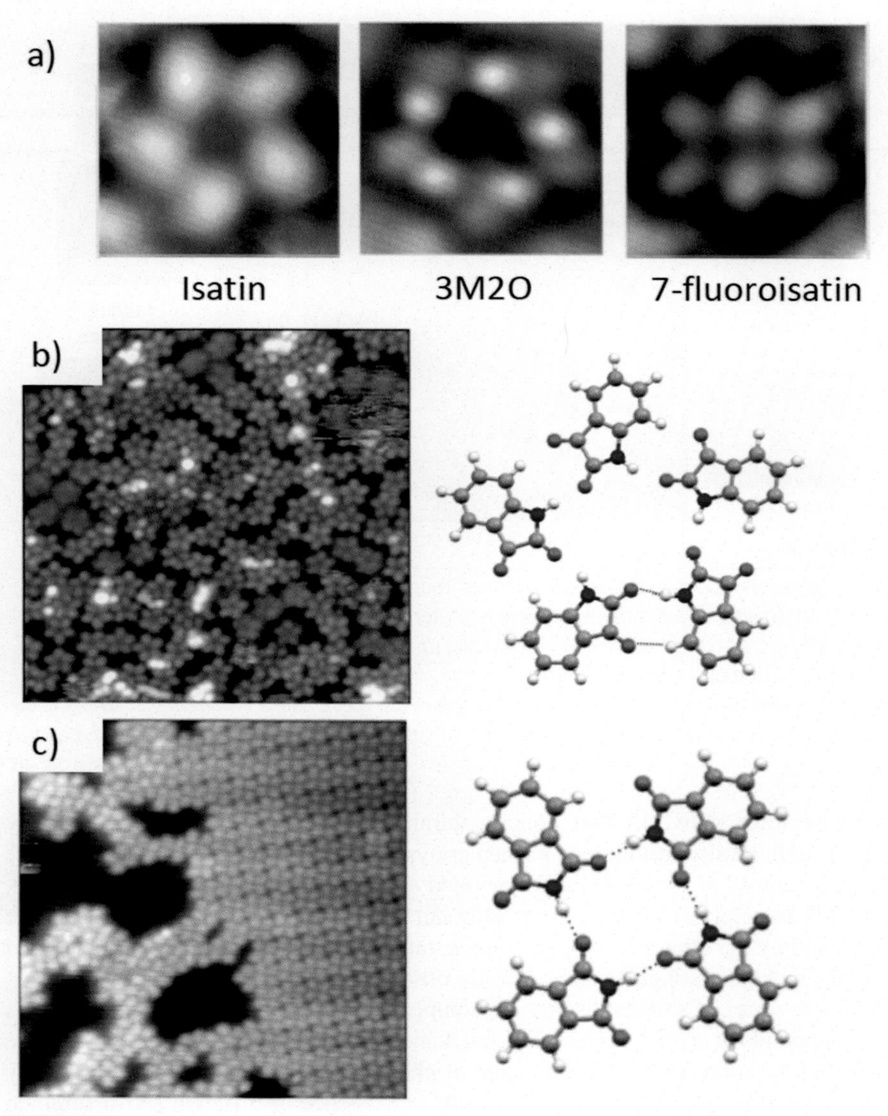

FIGURE 1.7

(A) STM images of pentamers of isatin, 3M2O, and hexamers of 7-fluoroisatin all on Au(111) [54]. (B) STM image of isatin on Au(111) (250 Å × 250 Å) with representative pentamers highlighted in blue and optimized geometry of the pentamer with N–H···O and C–H···O interactions shown. (C) STM image of phthalimide on Au(111) (220 Å × 220 Å) with representative tetramers highlighted in cyan and optimized geometry of the pentamer with N–H···O interactions shown [55].

Reproduced with permission from (A) A.M. Silski, R.D. Brown, J.P. Petersen, J.M. Coman, D.A. Turner, Z.M. Smith, S.A. Corcelli, J.C. Poutsma, S.A. Kandel. J. Phys. Chem. C 121 (2017) 21520–21526. Copyright 2017 American Chemical Society; (B) A.M. Silski, J.P. Petersen, R.D. Brown, S.A. Corcelli, S.A. Kandel. J. Phys. Chem. C 122 (2018) 25467–25474. Copyright 2018 American Chemical Society.

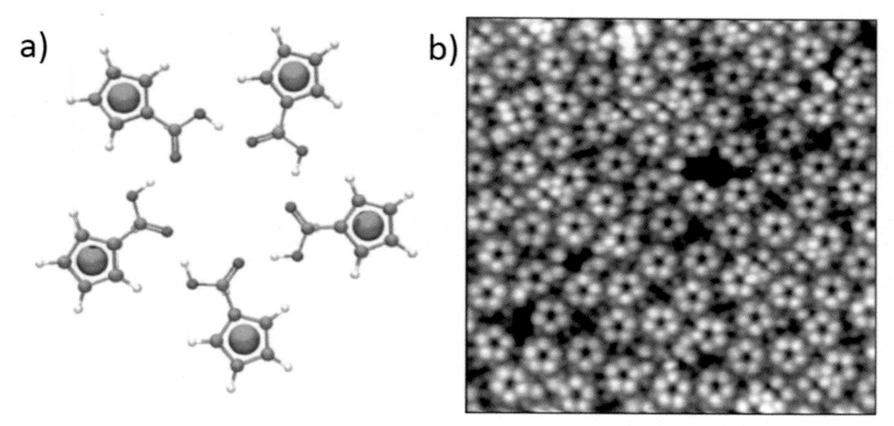

FIGURE 1.8

(A) Representation of the pentagonal arrangement formed by Fc(COOH) through intermolecular O—H···O and C—H···O hydrogen bonding; (B) pentagonal arrangements can be clearly seen in STM images of Fc(COOH) adsorbed on an Au(111) substrate [58].

Reproduced from N.A.Wasio, R.C. Quardokus, R.P. Forrest, C.S. Lent, S.A. Corcelli, J.A. Christie, K.W. Henderson, S.A. Kandel. Nature 507 (2014) 86–89, with permission from the Nature Publishing Group.

[54], discussed earlier, the pentamer was found to be more stable by DFT calculations than other arrangements, such as dimers. Interestingly, the observed pentamer arrangements are related to subunits of a Penrose P1 tiling, where Penrose tilings are related to quasicrystal structures in that they exhibit long-range, nonperiodic order, and unusual rotational symmetry [59].

Indeed, as surface self-assembled structures are most commonly characterized by scanning probe microscopies such as STM, it is possible to characterize structures that do not exhibit long-range order. This is in stark contrast to more traditional crystal engineering, which is more reliant on diffraction techniques that require long-range order. Thus, in addition to the Penrose tiling behavior observed for Fc(COOH), a number of unusual nonperiodic structures have been observed using coordination [60] or halogen bonds [61].

One of the first examples of such a structure is that formed by terphenyl-3,3″,5,5″-tetracarboxylic acid (TPTC) [62]. When TPTC is deposited onto a highly oriented pyrolytic graphite (HOPG) substrate, a two-dimensional hydrogen-bonded structure is formed that utilizes $R^2_2(8)$ intermolecular carboxylic acid—carboxylic acid hydrogen-bonding interactions. By using STM imaging, it is possible to identify the position of each molecule in the extended structure and directly visualize the random arrangement of the molecules within the array (Fig. 1.9). The intermolecular O—H···O hydrogen bonding leads to the formation of hexagonal junctions, which are formed from three, four, five, or six molecules as a result of the dimensions of the molecule (Fig. 1.9C); the distance from the centroids of the terminal phenyl

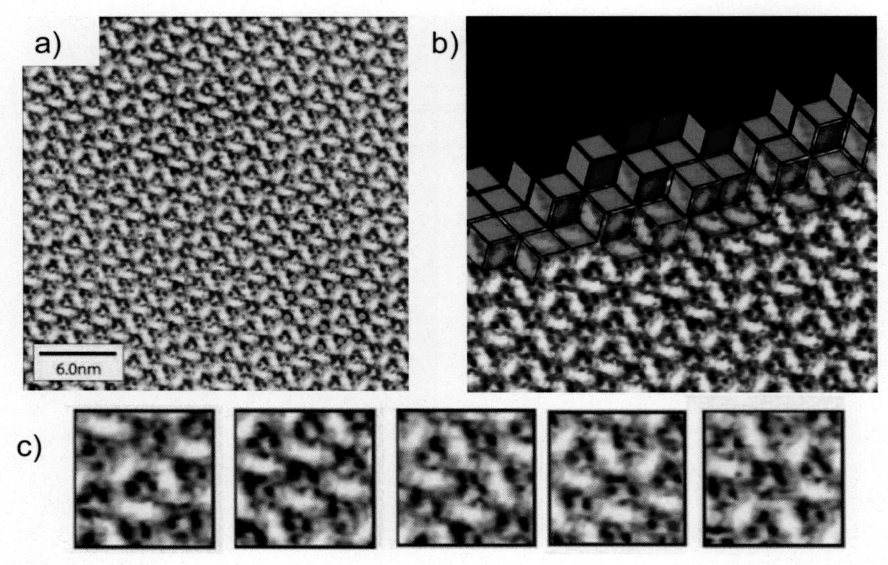

FIGURE 1.9

(A) STM image of a typical area of TPTC network at the nonanoic acid/HOPG interface. The group of three phenyl rings of the molecule backbone appear as bright features in the image. The hexagonal orientational order of the structure is indicated by the group of blue dots in the lower right-hand corner of the image. (B) Illustration of how the molecular arrangement, and each molecule, maps onto a rhombus tiling [65]. (C) Magnified STM images representing the five possible arrangements of TPTC molecules around a network pore. The locations of the magnified images are marked by blue dashed squares on (A) [62].

Parts (A) and (C) Reproduced from M.O. Blunt, J. Russell, M.C. Giménez-López, J.P. Garrahan, X. Lin, M. Schröder, N.R. Champness, P.H. Beton. Science 322 (2008) 1077–1081, with permission from the AAAS. Part (B) Reproduced from C. Pfeiffer, N.R. Champness. Two dimensional supramolecular chemistry on surfaces, in J. L. Atwood (Ed.), Comprehensive Supramolecular Chemistry II, vol. 2, Elsevier, Oxford, 2017, pp. 181–199.

groups of the TPTC molecule is similar to the analogous centroid–centroid distance across the carboxylic acid—carboxylic acid hydrogen bond linking adjacent molecules. As a result of this similarity in dimensions, the array forms an extremely rare example of a random, entropically stabilized, rhombus tiling. Detailed analysis of the rhombus tiling allows evaluation of the degree of randomness observed in the structure, and subsequent studies established that both solvent used for deposition and experiment temperature affect the degree of order, and hence randomness of the tiling [63]. It is noteworthy that the random rhombus tiling is only observed for molecules, which have the appropriate dimensions, i.e., those that are rhombus shaped, and analogous molecules such as quaterphenyl-3,3‴,5,5‴-tetracarboxylic acid form regular two-dimensional structures [64].

2. Conclusions

The parallels between surface self-assembly and crystal engineering are apparent, and it is clear that many of the principles of crystal engineering can be applied to creating molecular arrays on two-dimensional substrates. However, it is also apparent that the tenets of crystal engineering can only serve as a starting point for the design of surface-based structures. The major differences between traditional crystal engineering are twofold. Firstly, surface self-assembly is dominated by the two dimensions of the surface, whereas crystal engineering requires appreciation of the three dimensions of the crystal. It is also important to appreciate that the surface itself is far from an innocent bystander in the self-assembly process, just as nucleation is important in crystal growth and polymorph selection. Secondly, the role that the characterization technique of choice influences the field cannot be overstated. Crystal engineering is dominated by SCXRD, a technique that requires long-range order, i.e., crystallinity. In contrast, surface self-assembly relies upon scanning probe microscopies, such as STM, which are less adept at determining long-range order but allow the ready identification of local structure. The latter point is particularly important when studying arrangements that do not exhibit high symmetry such as rhombus or Penrose tilings. Ultimately, surface self-assembly continues to grow as a field, and many of the discoveries, which are made in the field, feed knowledge back into crystal engineering.

References

[1] K.S. Mali, N. Pearce, S. de Feyter, N.R. Champness, Chem. Soc. Rev. 46 (2017) 2520–2542.

[2] A.G. Slater, L.M.A. Perdigao, P.H. Beton, N.R. Champness, Acc. Chem. Res. 47 (2014) 3417–3427.

[3] S. De Feyter, A. Gesquière, M.M. Abdel-Mottaleb, P.C.M. Grim, F.C. De Schryver, Acc. Chem. Res. 33 (2000) 520–531.

[4] C.T. Seto, G.M. Whitesides, J. Am. Chem. Soc. 115 (1993) 905–916.

[5] Y. Liu, C. Hu, A. Comotti, M.D. Ward, Science 333 (2011) 436–440.

[6] S.K. Yang, S.C. Zimmerman, Isr. J. Chem. 53 (2013) 511–520.

[7] O.J.G.M. Goor, S.I.S. Hendrikse, P.Y.W. Dankers, E.W. Meijer, Chem. Soc. Rev. 46 (2017) 6621–6637.

[8] G. Gil-Ramirez, D.A. Leigh, A.J. Stephens, Angew. Chem. Int. Ed. 54 (2015) 6110–6150.

[9] J.D. Wuest, Chem. Commun. (2005) 5830–5837.

[10] J.D. Watson, F.H.C. Crick, Nature 171 (1953) 737–738.

[11] A.R. Fersht, Trends Biochem. Sci. 12 (1987) 301–304.

[12] J. Li, D.K. Ross, Nature 365 (1993) 327–329.

[13] G.R. Desiraju, Angew. Chem. Int. Ed. Engl. 34 (1995) 2311–2327.

[14] A.M. Sweetman, S. Jarvis, H. Sang, I. Lekkas, P. Rahe, Y. Wang, J. Wang, N.R. Champness, L. Kantorovich, P.J. Moriarty, Nat. Commun. 5 (2014) 3931.

[15] S. Aitipamula, R. Banerjee, A.K. Bansal, K. Biradha, M.L. Cheney, A.R. Choudhury, G.R. Desiraju, A.G. Dikundwar, R. Dubey, N. Duggirala, P.P. Ghogale, S. Ghosh, P.K. Goswami, N.R. Goud, R.K.R. Jetti, P. Karpinski, P. Kaushik, D. Kumar, V. Kumar, B. Moulton, A. Mukherjee, G. Mukherjee, A.S. Myerson, V. Puri, A. Ramanan, T. Rajamannar, C.M. Reddy, N. Rodriguez-Hornedo, R.D. Rogers, T.N. Guru Row, P. Sanphui, N. Shan, G. Shete, A. Singh, C.C. Sun, J.A. Swift, R. Thaimattam, T.S. Thakur, R.K. Thaper, S.P. Thomas, S. Tothadi, V.R. Vangala, N. Variankaval, P. Vishweshwar, D.R. Weyna, M.J. Zaworotko, Cryst. Growth Des. 12 (2012) 2147–2152.
[16] T. Gruene, J.T.C. Wennmacher, C. Zaubitzer, J.J. Holstein, J. Heidler, A. Fecteau-Lefebvre, S. De Carlo, E. Müller, K.N. Goldie, I. Regeni, T. Li, G. Santiso-Quinones, G. Steinfeld, S. Handschin, E. van Genderen, J.A. van Bokhoven, G.H. Clever, R. Pantelic, Angew. Chem. Int. Ed. 57 (2018) 16313–16317.
[17] A. Ciesielski, M. El Garah, S. Masiero, P. Samorì, Small 12 (2016) 83–95.
[18] N.C. Seeman, Nature 421 (2003) 427–431.
[19] T.G.W. Edwardson, K.L. Lau, D. Bousmail, H. Sleiman, Nat. Chem. 8 (2016) 162–170.
[20] E. Winfree, F. Liu, L.A. Wenzler, N.C. Seeman, Nature 394 (1998) 539–544.
[21] K. Hoogsteen, Acta Crystallogr. 16 (1963) 907–916.
[22] R. Otero, M. Schock, L.M. Molina, E. Laegsgaard, I. Stensgaard, B. Hammer, F. Besenbacher, Angew. Chem. Int. Ed. 44 (2005) 2270–2275.
[23] A. Ciesielski, R. Perone, S. Pieraccini, G. Piero Spada, P. Samorì, Chem. Commun. 46 (2010) 4493–4495.
[24] A. Ciesielski, S. Haar, M. El Garah, M. Surin, S. Masiero, P. Samori, L'Actualité Chimique 399 (2015) 31–36.
[25] M. El Garah, R.C. Perone, A. Santana Bonilla, S. Haar, M. Campitiello, R. Gutierrez, G. Cuniberti, S. Masiero, A. Ciesielski, P. Samorì, Chem. Commun. 51 (2015) 11677–11680.
[26] A.G. Slater, Y. Hu, L. Yang, S.P. Argent, W. Lewis, M.O. Blunt, N.R. Champness, Chem. Sci. 6 (2015) 1562–1569.
[27] M.O. Blunt, Y. Hu, C.W. Toft, A.G. Slater, W. Lewis, N.R. Champness, J. Phys. Chem. C 122 (2018) 26070–26079.
[28] N. Bilbao, I. Destoop, S. De Feyter, D. González-Rodríguez, Angew. Chem. Int. Ed. 55 (2016) 659–663.
[29] C. Montoro-García, J. Camacho-García, A.M. Lypez-Põrez, N. Bilbao, S. Romero-Põrez, M.J. Mayoral, D. González-Rodríguez, Angew. Chem. Int. Ed. 54 (2015) 6780–6784.
[30] C. Montoro- García, N. Bilbao, I.M. Tsagri, F. Zaccaria, M.J. Mayoral, C. Fonseca Guerra, D. González-Rodríguez, Chem. Eur. J. 24 (2018) 11983–11991.
[31] M.J. Mayoral, N. Bilbao, D. González-Rodríguez, ChemistryOpen 5 (2016) 10–32.
[32] J.A. Theobald, N.S. Oxtoby, M.A. Phillips, N.R. Champness, P.H. Beton, Nature 424 (2003) 1029–1031.
[33] L.M.A. Perdigão, E.W. Perkins, J. Ma, P.A. Staniec, B.L. Rogers, N.R. Champness, P.H. Beton, J. Phys. Chem. B 110 (2006) 12539–12542.
[34] P.A. Staniec, L.M.A. Perdigão, A. Saywell, N.R. Champness, P.H. Beton, ChemPhysChem 8 (2007) 2177–2181.
[35] R. Madueno, M.T. Räisänen, C. Silien, M. Buck, Nature 454 (2008) 618–621.
[36] C. Silien, M.T. Räisänen, M. Buck, Small 6 (2010) 391–394.

[37] L.M.A. Perdigão, A. Saywell, G.N. Fontes, P.A. Staniec, G. Goretzki, A.G. Phillips, N.R. Champness, P.H. Beton, Chem. Eur. J. 14 (2008) 7600–7607.
[38] M.T. Räisänen, A.G. Slater (née Phillips), N.R. Champness, M. Buck, Chem. Sci. 3 (2012) 84–92.
[39] A.G. Phillips, L.M.A. Perdigão, P.H. Beton, N.R. Champness, Chem. Commun. 46 (2010) 2775–2777.
[40] A.J. Pollard, E.W. Perkins, N.A. Smith, A. Saywell, G. Goretzki, A.G. Phillips, S.P. Argent, H. Sachdev, F. Müller, S. Hüfner, S. Gsell, M. Fischer, M. Schreck, J. Osterwalder, T. Greber, S. Berner, N.R. Champness, P.H. Beton, Angew. Chem. Int. Ed. 49 (2010) 1794–1799.
[41] R. Ricco, C. Pfeiffer, K. Sumida, C.J. Sumby, P. Falcaro, S. Furukawa, N.R. Champness, C.J. Doonan, CrystEngComm 18 (2016) 6532–6542.
[42] R.-B. Lin, Y. He, P. Li, H. Wang, W. Zhou, Chem. Soc. Rev. 48 (2019) 1362–1389.
[43] S.-Y. Ding, W. Wang, Chem. Soc. Rev. 42 (2013) 548–568.
[44] J.A. Theobald, N.S. Oxtoby, N.R. Champness, P.H. Beton, T.J.S. Dennis, Langmuir 21 (2005) 2038–2041.
[45] F. Silly, A.Q. Shaw, K. Porfyrakis, J.H. Warner, A.A.R. Watt, M.R. Castell, H. Umemoto, T. Akachi, H. Shinohara, G.A.D. Briggs, Chem. Commun. (2008) 4616–4618.
[46] L.M.A. Perdigão, P.A. Staniec, N.R. Champness, P.H. Beton, Langmuir 25 (2009) 2278–2281.
[47] A. Saywell, G. Magnano, C.J. Satterley, L.M.A. Perdigão, A.J. Britton, N. Taleb, M.C. Giménez-López, N.R. Champness, J.N. O'Shea, P.H. Beton, Nat. Commun. 1 (2010) 75.
[48] A. Lombana, N. Battaglini, G. Tsague-Kenfac, S. Zrig, P. Lang, Chem. Commun. 52 (2016) 5742–5745.
[49] A. Lombana, C. Rinfray, F. Volatron, G. Izzet, N. Battaglini, S. Alves, P. Decorse, P. Lang, A. Proust, J. Phys. Chem. C 120 (2016) 2837–2845.
[50] B. Karamzadeh, T. Eaton, I. Cebula, D. Munoz Torres, M. Neuburger, M. Mayor, M. Buck, Chem. Commun. 50 (2014) 14175–14178.
[51] B. Karamzadeh, T. Eaton, D. Munoz Torres, I. Cebula, M. Mayor, M. Buck, Faraday Discuss. 204 (2017) 173–190.
[52] C.J. Judd, N.R. Champness, A. Saywell, Chem. Eur. J. 24 (2018) 56–61.
[53] N.A. Wasio, R.C. Quardokus, R.D. Brown, R.P. Forrest, C.S. Lent, S.A. Corcelli, J.A. Christie, K.W. Henderson, S.A. Kandel, J. Phys. Chem. C 119 (2015) 21011–21017.
[54] A.M. Silski, R.D. Brown, J.P. Petersen, J.M. Coman, D.A. Turner, Z.M. Smith, S.A. Corcelli, J.C. Poutsma, S.A. Kandel, J. Phys. Chem. C 121 (2017) 21520–21526.
[55] A.M. Silski, J.P. Petersen, R.D. Brown, S.A. Corcelli, S.A. Kandel, J. Phys. Chem. C 122 (2018) 25467–25474.
[56] G.R. Desiraju, Acc. Chem. Res. 35 (2002) 565–573.
[57] M.K. Corpinot, D.-K. Bucar, Cryst. Growth Des. 19 (2019) 1426–1453.
[58] N.A. Wasio, R.C. Quardokus, R.P. Forrest, C.S. Lent, S.A. Corcelli, J.A. Christie, K.W. Henderson, S.A. Kandel, Nature 507 (2014) 86–89.
[59] R. Penrose, Eureka 39 (1978) 16–32.
[60] J.I. Urgel, D. Écija, G. Lyu, R. Zhang, C.-A. Palma, W. Auwärter, N. Lin, J.V. Barth, Nat. Chem. 8 (2016) 657–662.

[61] J. Shang, Y. Wang, M. Chen, J. Dai, X. Zhou, J. Kuttner, G. Hilt, X. Shao, J.M. Gottfried, K. Wu, Nat. Chem. 7 (2015) 389–393.
[62] M.O. Blunt, J. Russell, M.C. Giménez-López, J.P. Garrahan, X. Lin, M. Schröder, N.R. Champness, P.H. Beton, Science 322 (2008) 1077–1081.
[63] A. Stannard, J.C. Russell, M.O. Blunt, C. Sallesiotis, M.C. Gimenez-Lopez, N. Taleb, M. Schroder, N.R. Champness, J.P. Garrahan, P.H. Beton, Nat. Chem. 4 (2012) 112–117.
[64] M. Blunt, X. Lin, M.C. Gimenez-Lopez, M. Schröder, N.R. Champness, P.H. Beton, Chem. Commun. (2008) 2304–2306.
[65] C. Pfeiffer, N.R. Champness, Two dimensional supramolecular chemistry on surfaces, in: J.L. Atwood (Ed.), Comprehensive Supramolecular Chemistry II, vol. 2, Elsevier, Oxford, 2017, pp. 181–199.

CHAPTER

Crystal engineering of coordination networks: then and now

2

Daniel O'Nolan, Michael J. Zaworotko

Department of Chemical Sciences, Bernal Institute, University of Limerick, Limerick, Ireland

1. Then

The early 20th century was a golden era for structural chemistry thanks to the introduction and rapid development of analytical tools such as X-ray crystallography (Bragg, Laue, Ewald) [1–3], nuclear magnetic resonance (Rabi) [4], and electron microscopy (Ruska) [5]. These tools allowed researchers to "see" crystal and molecular structure and afforded the dream of not just observing structures but also controlling the arrangement of atoms and molecules from first principles. Indeed, Feynman's 1959 lecture series, which is considered by many to be the dawn of nanoscience [6,7], also addressed the importance to materials science of being able to create new materials by design. Today, over a century after the advent of X-ray crystallography, advances in computing, synthesis methodology, and theoretical chemistry have enabled the development of systematic approaches to control the arrangement of atoms/molecules in the solid state and, in some instances, to property advancements. In this chapter, crystal engineering is detailed in the context of a class of materials that are particularly amenable to design from first principles and relevant with respect to a number of functional properties: coordination networks.

1.1 What is crystal engineering?

Crystal engineering is the field of chemistry that studies the design, properties, and applications of crystals.

The aforementioned definition was recently introduced by us [8] and concisely captures how crystal engineering has evolved since the term was coined by Pepinsky [9]: Crystallization of organic ions with metal-containing complex ions of suitable sizes, charges, and solubilities results in structures with cells and symmetries determined chiefly by packing of the complex ions. These cells and symmetries are to a good extent controllable; hence crystals with advantageous properties can be "engineered."

G.M.J. Schmidt first demonstrated the utility of crystal engineering by introducing the concept of the topochemical principle to solid-state photodimerization [10]. He suggested that there were four phases of crystal engineering:

- Experimental determination of crystal structure and correlation with properties—the phase of the topochemical principle

Hot Topics in Crystal Engineering. https://doi.org/10.1016/B978-0-12-818192-8.00007-X

- Establishing the limits of the topochemical principle
- Development of empirical rules
- Systematic solid-state photochemistry

It is fair to assert that crystal engineering had reached phase 3 by 1970 in the context of photochemistry, but the application of crystal engineering concepts to coordination networks did not occur until later. It is no coincidence that supramolecular chemistry and crystal engineering have advanced concurrently [11,12]. This is because crystal engineering is in essence the application of supramolecular chemistry toward the understanding of existing crystals and the design of new crystalline materials [13,14]. The concept of molecular recognition was first hypothesized by Fischer in 1894 while working on enzymatic recognition of sugars [15].

Interest in crystal engineering grew rapidly in the late 1980s after Ermer [16], Desiraju [17], and Etter [18] laid important groundwork with respect to organic molecular crystals. The concepts of graph sets and supramolecular synthons [19–22] remain in widespread use. Ermer's report on the fivefold diamondoid (**dia**) network structure of adamantane-1,3,5,7-tetracarboxylic acid is particularly salient given the importance of **dia** topology and interpenetration in coordination networks [16]. It is perhaps ironic that it was in 1988 that the editor-in-chief of *Nature*, John Maddox, commented that the (perceived) inability to reasonably predict the simplest crystal structures by knowing their chemical composition was "one of the continuing scandals in the physical sciences…" [23]. This controversial statement at least partially inspired the growth in research into crystal engineering that started in the early 1990s (Fig. 2.1).

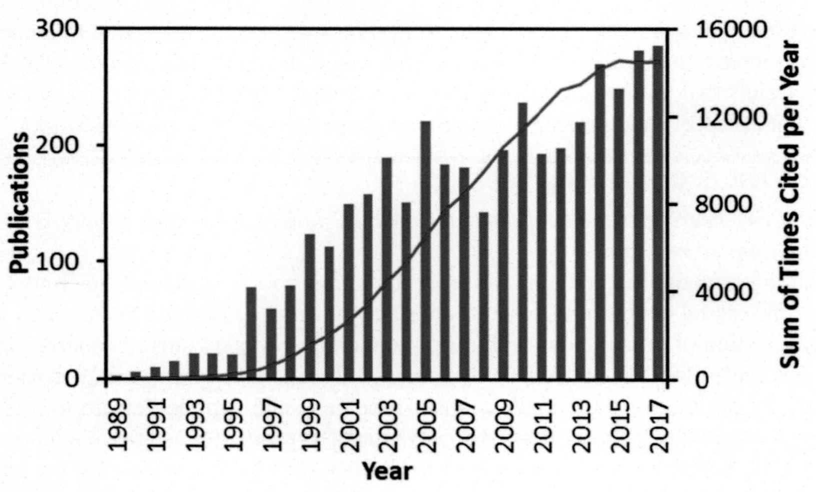

FIGURE 2.1

The number of publications per year (*blue*) between 1989 and 2017 on the topic "crystal engineering" and the sum of times cited (*red*) as found in the Web of Science core collection by Clarivate Analytics.

That certain molecules may be considered as suitable to serve as tectons [24] or molecular building blocks (MBBs) [25–27], for the a priori design of crystalline architectures is an elegant yet simple concept that remains an important general crystal engineering strategy. MBBs continue to represent an important concept with respect to the design of new coordination networks. However, when there are degrees of freedom in coordination networks, then supramolecular isomerism can occur [28–30], wherein the same ligands and metal components may pack in more than one topology. In addition, there remain classes of crystalline solid that are not amenable to crystal engineering, notably hydrates [31,32] and solvates of molecular compounds, for which even stoichiometry can be variable and difficult to predict. Nevertheless, hydrogen bonds and coordination bonds remain the most prevalent molecular interactions used in crystal engineering although halogen bonding [33], π-stacking [34], and ion-π interactions [35,36] have also been exploited.

Another powerful early concept that remains in use today is the reduction of network crystal structures to the connectivity of the nodes that can be defined by the MBBs. This approach to crystal structure analysis was first developed by Wells [37–40]. Wells' description of existing crystal structures as a combination of nodes and linkers that afford network structures is readily transferable to coordination networks. The use of known MBBs greatly simplifies a crystal engineering study since there is no need for new synthetic chemistry to be developed to enable the study. The inherent modularity of most coordination networks then enables systematic tuning of the structure and properties of a parent structure so that a family or platform of closely related materials can be studied.

It should be clear from the aforementioned that crystal engineering relies upon bottom-up design principles and that crystal engineering is not synonymous with "engineering crystals" (control of crystal size and morphology) and crystal structure prediction, both of which have also advanced in recent years [41–45]. Furthermore, whereas the ultimate goal of crystal engineering has always been driven by potential applications, it has only been in the past 10–15 years that the design principles discussed herein became reliable enough to move to systematic structure–function studies or "phase 4" crystal engineering as defined by Schmidt. Indeed, pharmaceutical cocrystals, which can offer an opportunity to create new intellectual property [46] as well as provide better physicochemical properties, have already resulted in better medicines reaching the market [47,48]. The design and synthesis of functional coordination networks represents another example of crystal engineering that has reached a degree of fruition and is the focus of the remainder of this contribution.

1.2 What is a coordination network?

A coordination network is a compound extending, through repeating coordination entities, in one dimension, but with cross-links between two or more individual chains, loops, or spirolinks, or a coordination compound extending through repeating coordination entities in two or three dimensions [49].

Coordinate covalent bonding and the directionality they typically afford were first described by Werner [50]. The directionality of coordination bonds enabled the first generation of coordination polymers reported in the early 1960s by Bailar [51]. Coordination networks existed before this time with notable examples being the pigment Prussian Blue [52], ($Fe_4[Fe(CN)_6]_3$), which exhibits primitive cubic (**pcu**) topology, and the mineral, stepanovite ($[NaFe(oxalate)_3]\cdot(Mg(H_2O)_6)$) [53]. Guest inclusion compounds such as Hofmann's clathrate ($[Ni(NH_3)_2Ni(CN)_4]\cdot$ guest) are in effect square lattice (**sql**) and variants of Prussian Blue are to a certain extent modular [54,55]. The synthesis of **dia** coordination networks was reported in 1959 by Kinoshita et al. [56] and by Aumueller et al. [57] in the 1980s. However, Robson and coworkers were the first group to recognize and enable the scope of coordination networks possible using the concept of nodes and linkers. By 1990, Robson's group had published papers outlining the prospects (1) for combining octahedral and tetrahedral metal centers (nodes) with "rodlike connecting units" (i.e., linkers) to afford network materials and (2) that the underlying components could be substituted to rationally afford "unusual and useful properties" [58—60]. Seminal papers reported the design, synthesis, and structures of coordination networks such as $[Cu(4,4',4'',4'''\text{-tetracyanotetraphenylmethane})]_n\cdot BF_4\ xC_6H_5NO_2$ and $[Zn(4,4'\text{-bipyridyl})_2(H_2O)_2]_n\cdot SiF_6$ (Fig. 2.2A and B). These papers introduced

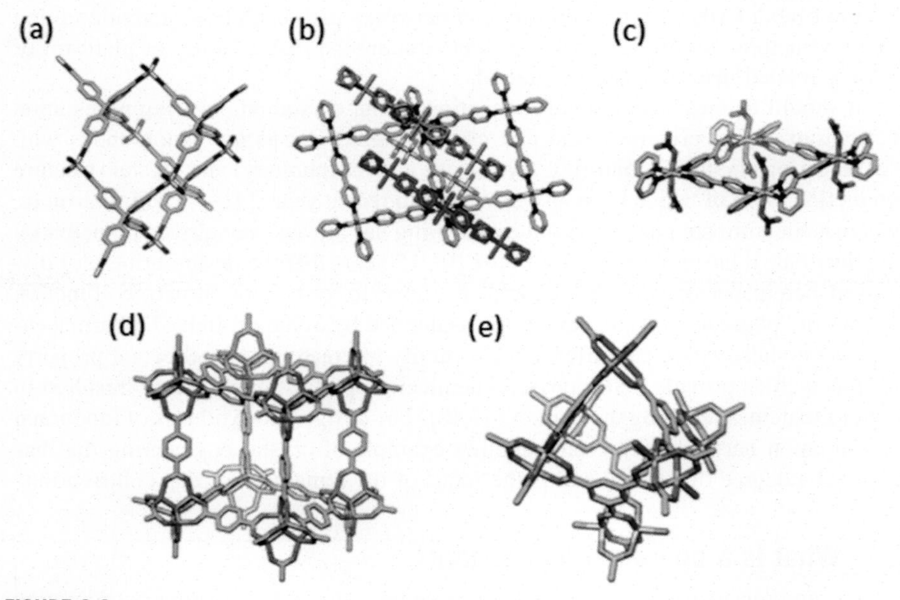

FIGURE 2.2

Crystal structures of (A) $[Cu(4,4',4'',4'''\text{-tetracyanotetraphenylmethane})]BF_4\cdot xC_6H_5NO_2$, (CSD code JARMEU), (B) $[Zn(4,4'\text{-bipyridyl})_2(H_2O)_2]$ $OSiF_6$, (CSD code JEZRUB), (C) $[Cd(4,4'\text{-bipyridyl})_2]$ $(NO_3)_2$ (CSD code YECFAN), (D) MOF-5, $[Zn_4O(1,4\text{-benzenedicarboxylate})_3]_n$ (CSD code SAHYIK), (E) HKUST-1, $[Cu3(1,3,5\text{-benzenetricaboxylate})_2]_n$ (CSD code FIQCEN). *CSD*, Cambridge Structural Database.

two important aspects of coordination networks: (1) that coordination networks can be synthesized with large cavities to offer applications similar to those offered by zeolites [60–65]; (2) that interpenetration would be a commonly encountered but hard to control phenomenon [66–68].

In 1994, Fujita's group reported the noninterpenetrated **sql** coordination network $[Cd(4,4'\text{-bipyridine})_2]_n\cdot(NO_3)_2$ (Fig. 2.2C) [69], and addressed its catalytic properties. Almost concurrently, Moore and Lee reported sixfold interpenetration of honeycomb networks and suggested the possibility of a crystal engineering approach to auxetic materials [26]. Indeed, Robson (and later Batten) continued to focus upon interpenetration, its control, and its utility [70]. By 1997 and 1998, Kitagawa [71] and Yaghi [72], respectively, had demonstrated that permanent porosity could exist in coordination networks. These papers introduced the terms porous coordination polymers (PCPs) and metal–organic frameworks (MOFs), respectively. Interest in coordination networks surged shortly thereafter in 1999 when Williams and Yaghi published HKUST-1, $[Cu_3(1,3,5\text{-benzenetricaboxylate})_2]_n$ (Fig. 2.2E) [73], and MOF-5, $[Zn_4O(1,4\text{-benzenedicarboxylate})_3]_n$ (Fig. 2.2D) [74], respectively. These coordination networks were found to exhibit extralarge gravimetric surface area values (>3000 m^2/g for MOF-5) well beyond that observed for any existing type of porous material. The study of porous materials with extralarge surface area remains a topic of interest; Hupp and Kaskel have recently pushed the upper range of Brunauer–Emmett–Teller (BET) surface areas to *ca.* 7000 m^2/g and 7800 m^2/g, respectively [75,76].

Given the explosive growth rate of this field, it is perhaps unsurprising that nomenclature has at times been a subject of confusion and/or disagreement. However, an IUPAC panel came to agreement on terminology for MOFs in 2013 [49], and it would be pertinent herein to address how MOFs fit under the broader umbrella of coordination networks and other classes of network material [77]. A MOF was defined as "a coordination network with organic ligands containing potential voids." This definition differentiates MOFs from zeolites and other types of coordination networks such as those that are sustained by combinations of inorganic and organic linkers. The former are well studied and, of course, industrially relevant, but they are not as diverse in terms of composition and pore chemistry as MOFs. The latter, which could be called hybrid coordination networks, remain relatively understudied but are also modular in nature and therefore amenable to systematic crystal engineering studies.

1.3 The role of the Cambridge Structural Database

1,000,000 answers … but what are the questions?

Successful crystal engineering of network structures typically requires knowledge of the chemistry and geometry of the underlying molecular components or "building blocks," the interactions that direct crystal packing and whether one can controllably fine-tune the structure to optimize properties. For this, the crystal engineer not only needs to understand the underlying chemistry but must also have a handle on statistics to predetermine whether or not the desired compound is reasonably expected [78,79]. For this reason, the Cambridge Structural Database (CSD) was and continues to be an indispensable resource for crystal engineers [80]. Bernal was first to propose such a database [81], and at the time of writing, the CSD had archived >1,000,000 crystal structures. Given the number of crystal structures in the CSD, others have noted its importance in understanding trends in crystal chemistry with each structure being an answer to a question waiting to be asked [82]. Herein, we use the CSD to survey MBBs and the propensity of various metals to enable their construction.[1] Analysis of this survey is tabulated in Tables 2.1 (O-donor ligands), 2.2 (N-donor ligands), and 2.3 (mixed-donor ligands).

1.4 What is a molecular building block?

An MBB is a discrete coordination complex or multifunctional organic ligand that is suited to serve as a node for the generation of a coordination network.

MBBs, a term coined by Moore and Lee in 1993 [26], are typically based upon known discrete coordination complexes or organic molecules with appropriate geometry and functionality. Their inherent geometry and connectivity are key to enabling a crystal engineering or bottom-up design of coordination networks. The node-and-linker design strategy introduced by Robson's group was and remains an effective approach to design families of structurally related coordination networks. The process of selecting an MBB(s) entails both an understanding of the coordination chemistry of the metal cation/cluster of the MBB(s) and the suitability of the coordinating ligands to serve as linkers (or additional nodes). The electronic structure of the metal moiety will dictate what ligand types will form robust coordination bonds. In turn, the types of ligands that can coordinate to the metal nodes will influence what complexes may form and their potential connectivity. First-row transition metals have been widely used in coordination networks and Pearson's hard—soft acid—base theory is consistent with the widespread use of linker ligands with pyridyl, carboxylate, or imidazolate moieties [83]. Importantly, the node-and-linker approach can be based upon existing MBBs. For further reading on the underlying chemistry of MBBs, the reader is directed elsewhere [84].

1.4.1 O-donor only molecular building blocks

3-connected nodes. Octahedral metal cations or clusters can readily form three-connected (3-c) trigonal MBBs (n-c where $n =$ the connectivity) when complexed by three bidentate chelating O-donor ligands (Fig. 2.3A). CSD statistics (5.39, Nov 2017) reveal that this MBB can be sustained by chromium (27% of 686 entries),

[1] CSD version 5.39 (November 2017) as applied in ConQuest 1.20 was used for the MBB survey. Ambiguous bond types were assigned as "Any." The following metal centers were surveyed: Mg, Ca, Sc, Ti, V, Cr, Mn, Fe, Co, Ni, Cu, Zn, Sr, Y, Zr, Nb, Mo, Tc, Ru, Rh, Pd, Ag, and Cd. Where necessary, the number of atoms bonded to the metal centers was assigned, and the torsional angle between coordinating atoms was assigned to control geometry. A filter requiring 3D coordinates was applied.

Table 2.1 O-donor molecular building blocks.

MBB	Trigonal mononuclear (3-c)	Trigonal paddlewheel (3-c)	Square paddlewheel (4-c)	Basic zinc acetate (6-c)	Basic iron acetate (6-c)	Cubic (8-c)	Cuboctahedron (6-c, 8-c, 12-c)
Mg	–	–	1 (<1%)	–	–	–	–
Ca	1 (<1%)	–	–	–	–	–	–
Sc	–	–	–	–	1 (<1%)	–	–
Ti	51 (7%)	–	1 (<1%)	–	–	–	–
V	12 (2%)	–	–	–	16 (3%)	–	–
Cr	186 (27%)	–	34 (1%)	–	130 (20%)	–	–
Mn	95 (14%)	–	7 (<1%)	–	55 (9%)	–	–
Fe	198 (29%)	–	45 (2%)	–	252 (39%)	–	–
Co	31 (5%)	–	37 (1%)	5 (5%)	11 (2%)	–	–
Ni	5 (1%)	–	38 (1%)	–	16 (3%)	–	–
Cu	52 (8%)	–	1357 (52%)	2 (2%)	–	4 (100%)	–
Zn	32 (5%)	40 (100%)	163 (6%)	89 (89%)	1 (<1%)	–	–
Zr	–	–	–	–	–	–	62 (93%)
Mo	8 (1%)	–	157 (6%)	–	–	–	–
Tc	2 (<1%)	–	5 (<1%)	–	–	–	–
Ru	19 (3%)	–	329 (13%)	–	52 (8%)	–	–
Rh	12 (2%)	–	434 (17%)	–	7 (1%)	–	–
Cd	4 (1%)	–	4 (<1%)	–	–	–	–
Total	686 (100%)	40 (100%)	2614 (100%)	100 (100%)[a]	638 (100%)	4 (100%)	67 (100%)[b]

MBB, *molecular building block.*
[a] *Four additional structures were observed with Be metal centers.*
[b] *Five additional structures were observed with Hf metal centers.*

Table 2.2 N-donor molecular building blocks.

MBB	Trigonal trinuclear (3-c)	Trigonal mononuclear (3-c)	Square planar (4-c)	Square paddlewheel (4-c)	Tetrahedral (4-c)	Basic zinc pyrazolate (6-c)	Basic iron pyrazolate (6-c)	Cubic (8-c)
Mg	–	11 (<1%)	11 (<1%)	–	115 (3%)	1 (20%)	–	–
Ca	–	2 (<1%)	–	–	23 (1%)	–	–	–
Sc	–	–	–	–	17 (<1%)	–	–	–
Ti	–	5 (<1%)	–	–	60 (1%)	–	–	–
V	–	6 (<1%)	–	–	34 (1%)	–	–	–
Cr	–	96 (2%)	49 (1%)	–	32 (1%)	–	–	–
Mn	–	219 (4%)	21 (<1%)	–	87 (2%)	–	1 (11%)	4 (20%)
Fe	–	1381 (26%)	89 (1%)	–	197 (5%)	–	6 (67%)	3 (15%)
Co	–	1100 (20%)	180 (3%)	–	315 (7%)	3 (60%)	2 (22%)	5 (25%)
Ni	–	873 (16%)	2338 (37%)	–	124 (3%)	–	–	2 (10%)
Cu	19 (100%)	250 (5%)	1759 (28%)	–	1400 (32%)	–	–	2 (10%)
Zn	–	368 (7%)	387 (6%)	–	781 (18%)	1 (20%)	–	1 (5%)
Sr	–	–	–	–	6 (<1%)	–	–	–
Y	–	4 (<1%)	–	–	40 (1%)	–	–	–
Zr	–	3 (<1%)	–	–	30 (1%)	–	–	–
Nb	–	1 (<1%)	–	–	27 (1%)	–	–	–
Mo	–	5 (<1%)	–	–	43 (1%)	–	–	–
Tc	–	3 (<1%)	–	–	–	–	–	–
Ru	–	928 (17%)	2 (<1%)	–	–	–	–	–
Rh	–	29 (1%)	11 (<1%)	5 (100%)	1 (<1%)	–	–	–
Pd	–	–	1250 (20%)	–	–	–	–	–
Ag	–	18 (<1%)	160 (3%)	–	935 (22%)	–	–	–
Cd	–	123 (2%)	11 (<1%)	–	46 (1%)	–	–	3 (15%)
Total	19 (100%)	5404 (100%)	6253 (100%)	5 (100%)	4309 (100%)	5 (100%)	9 (100%)	20 (100%)

MBB, *molecular building block.*

Table 2.3 Mixed donor molecular building blocks.

MBB	Octahedral (*cis*-chelate A)[a] (4-c)	Octahedral (*cis*-chelate B)[a] (4-c)	Octahedral (*trans*-chelate A)[a] (4-c)	Octahedral (*trans*-chelate B)[a] (4-c)	Octahedral (capped) (4-c)
Mg	–	–	–	–	16 (<1%)
Sc	–	–	–	–	1 (<1%)
Ti	–	1 (<1%)	–	–	2 (<1%)
V	1 (<1%)	–	1 (1%)	–	40 (1%)
Cr	72 (33%)	1 (<1%)	1 (1%)	–	175 (4%)
Mn	15 (7%)	9 (2%)	22 (14%)	–	288 (7%)
Fe	10 (5%)	8 (1%)	2 (1%)	1 (1%)	167 (4%)
Co	76 (35%)	85 (15%)	95 (62%)	9 (5%)	1066 (26%)
Ni	9 (4%)	75 (13%)	4 (3%)	7 (4%)	831 (20%)
Cu	11 (5%)	84 (15%)	14 (9%)	159 (87%)	874 (21%)
Zn	22 (10%)	150 (27%)	4 (3%)	1 (1%)	376 (9%)
Zr	–	1 (<1%)	–	–	–
Mo	1 (<1%)	–	–	–	3 (<1%)
Tc	–	–	–	–	1 (<1%)
Ru	2 (1%)	–	8 (5%)	–	71 (2%)
Rh	2 (1%)	–	–	–	15 (<1%)
Pd	–	–	–	–	1 (<1%)
Ag	–	–	–	–	1 (<1%)
Cd	1 (<1%)	144 (26%)	5 (3%)	5 (3%)	221 (5%)
Total	220 (100%)	558 (100%)	182 (100%)	182 (100%)	4142 (100%)
MBB	**DMOF (6-c)**	**MN_4X_2 (6-c)**	**Trigonal prismatic (6-c, 9-c)**	**Tricapped trigonal prismatic (3-c, 9-c)**	**Tetrahedral (4-c)**
Mg	1 (<1%)	142 (1%)	–	–	–
Ca	–	54 (<1%)	–	–	–
Sc	–	18 (<1%)	–	–	–
Ti	1 (<1%)	74 (<1%)	–	–	–
V	–	77 (<1%)	1 (25%)	2 (1%)	–
Cr	26 (1%)	591 (2%)	–	24 (10%)	–
Mn	7 (<1%)	1497 (5%)	–	33 (13%)	1 (<1%)
Fe	45 (2%)	4786 (17%)	–	92 (37%)	8 (1%)
Co	37 (1%)	6680 (24%)	3 (75%)	7 (3%)	76 (14%)
Ni	38 (2%)	4482 (16%)	–	15 (6%)	1 (<1%)
Cu	1305 (53%)	3412 (12%)	–	–	–
Zn	163 (7%)	1423 (5%)	–	1 (<1%)	441 (82%)

Continued

Table 2.3 Mixed donor molecular building blocks.—*cont'd*

MBB	DMOF (6-c)	MN_4X_2 (6-c)	Trigonal prismatic (6-c, 9-c)	Tricapped trigonal prismatic (3-c, 9-c)	Tetrahedral (4-c)
Sr	–	13 (<1%)	–	–	–
Y	–	17 (<1%)	–	–	–
Zr	–	34 (<1%)	–	–	–
Nb	–	19 (<1%)	–	–	–
Mo	86 (3%)	126 (<1%)	–	–	–
Tc	4 (<1%)	23 (<1%)	–	–	–
Ru	327 (13%)	3237 (11%)	–	14 (6%)	–
Rh	424 (17%)	390 (1%)	–	–	–
Pd	–	22 (<1%)	–	–	–
Ag	–	22 (<1%)	–	–	6 (1%)
Cd	4 (<1%)	1334 (5%)	–	–	2 (<1%)
Total	2468 (100%)	28,343 (100%)	4 (100%)	251 (100%)	535 (100%)

Where all MBBs within a table did not form with a specific metal center, that metal center was excluded from the list. MBB, *molecular building block.*
[a] A, *catecholate/oxalate;* B, *carboxylate.*

manganese (14%), and iron (29%). The proposed mechanism for the racemization of such enantiomers is named after Bailar [85]. Tetrahedral metal centers can form 3-c trigonal "paddlewheel" MBBs when two metals are bridged by three carboxylate ligands (Fig. 2.3B). The trigonal paddlewheel MBB is understudied because it is not widely accessible; the tendency of Zn(II) cations to exhibit tetrahedral geometry means that Zn(II) is the only metal to form this MBB, and there are only 40 observed entries.

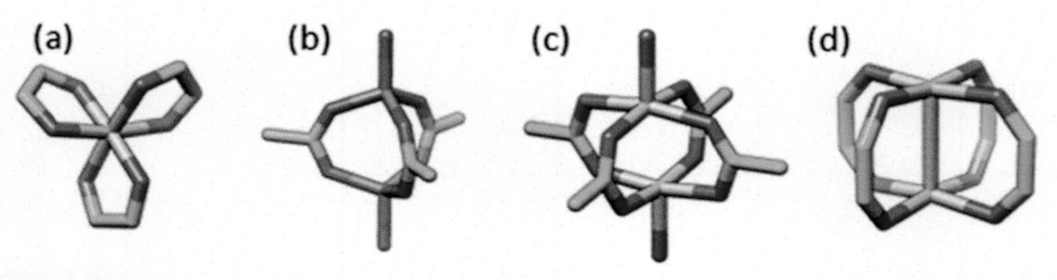

FIGURE 2.3

Commonly used MBBs based upon O-donor groups: (A) trigonal mononuclear, (B) trigonal paddlewheel, (C) square paddlewheel (carboxylate), and (D) square paddlewheel (catecholate/oxalate). *MBBs*, molecular building blocks.

4-connected nodes. The most widely used 4-c node based upon carboxylate ligands is the "square paddlewheel" MBB. Two metal centers are bridged by four carboxylate ligands to form a binuclear complex of formula $M_2(carboxylate)_4(L)_2$ (Fig. 2.3C). As a 4-c node, the square paddlewheel MBB is usually capped by a terminal ligand, and the resulting coordination network is generated by propagating the 4-c node through dicarboxylate linker ligands (e.g., 1,4-benzenedicarboxylate, 1,4,-bdc) or polycarboxylate nodes (e.g., 1,3,5-benzenetricarboxylate, 1,3,5-btc). In the case of Cu(II), there is weak coupling of the unpaired electrons; however, there is no metal–metal bond. This is not the case for other metal cations that sustain the square paddlewheel MBB such as Cr^{2+}, Mo^{2+}, Ru^{2+}, Ru^{3+}, and Rh^{2+} [84]. The CSD reveals that Cu(II) is the most prevalent metal center found in square paddlewheel MBBs with >50% of the 2614 reported paddlewheel structures. Rhodium (17%) and ruthenium (13%) metal centers are prevalent. Catecholate or oxalate-type ligands might be used in a similar manner, but $Pt_2(squarate)_4$ square paddlewheel MBBs are the only such examples studied thus far (Fig. 2.3D) [85].

The angular nature of 1,3-bdc linker ligands with the square paddlewheel MBB has afforded several supramolecular isomers including the small rhombihexahedron (nanoball), a discrete polyhedron [86,87] that has been used as a supermolecular building block (SBB) (vide infra). Undulating **sql** [88] and **kgm** topology networks have also been generated (Fig. 2.4) [89]. The square paddlewheel MBB can also be used to form 3D coordination networks as exemplified by MOF-101 (**nbo** topology, Fig. 2.5A) [90], HKUST-1 (**tbo** topology, Fig. 2.5B) [73], NOTT-100 (**stx** topology, Fig. 2.5C) [91], and PCN-61 (**rht** topology, Fig. 2.5D) [92,93]. Each structure is comprised of the same MBB; however, the structure directing nature of di-, tri-, tetra-, and hexatopic carboxylate linkers can afford a range of topologies.

6-connected nodes. The prototypal 6-c MBB built entirely of carboxylate anions is based upon tetranuclear basic zinc acetate, $Zn_4O(acetate)_6$. An oxide anion exhibits tetrahedral coordination to four metal centers with general formula $M_4O(carboxylate)_6$. The metal centers are coordinated to four oxygen atoms; one being the central oxygen anion and the other three belonging to bridging carboxylate anions (Fig. 2.6A). When the terminal carboxylate ligands are substituted for carboxylate-containing organic linker ligands (e.g., 1,4-bdc) to afford coordination networks, Zn^{2+} was the first and remains the most commonly used metal to form this MBB [94]. However, Be^{2+} is also known to afford this complex [95]. CSD statistics reveal that *ca.* 89% of basic zinc acetate MBBs are constructed from zinc metal centers. Only 100 such MBBs were retrieved by our CSD survey so it is one of the least commonly used MBBs. Nevertheless, Yaghi and coworkers exploited this MBB to afford an extensive family of "MOF-*n*" and "IRMOF-*n*" coordination networks (Fig. 2.6B and C) [96]. This body of work exemplifies the concept of reticular chemistry and underlines (1) the importance of linker ligands in directing the desired structure, (2) the reliability that certain MBBs can exhibit, and (3) the diversity of pore size in the family of coordination networks that can thereby be afforded [97].

Another important 6-c node based upon carboxylate ligands is the trinuclear trigonal prismatic basic iron acetate MBB [98]. A central oxygen anion exhibits

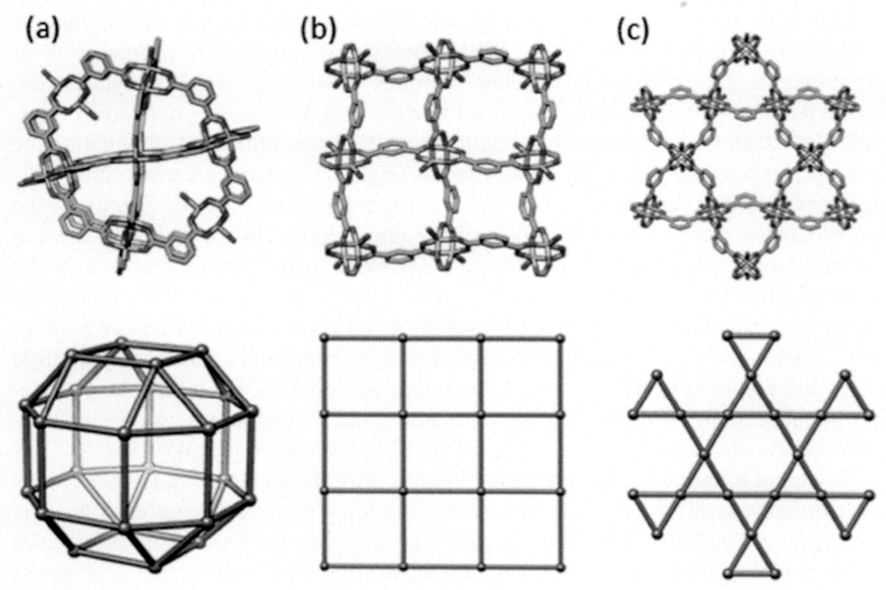

FIGURE 2.4

Chemical structure (top) and connectivity (bottom) of supramolecular isomers of [Cu(1,3-benzenedicarboxylate) $(H_2O)]_n$ (A) small rhombihexahedron (e.g., CSD code XEYBEI). (B) Square lattice network (sql) (e.g., CSD code HATGUF), and (C) kagomé lattice network (kgm) (e.g., CSD code PACFOP). *CSD*, Cambridge Structural Database.

trigonal geometry and is bound by three (typically trivalent) metal centers. The metal centers are doubly bridged with six carboxylates and capped by terminal ligands (typically water) to form $M_3O(carboxylate)_6(H_2O)_3$ MBBs (Fig. 2.7A). When the terminal carboxylate ligands are substituted by organic linkers, then coordination networks are generated. Cr^{3+} and Fe^{3+} are commonly found metal centers for this MBB, but V^{3+}, Mn^{3+}, Co^{3+}, Ru^{3+}, and Rh^{3+} have also been used. Our survey of the CSD indicates that iron and chromium are the most widely studied examples, with 39% and 20% of the 638 observed MBBs, respectively. Férey used this MBB to form MIL-100 and MIL-101 (**mtn** topology) 1,3,5-btc and 1,4-bdc, respectively (Fig. 2.7B and C) [99–101].

1.4.2 N-donor only molecular building blocks

3-connected nodes. Pyrazole-type ligands can afford a 3-c trigonal trinuclear MBB. A tridentate oxygen atom or hydroxyl group can be coordinated by three square planar/octahedral metal centers bridged by pyrazolate ligands (Fig. 2.8A). The metal centers may be a site for other linker ligands to bind. This MBB was used as the 3-c node in the first example of an **rht**-MOF (Fig. 2.8B) [102]. Substitution of the pyrazolate ligand for triazolate afforded a self-penetrating coordination network in which the MBB acts as a 6-c node (Fig. 2.8C) [103]. Of the 19 entries in the

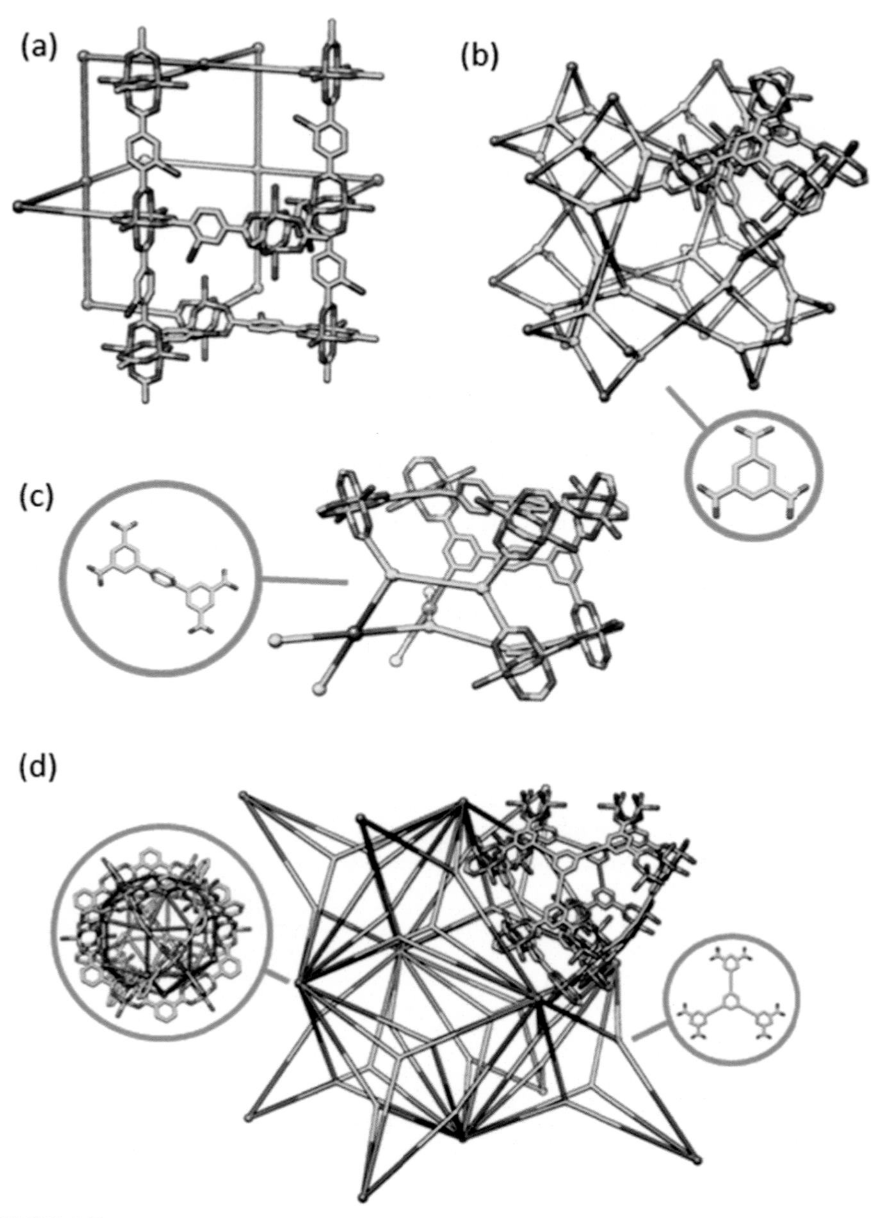

FIGURE 2.5

Examples of 3D coordination networks sustained by the square paddlewheel MBB and organic linkers: (A) MOF-101 (CSD code YIXBIQ), (B) HKUST-1 (CSD code FIQCEN), (C) NOTT-100 (CSD code CESFOW), and (D) PCN-61 (CSD code VUJBIM). *CSD*, Cambridge Structural Database; *MBB*, molecular building block; *MOF*, metal–organic framework.

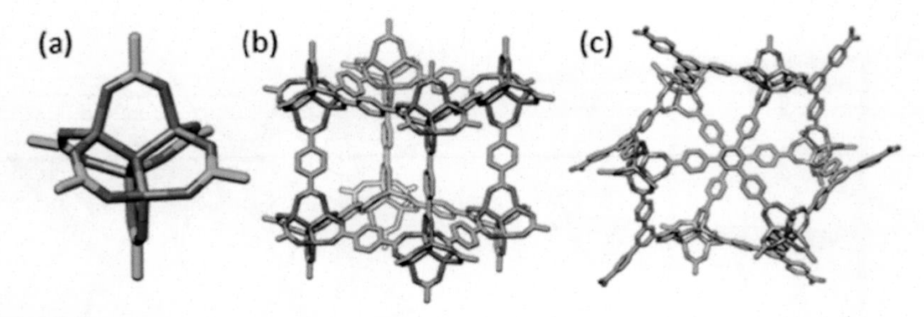

FIGURE 2.6

(A) The basic zinc acetate MBB. (B) MOF-5 (CSD code SAHYIK), (C) MOF-177 (CSD code BABRII). *CSD*, Cambridge Structural Database; *MBB*, molecular building block; *MOF*, metal–oganic framework.

CSD, all were sustained by Cu(II). Alternatively, three 2,2′-bipyridine or diethylamine-type linkers can chelate an octahedral metal center to afford a trigonal mononuclear MBB (Fig. 2.3A). Iron (26% of 5404 entries) and cobalt (20%) are the most commonly observed metal centers in such MBBs.

4-connected nodes. The most commonly used 4-c node based upon N-donor ligands is the mononuclear square planar MBB first used by Robson [60] and Fujita [69]. This MBB is typically based upon an octahedral metal center coordinated in the equatorial positions by four N-donor ligands (e.g., pyridyls, pyrroles) that are propagated using suitable organic linkers such as 4,4′-bipyridine. The axial positions are typically occupied by terminal aqua ligands or counterions (Fig. 2.9A). With >34,000 structures in the CSD exhibiting square planar or octahedral geometry with four N-donor ligands in the equatorial plane, this MBB is the most prevalent in our survey. Its ubiquity might be attributable to the number of metal cations that exhibit this MBB. Nickel and cobalt (both 20%) are the most common metal centers for this MBB, with iron (14%), copper (15%), ruthenium (9%), zinc (5%), and manganese (4%) also being prominent. The relative ease of synthesis of this

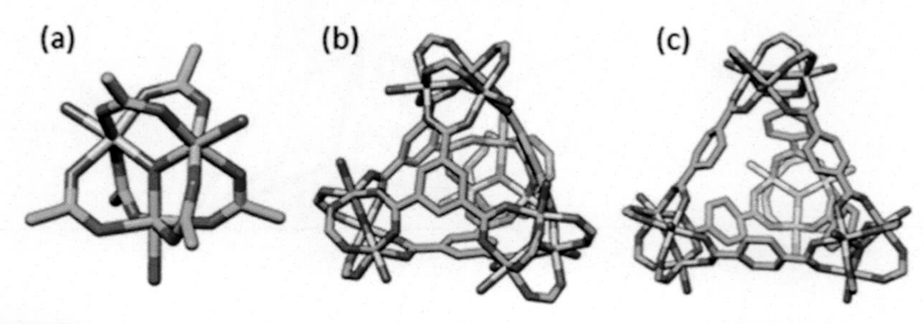

FIGURE 2.7

(A) The basic iron acetate MBB. As found in (B) MIL-100 (CSD code UDEMEW) and (C) MIL-101 (CSD code OCUNAC). *CSD*, Cambridge Structural Database; *MBB*, molecular building block.

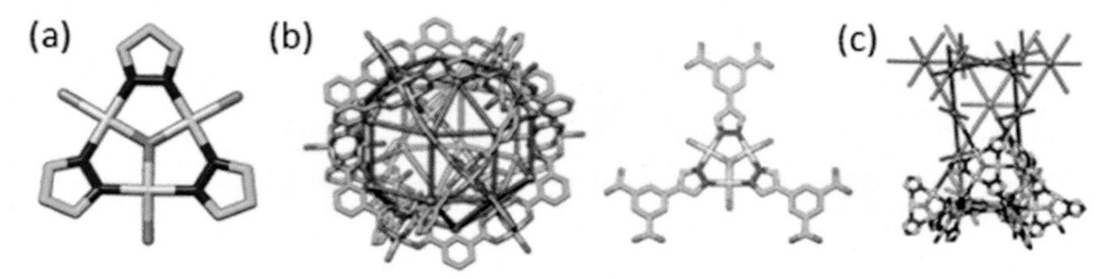

FIGURE 2.8

(A) A trigonal trinuclear MBB can act as both (B) a 3-c node (CSD code UGOCEA) and (C) a 6-c node (CSD code DEGPEL). *CSD*, Cambridge Structural Database; *MBB*, molecular building block.

MBB has enabled crystal engineering studies including those focused upon interpenetration. Macrocycles such as porphyrins and cyclams also help to explain the occurrence of this MBB. Another type of 4-c square planar MBB based upon N-donor ligands is exemplified by pyrazolate paddlewheel MBBs with rhodium (Fig. 2.9B).

Another common 4-c MBB based upon N-donor ligands utilizes tetrahedral metal cations. Cu(I), Ag(I), and Zn(II) metal centers are the most commonly observed metal centers (32%, 22%, and 18%, of 4309 entries, respectively) for this tetrahedral MBB. There are more than 4000 entries in the CSD (Fig. 2.9C and D). The use of linear N-donor ligands to afford **dia** topology coordination networks was reported as far back as the early 1990s [104]. **dia** coordination networks are also of interest for study of interpenetration [105]. Imidazolate-type linkers have been used to generate so-called zeolitic−imidazolate frameworks (ZIFs) or zeolite-like metal−organic frameworks (ZMOFs). The M-Im-M bridges in such structures exhibit similar angles to the Si−O−Si links of zeolites (Fig. 2.9D) [106−108].

6-connected nodes. The basic zinc acetate MBB discussed earlier can also be prepared from pyrazole-type ligands (basic zinc pyrazolate MBB). These MBBs are as described earlier, but carboxylate bridges are substituted by N-donor groups (Fig. 2.10A). Such MBBs are less commonly studied than their carboxylate counterparts and are exemplified by coordination networks such as the pcu networks

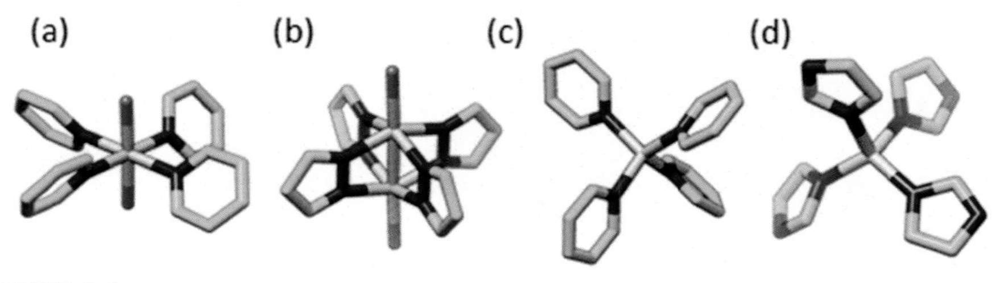

FIGURE 2.9

(A) A square planar 4-c MBB built from an octahedral metal center capped by terminal ligands. (B) A pyrazolate paddlewheel 4-c MBB. (C) and (D) tetrahedral 4-c MBBs. *MBBs*, molecular building blocks.

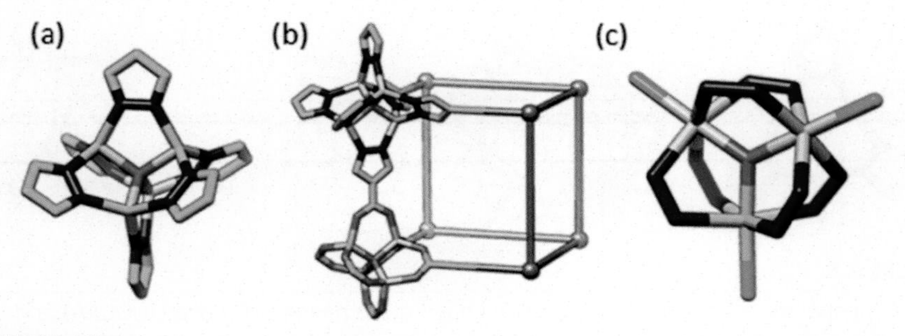

FIGURE 2.10

(A) A basic zinc pyrazolate MBB was used by Janiak to produce (B) A pcu coordination network isostructural to MOF-5 (CSD code LILNUR). (C) A basic iron pyrazolate MBB (CSD code RIVYAX). *CSD*, Cambridge Structural Database; *MBB*, molecular building block; *MOF*, metal–organic framework.

reported by Janiak and coworkers (Fig. 2.10B) [109]. Similarly, the carboxylate groups of the basic iron acetate MBB can be substituted by pyrazolate-type ligands (basic iron pyrazolate MBB). This MBB is less commonly observed, and no coordination network has thus far been constructed from it (Fig. 2.10C).

1.4.3 Mixed-donor molecular building blocks

3-connected nodes. The trigonal prismatic basic iron acetate MBB (Fig. 2.7A) can also be used as a 3-c trigonal MBB if the carboxylate anions are terminal and the aqua ligands are replaced by N-donor ligands. This tricapped trigonal prismatic MBB was present in 39% of the 251 relevant structures in the CSD. Iron (37%), manganese (13%), and chromium (10%) are the most commonly used metal centers. This MBB can form a 12-c octahemioctahedron with a 1,3-substituted ditopic pyridyl linker (Fig. 2.11A) [110] and a twofold interpenetrated honeycomb coordination network with 4,4′-bipyridine (**hcb** topology) (Fig. 2.11B) [111].

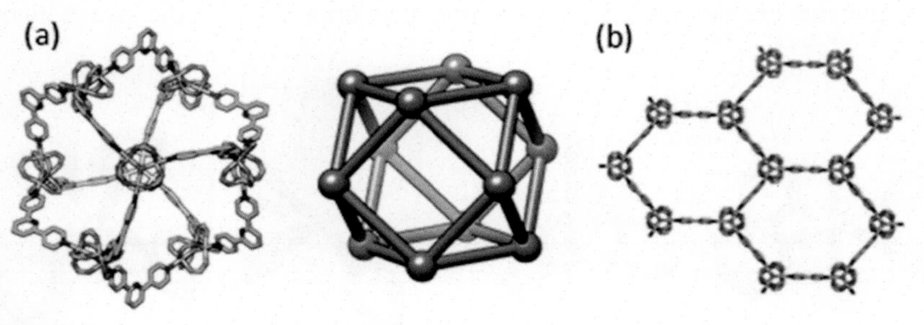

FIGURE 2.11

(A) As a 3-c node, the tricapped trigonal prismatic MBB has been found to form a zero-dimensional octahemioctahedron (CSD code CANZAU) and (B) A **hcb** coordination network (CSD code GIXGUR). *CSD*, Cambridge Structural Database; *MBB*, molecular building block.

4-connected nodes. The combination of N-donor and carboxylate ligands with either octahedral or tetrahedral metal centers can afford a variety of 4-c MBBs exhibiting either square planar or tetrahedral geometry. For tetrahedral metal centers, the geometry of the MBB is dependent on the organic linker ligands used. MBBs comprised from octahedral metal centers are more diverse in scope. Square planar MBBs can be derived from octahedral metal centers bound by two N-donor ligands and two carboxylate ligands with two terminal ligands (e.g., water). Indeed, the carboxylate ligand could be monodentate with a terminal or additional chelating ligand coordinating (Fig. 2.12B). Without terminal ligands, octahedral metal centers may still afford square planar or tetrahedral MBBs if the carboxylate moieties are bidentate. Such coordination can result in structural isomerism whereby the *trans*-isomer affords a square planar MBB (Fig. 2.12A) and the *cis*-isomer results in a tetrahedral MBB (Fig. 2.12C). Catechol and oxalate-type chelate ligands may also sustain this MBB. With respect to tetrahedral metal centers, the tetrahedral MBB in Fig. 2.12D is the only option.

Square planar (mixed-donor) MBBs as shown in Fig. 2.12A are dominated by copper (87% of 182). The square planar MBB shown in Fig. 2.12B forms with cobalt (26% of 4142 entries), nickel (20%), and copper (21%). Where catechol/oxalate-type ligands chelate the metal, cobalt was most commonly observed (62% of 182 entries). Tetrahedral (mixed-donor) MBBs constructed from octahedral metal centers were observed to be sustained by cobalt (15% of 558 entries), nickel (13%), copper (15%), zinc (27%), and cadmium (26%) metal centers. For tetrahedral metal centers, zinc (82% of 535 entries) was dominant. With respect to catechol- or oxalate-type ligands, chromium (33% of 220 entries) and cobalt (35%) were found to be the most frequent.

6-connected nodes. Square paddlewheel MBBs in which the terminal ligand is substituted by an N-donor ligand to generate a 6-c octahedral node are well studied. The most widely used linker is 1,4-diazabicyclo[2.2.2]octane (DABCO), introduced by Kim and coworkers in the **pcu** structure DMOF-1, $[Zn_2(1,4\text{-benzendicarboxylate})_2(DABCO)]_n$ (Fig. 2.13B) [112]; however, many pyridyl-type linker ligands have also been used in this context [113]. CSD statistics reveal that this MBB is versatile and readily constructed from copper (53% of 2468

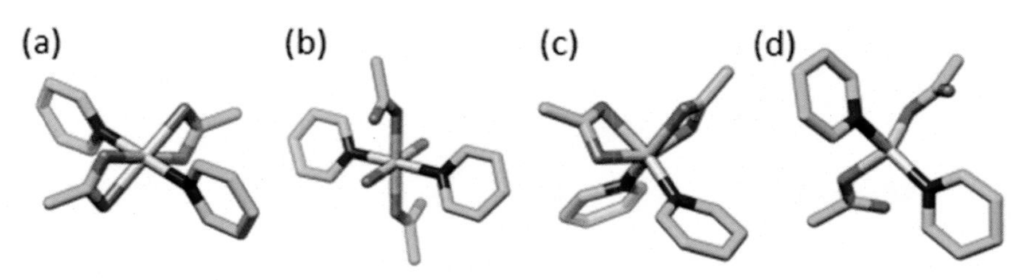

FIGURE 2.12

(A) A square planar MBB. (B) A square planar MBB from an octahedral metal center axially capped by terminal ligands. A tetrahedral MBB from (C) octahedral and (D) tetrahedral metal centers. *MBB*, molecular building block.

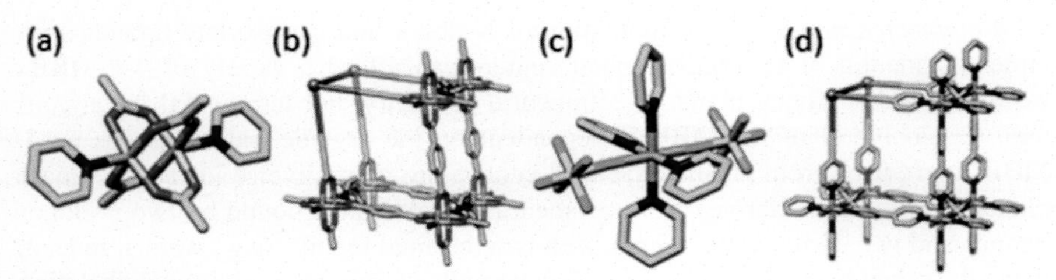

FIGURE 2.13

(A) A pillared square paddlewheel MBB can afford (B) DMOF-1. (C) A MN_4X_2 MBB can afford (CSD code WAFKIY) (D) SIFSIX-1-Cu (CSD code ZESFUY). *CSD*, Cambridge Structural Database; *MBB*, molecular building block.

entries), rhodium (17%), and ruthenium (13%). The wide variety of carboxylate, N-donor, and carboxylate/N-donor linker ligands that are readily available means that DMOF-1 and the structures reported by Lee et al. are prototypal of a large family of related structures. Interpenetration and chemical/hydrolytic stability have been addressed using such coordination networks [113−115].

MBBs constructed from octahedral metal centers and N-donor ligands that afford 4-c square planar nodes can be utilized to create 6-c nodes by substitution of the terminal axial ligands with linker ligands. These new linker ligands may be counterions or neutral ligands and are not necessarily organic (Fig. 2.13C). Within the CSD, 28,343 structures exhibit this MBB with cobalt (24%), iron (17%), nickel (16%), ruthenium (11%), and copper (12%) being the most commonly encountered metals. Pillared square grid networks that use this MN_4X_2 6-c MBB include hybrid ultramicroporous materials and hybrid pillared square grid materials with **pcu** (Fig. 2.13D) or **mmo** network topologies [116,117].

If the basic iron acetate MBB is modified by replacing half of the carboxylate anions with pyrazole-type linkers, a mixed-linker trigonal prismatic MBB can be generated (Fig. 2.14A). Although uncommon in the CSD, coordination networks and metal−organic cages used this MBB (Fig. 2.14B and C) [118,119].

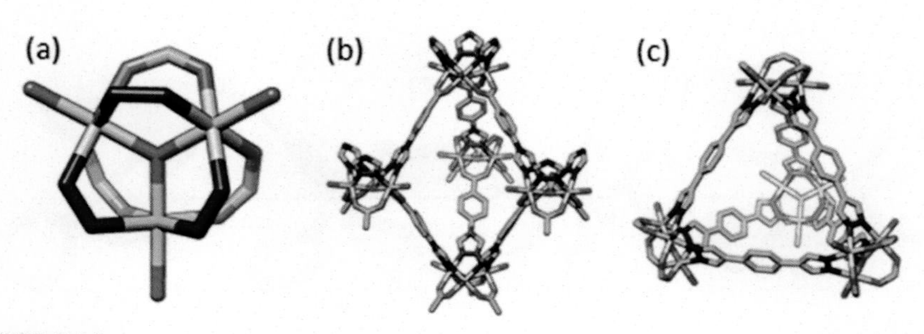

FIGURE 2.14

(A) A mixed-linker trigonal prismatic MBB can afford both (B) coordination networks (CSD code VIDJID) and (C) metal−organic cages (CSD code PULWEA). *CSD*, Cambridge Structural Database; *MBB*, molecular building block.

1.4.4 High-connectivity molecular building blocks or supermolecular building block (≥8-c)

Although 4-c and 6-c nodes are prominent in coordination networks, there are MBBs that are suited to serve as high connectivity 8-c, 9-c, 12-c, or 24-c nodes. Some such MBBs have been termed SBBs [120]. The SBB approach affords coordination networks that are less prone to supramolecular isomerism, thanks in part to the higher connectivity and symmetry of these structures, which in turn reduces the degrees of freedom.

With respect to 8-c nodes, two types of cubic MBBs are known. The first is a tetranuclear complex comprised of a four-coordinate square planar moiety (typically a halogen or hydroxyl group) coordinated to four metal centers that exhibit square pyramidal geometry. Each metal center is doubly bridged to neighboring metal centers by either two carboxylate anions or two pyrazolate-type anions affording the cubic MBB (Fig. 2.15A–D) [121]. When solely constructed from carboxylate linker ligands, only copper metal centers have been shown to afford coordination networks wherein the MBB is cross-linked by 1,3,5-btc ligands to generate a (3,8)-c network [122]. The use of pyrazolate-type ligands is more common, and the metal centers used include manganese (20% of 20 entries), iron (15%), and cobalt (25%). Similar coordination networks were afforded by tris-triazolate linkers [123–125].

A 9-c node can be derived via propagation of all peripheral ligating sites in the tricapped trigonal prismatic MBB. By substituting terminal aqua ligands for, e.g., N-donor linker ligands, as well as propagating structures through carboxylate linker anions, a 9-c node is obtained (Fig. 2.16A). Chen has used this MBB to design coordination networks with interesting network topologies (Fig. 2.16B) and sorption properties. Similarly, the mixed-linker trigonal prismatic MBB can also be exploited [126].

A hexanuclear cuboctahedron MBB can be constructed from six square antiprismatic metal centers (typically Zr^{4+} or Hf^{4+}) that sit at the vertices of an octahedron. Tridentate oxygen atoms are positioned at the faces of this octahedron, coordinating to the three neighboring metal centers. Metal centers are also bridged along the

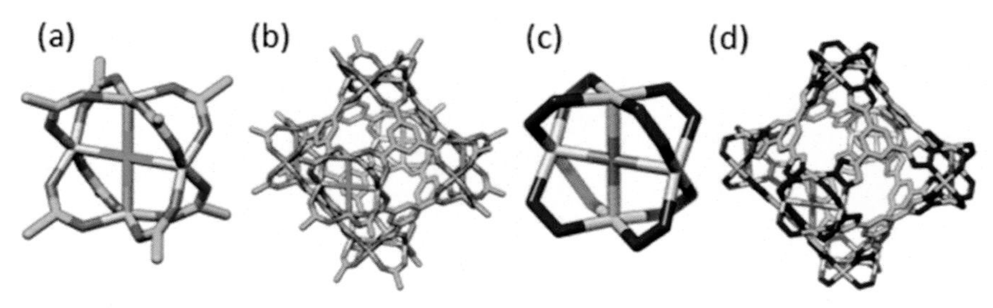

FIGURE 2.15

(A) A carboxylate cubic MBB and (B) a derived coordination networks (CSD code ABEMIF), (C) and (D) the pyrazolate analogue (CSD code JEWYAM). *CSD*, Cambridge Structural Database; *MBB*, molecular building block.

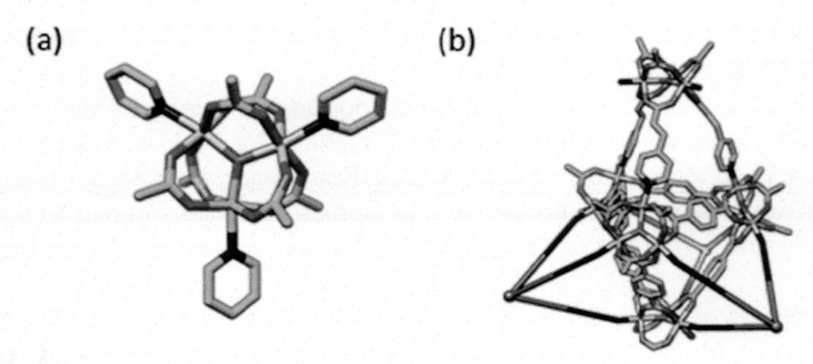

FIGURE 2.16

(A) A tricapped trigonal prismatic MBB can afford (B) a coordination network with ncb topology (CSD code KARLEW). *CSD*, Cambridge Structural Database; *MBB*, molecular building block.

edges of this octahedron by carboxylate anions to afford the cuboctahedron MBB. This MBB sustains the well-studied coordination network UiO-66 (Fig. 2.17A and B) and its derivatives [127]. The CSD survey of this MBB indicates that zirconium metal centers are most prominent with 93% of 67 entries. Substituting four carboxylates with hydroxyl groups within one plane affords a second cubic MBB. This

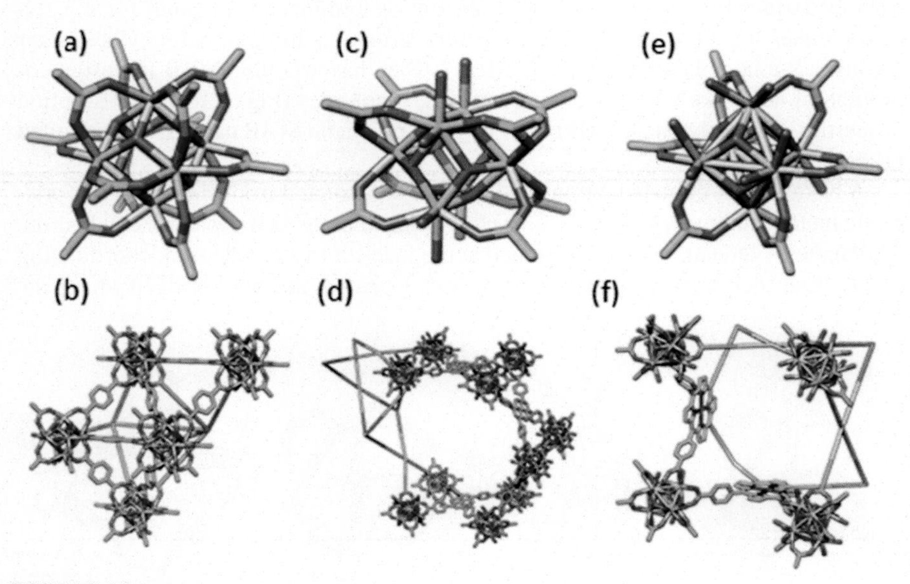

FIGURE 2.17

(A) A 12-c cuboctahedron MBB affords (B) UiO-66 (CSD code RUBTAK). (C) As an 8-c node, this MBB affords (D) NU-1000 (CSD code FIFFUX), and as a (E) 6-c node, it affords (F) PCN-224 (CSD code DOGBEI). *CSD*, Cambridge Structural Database; *MBB*, molecular building block.

FIGURE 2.18

The square paddlewheel MBB can assemble to form an octahedral node that can be pillared in a similar manner to DMOF-1 to afford a **pcu-a** coordination network (CSD code SUPQAW). *CSD*, Cambridge Structural Database; *MBB*, molecular building block.

MBB sustains ultrahigh surface area coordination networks such as NU-1000 (Fig. 2.17C and D) [128], PCN-222 [129], and DUT-67 [130]. Further substitution with hydroxyl groups affords the six-connected node of PCN-224 (Fig. 2.17E and F) and MOF-808 [131,132].

A 24-c node may be derived from the small rhombihexahedron SBB that is constructed from 12 square paddlewheel MBBs cross-linked by 24 1,3-bdc linker ligands (Fig. 2.4A). The bdc ligands may be substituted at the 5-position to enable cross-linking of the SBBs which then can serve as a 24-c node. This is why **rht**-MOF coordination networks (Figs. 2.5D and 2.8B) can be regarded as a platform. Similarly, an octahedral MBB can be pillared in a manner equivalent to DMOF-1 and become a 6-c node in a twofold interpenetrated **pcu-a** network topology (Fig. 2.18) [133]. The use of such SBBs has been discussed more thoroughly elsewhere [120].

There are 57,025 structures sustained by the metal centers and MBBs surveyed herein; iron (13%), cobalt (17%), nickel (16%), and copper (19%) are the most commonly used metals. There are >360,000 metal complexes in CSD, and cobalt (9%) and nickel (9%) are less commonly studied in general (Fig. 2.19). This may be a reflection of the much broader scope of metal complexes in general, which include many organometallic compounds and ligands such as phosphates and sulfonates that have not been well studied in the context of coordination networks.

1.5 The architecture of network structures

Although the previous section focuses upon coordination networks that may be propagated directly from their MBBs using organic linkers or nodes, well-studied coordination networks such as MIL-53 [134] and CPO-27 (MOF-74) [135] exhibit

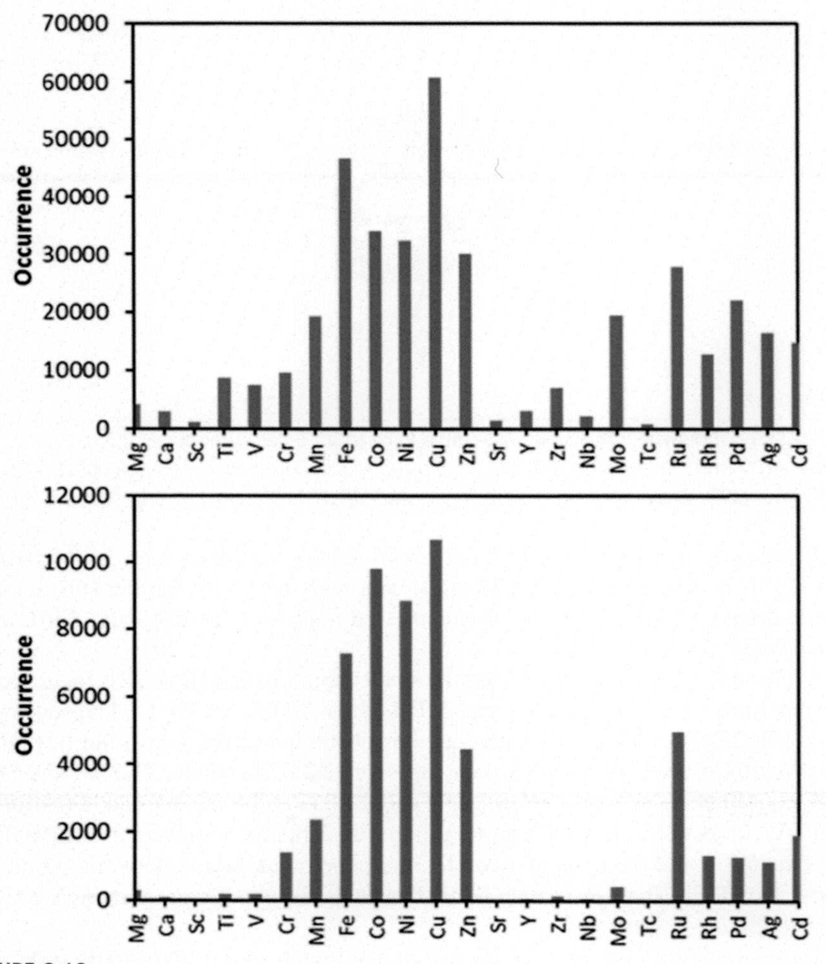

FIGURE 2.19

(top) Histogram depicting the occurrence of different metal centers observed within the CSD. (bottom) Histogram depicting the occurrence of different metal centers in MBBs observed in the CSD. *CSD*, Cambridge Structural Database; *MBBs*, molecular building blocks.

structures afforded by cross-linked one-dimensional coordination entities or rod building blocks (RBBs). RBBs can be derived from vertex-sharing and edge-sharing octahedral metal centers, respectively, and not by discrete MBBs/SBBs (Fig. 2.20). Nonetheless, the nature of these compounds would make them amenable to crystal engineering principles. For the sake of brevity, only coordination networks built from one kind of MBB are addressed herein. Nevertheless, it should be noted that cross-linking of multiple different MBBs within one network is an active area of

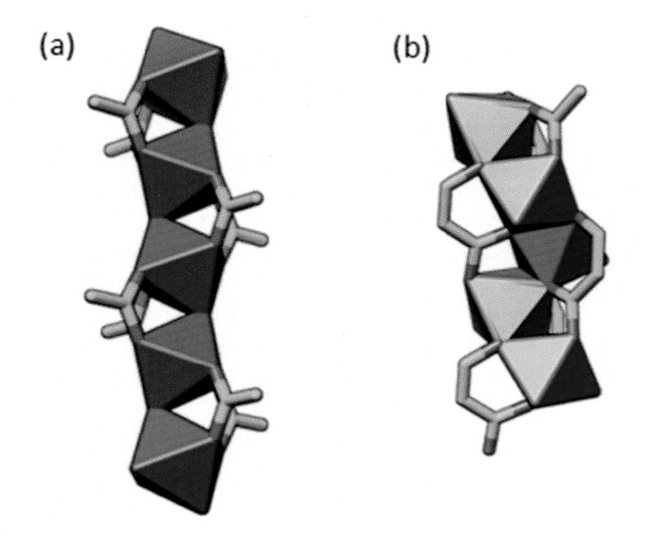

FIGURE 2.20
(A) MIL-53 (CSD code GUSNEN) and (B) CPO-27 (MOF-74) (CSD code LEJRIC) are constructed from vertex- and edge-sharing octahedra, respectively. *CSD*, Cambridge Structural Database.

crystal engineering research that has afforded coordination networks with new topologies (e.g., two-step crystal engineering) [136].

1.5.1 Nodes, linkers, and topological networks

The design of a functional coordination network with exact control over packing and symmetry remains elusive. However, with a combination of the discussed MBBs and appropriate ligands, sophisticated coordination networks can be derived whose packing and symmetry are the result of a combination of the packing and symmetry of its components. It is the strength and directionality of coordinate covalent bonding that allows for reasonable fidelity and the ability to crystal engineer functional coordination networks.

The broad application of these principles coupled with the node-and-linker approach of Wells led to the development of an aspect of crystal engineering, reticular chemistry, which focuses specifically on the linking of MBBs into network materials. Reticular chemistry can be applied to not only coordination networks but also covalently bonded networks, e.g., covalent organic frameworks (COFs) [137].

Linker ligands. The linker ligands used in the design and synthesis of functional coordination networks are of equal importance to the metal center used. The linker ligands can propagate the symmetry of the MBBs or generate new networks based on their own symmetry and the number of MBBs being bridged. From a crystal engineering perspective, the ligands are also important because symmetry can be maintained while size and functionality can be altered. For instance, a linear ditopic linker ligand such as 1,4-benzenedicarboxylate could cross-link square paddlewheel MBBs to form a square grid coordination network. This network topology may be retained but expanded by using biphenyl-4,4′-dicarboxylate. A growing area of research in crystal engineering involves postsynthetic modification of coordination networks wherein new functional groups can be added to linker ligands or MBBs after the coordination network has been obtained [138−140]. The alteration of the linker ligand thereby allows the pore chemistry and pore size of a coordination network to be manipulated. Given that they are readily available and cheap, 4,4′-bipyridyl and 1,4-benzenedicarboxylate have historically been the most widely utilized organic linker ligands used in the crystal engineering of coordination networks. To that end, there are currently over 4984 structures in the CSD where 4,4′-bipyridyl links metal centers and over 5456 structures wherein 1,4-benzenedicarboxylate links metal centers.

Topological analysis. Topology is a mathematical construct that can be used to classify coordination networks. Similar to the concept of ribbon diagrams in structural biology, topological analysis of crystal structures (especially coordination networks) allows researchers to more clearly understand the connectivity of complex structures. Beyond Wells' work, more recent efforts in the understanding of coordination network topology and catenation have been provided [141−149]. Indeed, the chemistry involved in coordination networks can often be disparate; however, the underlying network topology can be the same. Topology can allow a crystal engineer to more easily visualize and design new materials. Therefore, the use of the Reticular Chemistry Structural Resource (RCSR) [150] as well as the use of programs such as ToposPro [147] in understanding and visualizing underlying network topologies is highly recommended.

However, to the crystal engineer, chemist, and crystallographer, such mathematical treatment of coordination networks may not be intuitive as topology provides no information on shape or size and further confusion can result from the notation and terminology used (e.g., transitivity, Schlaefli symbols, vertex symbols, point symbols, face symbols, or Delaney symbols). A brief description is given herein.

A coordination network's transitivity is defined by the quantity of unique vertices (p), edges (q), faces (r), and tiles (s) that sustain the network. The network is defined as regular when the transitivity (pqrs) equals 1111, i.e., there is one unique vertex, edge, face, and tile. Using Schlaefli symbols, a network may be described by the connectivity of the vertices (p) and the edges (q, polygons) that pattern the network. For example, the notation [q,p] may be used to describe a discrete cube as a structure where three squares meet at each vertex [4,3] (Fig. 2.21A); the **sql** network topology can be described as four squares meeting at each vertex, i.e., [4,4] (Fig. 2.21B) and similarly, a honeycomb (**hcb**) network topology can be described as three hexagons meeting at each vertex: [6,3] (Fig. 2.21C). However, considering three-dimensional (three-periodic) structures, we must consider the amount of polygons that meet at each vertex [q,p] and the numbers of faces (r) that coincide at each edge and is described with the notation [q,p,r]. Considering the primitive cubic (**pcu**) network

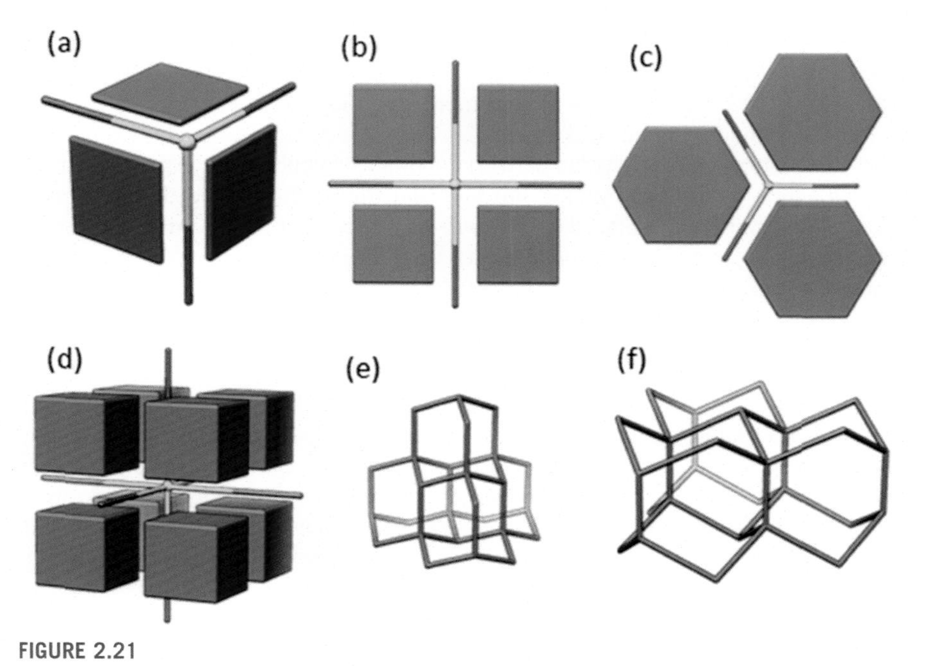

FIGURE 2.21

(A) The [4,3] packing of a cube. (B) The [4,4] packing of the square lattice (sql) network topology. (C) The [6,3] packing of the honeycomb (hcb) network topology. (D) The [4,3,4] packing of the primitive cubic (pcu) network topology. (E) The diamondoid (dia) network topology. (F) The lonsdaleite (lon) network topology.

topology, three squares meet at a vertex and are repeated fourfold about each edge: [4,3,4] (note that a discrete cube is described as [4,3] and the pcu network is the simple tiling of cubes wherein four cubes meet at each edge, thus [4,3,4] (Fig. 2.21D). Conversely, the **pcu** network topology may be considered based upon the number of faces (r) constructed from polygons that define the cage/tiles that construct the network topology, [q^r] (face symbol): six squares (tetragon) pack to form a cubic tile, [4^6].

The vertex symbol can be used to list the shortest rings that involve a vertex at a specific angle. This notation follows as $A_a.B_b.C_c$ ... where a is the number of rings of size A at the first angle. For example, the diamondoid (**dia**) network topology would have the vertex symbol of $6_2.6_2.6_2.6_2.6_2.6_2$ (Fig. 2.21E). Indeed, the network topology lonsdaleite (**lon**) exhibits the same vertex symbol (Fig. 2.21F). To discern the two network topologies, we calculate the sum of the topological density at the 10th nearest neighbor (TD10). To calculate TD10, we count the length between vertices, with the closest connecting vertices being the first nearest neighbors, and each unique vertex connecting to the first nearest neighbors being part of the second nearest neighbors and so on until we have reached the *k*th nearest neighbor. The quantity of neighbors at each step can be written as a series known as the coordination sequence (CS), the sum of which at the 10th nearest neighbor is TD10. For **dia** topology, TD10 is 981, whereas for **lon** topology, it is 1027.

The three underlying network topologies may be modified by an additional letter to the three letter code assigned to network topologies by the RCSR, e.g., **lon-e** (new vertices added to the edge of the **lon** net) and **dia-a** (the **dia** net is augmented wherein each vertex is replaced by a group of vertices but the coordination figure shape of the original vertex is maintained). Emboldened lower-case letters are designators assigned to topologies by the RCSR and allow for network topologies of coordination networks to be easily recognized by readers.

1.5.2 Supramolecular isomerism

A crystal engineer may be judicious in their selection of MBBs, linker ligands, and the network topology that is sought and could have used statistical data from the CSD to suggest that a certain outcome is reasonably expected. However, nodes and linkers may pack in more ways than one; supramolecular isomerism can then occur. The topic of supramolecular isomerism has been covered in reviews and research articles [13,29,151,152]; however, given its importance to crystal engineering and the improved understanding that is available today, a brief summation is given herein.

There are four recognized types of supramolecular isomerism: structural, conformational, catenane, and optical. Structural supramolecular isomerism is a result of the same molecular components exhibiting different crystal packing resulting in different network topologies. An early example of structural supramolecular isomerism in coordination networks was exhibited by coordination networks of formula $[Cu(1,3\text{-benzenedicarboxylate})_2]_n$ (Fig. 2.4). Conformational supramolecular isomerism involves changes in the conformation of a component of a coordination network that results in a different network topology [153].

Catenation/interpenetration (e.g., **pcu-c**) can result in catenane supramolecular isomerism. Catenane supramolecular isomerism involves at least two distinct networks that are mechanically bonded as two separate networks that would require disruption of at least one network's connectivity to become unlinked. Catenation may be difficult to enumerate topologically especially when one network topology can exhibit multiple modes of interpenetration or polycatenation. Indeed, the study of catenation with respect to topology has recently been of growing interest [68,70,148,149]. More recently, this phenomenon has also been dubbed entanglement isomerism [154]. Terminology in the study of catenane-type structures has been often misused due to misunderstandings in definitions, and some basic guidelines are given herein. The term interpenetration would refer to a finite number of networks that are entangled but wherein overall dimensionality/periodicity remains the same. An example of interpenetration would be the twofold interpenetration of three-dimensional networks observed in SIFSIX-2-Cu-i (Fig. 2.22A) [155]. Conversely, polycatenation would refer to an infinite number of networks whose entanglement results in an increase in dimensionality. Robson's structure

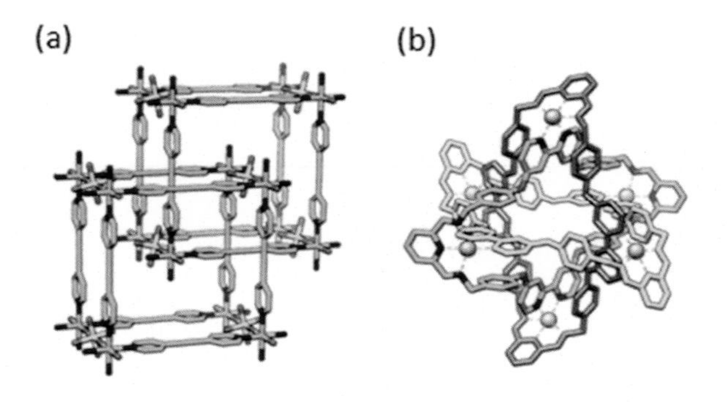

FIGURE 2.22

(A) The twofold interpenetrated **pcu** HUM, SIFSIX-2-Cu-i (CSD code YEMTIV) (B) Stoddart's Borromean rings (CSD code IYOQOC). *CSD*, Cambridge Structural Database; *HUMs*, hybrid ultramicroporous materials.

(Fig. 2.2B) is a good example of polycatenation as square lattice (**sql**) coordination networks exhibit inclined interpenetration such that a three-dimensional structure is afforded [60,156].

Both interpenetration and polycatenation are often realized by Hopf links wherein the polygons (or rings) that define each network link once. Partial interpenetration, wherein the level of interpenetration is not an integer (i.e., the degree of interpenetration is not consistent across the entire coordination network), is also possible. In contrast to mixtures of interpenetrated and noninterpenetrated coordination networks, partially interpenetrated networks are solid solutions wherein the inconsistent interpenetration is distributed across one crystal lattice [157–159]. Alternatively, networks may be bound together by Brunnian (e.g., Borromean) links. In this instance, at least three networks are entangled; however, should one of the networks be removed, the remaining two networks would no longer be entangled. Stoddart's Borromean rings are an excellent discrete example of such knotting (Fig. 2.22B) [160]. With respect to the topological understanding of interpenetration and polycatenation, the use of Hopf-ring nets and extended ring nets have recently been reported [154]. When it comes to Hopf-ring nets, the network topologies of the coordination networks are drawn. The point at which each ring entangles another ring of another network, a new topological node is drawn; the old network topologies are removed, and the new network topology is drawn. Indeed, for two-dimensional structures, the Hopf-ring net may well just lead to a one-dimensional chain. However, by creating a node at the point at which a network is entangled, a fused-ring net is formed. The addition of the Hopf-ring net and fused-ring net affords the extended ring net.

Chirality of a coordination network may be the result of the molecules used and/or their packing and is the topic of optical supramolecular isomerism [161,162]. The MBB, ligand, their packing, and interpenetration can all cause chirality to occur within a coordination network. The chirality of a coordination network may not only be induced by the orientation of the MBBs and ligands but also by the guest molecules which incorporate the voids [162].

2. Now

2.1 Properties and potential applications

In our preoccupation with the detailed problems of our daily work of finding out how atoms are arranged in space, we are in danger of losing sight of the whole picture….

[163].

These words of Mackay (a student of J.D. Bernal) in 1975 ring true even today. In this chapter, the focus is the design of coordination networks using the principles of crystal engineering. However, a crystal engineer ultimately seeks to understand and control structure—property relationships. Ostensibly, the properties that may be exhibited by coordination networks are as varied as the underlying molecules, their combination, and packing thereof. To discuss the crystal engineering of functional coordination networks, just a few of the properties and potential applications observed are addressed herein. Indeed, many studies into coordination networks focus on their properties, and several reviews have been written regarding their properties and potential applications [61—63]. With respect to functional coordination networks, Kitagawa has likely explored more functions than any other active researcher [164]. Kitagawa has classified coordination networks (specifically, PCPs) into different "generations" based upon their properties [165]. The term generation to classify coordination networks may be a misnomer, as the properties and characteristics of different generations of coordination networks may not have been discovered nor developed in a chronological manner. However, as mentioned earlier, properties of coordination networks are derived from their underlying components and isostructural compounds may exhibit vastly different properties. Herein, the properties observed in different generations of coordination networks and their potential applications are discussed.

2.1.1 First-generation coordination networks

The first generation of coordination networks enclathrate guest molecules; which, when removed, results in decomposition of the network.

That guest removal results in the decomposition of a coordination network undermines their practical utility. Nonetheless, their study and design afforded important principles with regard to the design aspects of crystal engineering. A prototypal first-generation coordination network is MOF-101 (Fig. 2.5A) [90]. MOF-101 revealed that symmetry of the MBB and ligand connectivity are not always controlling; noncoordinating functional groups attached to the ligand may also influence the ligand conformation and crystal packing.

The stability in coordination networks is a relatively understudied area, and as such, it is difficult to determine all the coordination networks that may be categorized by this generation [166–168]. However, some properties and potential applications of coordination networks may be addressed wherein the removal of the guest/porosity is not required or will not influence the resulting properties, i.e., nonporous coordination networks. For example, the use of appropriate MBBs and ligands can afford a coordination network in a noncentrosymmetric space group, leading to nonlinear optical properties [169]. Indeed, control of interpenetration is a crucially important aspect herein as n-fold interpenetration likely requires $n =$ odd for noncentrosymmetry to be maintained [170]. Alternatively, a polar guest molecule may pack unsymmetrically within the void space of a coordination network also affording nonlinear optical properties. Coordination networks that exhibit spin-crossover properties may also be studied in this regard as the origin of the properties is found within the MBB itself. Guest molecules have also been found to have an effect on spin-crossover properties [171]. Similarly, conductivity and luminescence could be based upon the inherent nature of the coordination network and its interactions with guests or other interpenetrated networks [172–174].

Studies that address the stability of coordination networks have grown in recent years and revealed that both the supramolecular interactions of the ligands and coordination bonding can be important facets [175,176].

2.1.2 Second-generation coordination networks

The second generation of coordination networks may have guest molecules removed without the structure decomposing and thereby exhibiting "permanent porosity."

That the second generation of coordination networks may retain structure upon removal of guest molecules and still exhibit accessible void space has perhaps been the largest achievement to date in this area of chemistry. Indeed, Robson's recognition was that such properties could lead to applications like zeolites, but with the chemist having much greater control over design and synthesis.

High-surface-area coordination networks. As mentioned earlier, Kitagawa's and Yaghi's work in the late 1990s revealed permanent porosity in coordination networks. Kitagawa recently noted that society has moved from the use of solid fuels, to liquid fuels, and now to gaseous fuels as a major source of energy. In this capacity, the high gravimetric gas uptake of coordination networks means they are poised to find utility for gas storage in commercial and industrial settings. Indeed, both the storage and upgrading of gases to higher purity commodities coupled with the challenge of designing materials to exhibit the necessary properties create a subject of technological and scientific importance applicable to both industrial and basic chemistry.

For example, the US Department of Energy set fuel cell technology goals in the context of onboard hydrogen storage systems, which led to extensive research into hydrogen storage by porous coordination networks [177,178]. The targets set out for hydrogen storage, to be achieved by 2020, were to obtain physical or material means to store 4.5 wt.% hydrogen onboard, for the storage of hydrogen to be at least

0.03 kg/L, and for the cost to be $333/kg of stored hydrogen capacity. Similarly, ARPA-E created the MOVE (Methane Opportunities for Vehicular Energy) project which set a target of >9.2 MJ/L of stored methane with a lifetime of 100 cycles also leading to extensive research in methane storage in porous coordination networks [179–181]. It has therefore often been the goal of researchers to design and synthesize porous coordination networks with extralarge gravimetric surface area.

The isoreticular metal–organic framework (IRMOF, isoreticular = of the same net) series described by Yaghi in the early 2000s exhibits the use of basic zinc acetate MBBs for formation of same topology nets with extended ditopic ligands. The free volume of such structures can range from *ca.* 56% in IRMOF-1 (MOF-5, 1,4-benzenedicarboxylate) to ca. 91% in IRMOF-16 (1,4-bis(4-carboxyphenyl)benzene) [97].

That the coordination network's topology and packing upon expansion of the ligand is predictable enabled molecular simulations to be conducted before the chemist had even synthesized the materials. NU-100 was designed in silico to have long hexa-topic ligands that could form the square paddlewheel MBB, affording an **rht**-MOF with ultrahigh surface area [75,182].

Selective gas sorption. The design of coordination networks for gas separations is ultimately the design of specific binding sites wherein a component of a gas mixture is preferentially adsorbed over others. This area of study has received wide attention within coordination networks and crystal engineering. In contrast to gas storage, for which high surface area is important, selective separation may be based upon (1) porosity (size exclusion) wherein microporous cavities may adsorb smaller gas molecules while excluding larger gas molecules (sieving) or (2) favourable binding energy for one gas molecular over another.

The **bnn**-MOF CPO-27 (MOF-74) exhibits different gas sorption properties based upon the metal center used (Fig. 2.23) [135,183]. In the adsorption of H_2 or CO_2, a combination of polarisability of the unsaturated metal center and the steric

FIGURE 2.23

The porous coordination network CPO-27 (MOF-74) (CSD code YEMTIV). *CSD*, Cambridge Structural Database; *MOF*, metal–organic framework.

effects of metal—ligand bond length results in a distinct ranking of metal centers: $Ni^{2+} > Co^{2+} > Mg^{2+} > Zn^{2+}$ for H_2 sorption and $Mg^{2+} > Ni^{2+} > Co^{2+} > Zn^{2+}$ for CO_2 sorption. The differences found indicate the importance of a network's constituents with respect to the targeted sorbate molecule [184]. However, large pores result in poorer selectivity after open metal sites are filled. Furthermore, although Mg-MOF-74 has the strongest interactions with CO_2 within the series, in the presence of water, affinity for water by complexation (aqua ligands) is invariably higher [185].

CPO-27 and expanded analogues have been used to demonstrate postsynthetic functionalization with amine-type ligands [186,187]. In the context of selective CO_2 capture, this added functionalization improves the separation performance of the materials. However, the chemisorptive nature of amine ligands implies that higher regeneration energies are required and a physisorptive approach may ultimately be more desirable in a practical setting.

Catalytic activity. Heterogeneous catalytic reactions are a growing part of green chemistry wherein the catalytically active site is bound within (or is a part of) the coordination network. As mentioned earlier, Fujita's work from 1994 used the open metal sites of a cadmium **sql** coordination network to catalyze a cyanosilylation reaction and may well be the first example of coordination networks used in catalysis [69]. Chen recently demonstrated that the open metal sites of MIL-88B can template [2+2+2] cyclic trimerization using 4-pyridyl compounds. MIL-88B is sustained by the basic chromium acetate MBB, and the open metal site was used to coordinate the reactants (forming the tricapped trigonal prismatic MBB) in a manner amenable to cyclization [188].

Kim combined the tricapped trigonal prismatic MBB with a chiral pyridyl/carboxylate ligand to afford POST-1, a two-dimensional coordination network wherein the uncoordinated pyridyl groups promoted a transesterification reaction with 8% enantiomeric excess (Fig. 2.24A) [189]. Furthermore, POST-1 was shown

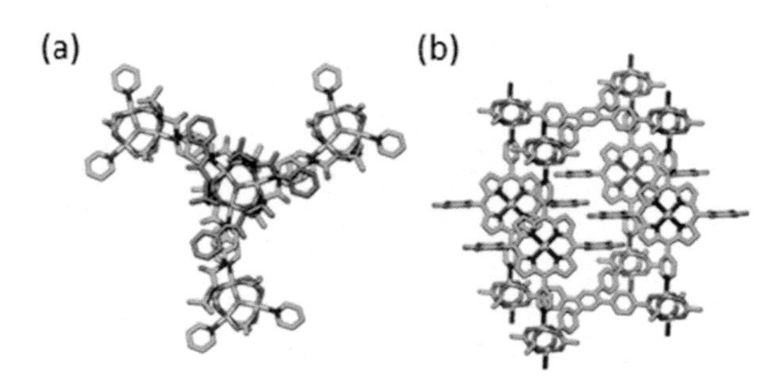

FIGURE 2.24

The catalytically active (A) POST-1 by Kim (CSD code UHOPUC) and (B) ZnPO-MOF by Hupp (CSD code XUGSEY). *CSD*, Cambridge Structural Database; *MOF*, metal—organic framework.

to be able to undergo postsynthetic modification through N-alkylation with CH_3I. Cohen studied postsynthetic modification suggesting that any robust coordination network with suitable linker ligands might be altered *in situ* to afford catalytically active sites [138–140].

Hupp used the pillared paddlewheel MBB with tetratopic carboxylate ligands and ditopic dipyridyl–porphyrin ligands to afford ZnPO-MOF. The incorporation of a porphyrin unit within the coordination network was found to result in a 2420-fold enhancement in the rate of an acyl transfer reaction between *N*-acetylimidazole and 3-pyridylcarbinol in comparison with the reaction without catalyst (Fig. 2.24B) [190]. Conversely, the guest molecule may be the catalytically active site. Eddaoudi revealed that a Mn-metallated porphyrin molecule could be trapped within the pores of a coordination network constructed from In^{3+} and imidazoledicarboxylate affording the catalytic oxidation of cyclohexane [191].

Crystalline sponge. A recently introduced function of second-generation coordination networks is their application in what is known as the crystalline sponge method [192,193]. Several important small molecules do not easily crystallize, and understanding of their structure typically comes from nuclear magnetic resonance studies. However, important aspects of such molecules (particularly absolute configuration) are difficult to determine without crystallographic data. First developed by Fujita, the crystalline sponge method uses an electron-rich coordination network such as $[(ZnI_2)_3(1,3,5\text{-tris(4-pyridyl)triazine})]_n$ (Fig. 2.25) [194]. That guest molecules may be exchanged means that targeted compounds can be enclathrated. High-quality diffraction studies allows for high-quality diffraction data to be collected, and the structure of the guest may be determined, including absolute configuration, if appropriate.

2.1.3 Third-generation coordination networks

The third generation of coordination networks are described as being flexible or dynamic; they may exhibit chemical and/or physical responses to stimuli that are reversible.

Kitagawa noted that drawbacks of first-generation coordination networks could become advantageous if collapse of a coordination network upon guest removal is

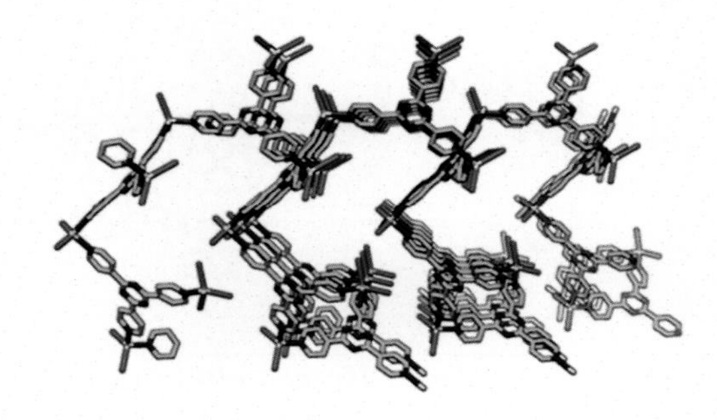

FIGURE 2.25

A crystalline sponge coordination network such as $[(ZnI_2)_3(1,3,5\text{-tris(4-pyridyl)triazine})]_n$ by Fujita (CSD code IJAWUN). *CSD*, Cambridge Structural Database.

reversible and if crystallinity is maintained throughout the process. In this respect, a coordination network could be envisaged wherein the removal of guest molecules would lead to a phase transformation that is reversible by reintroducing those guest molecules (by extension, at least one phase of the coordination network must have void space).

Forms of flexibility. Férey and Kitagawa pioneered the development of flexible coordination networks in the early 2000s, and since then, there have been a number of distinct mechanisms recognized [164,195]. MIL-53 exhibits a breathing-type flexibility wherein the three-dimensional network distorts and the one-dimensional void space decreases upon adsorption of water molecules (Fig. 2.26A and B) [134]. This action was associated with the hingelike behavior of the carboxylates about the metal node [196]. The use of the basic iron acetate MBB led to the construction of MIL-88, which exhibits swelling-type flexibility wherein the pores of the structure expand upon addition of guest molecules [197]. Both breathing and swelling mechanisms can be associated with the interactions between metal centers and ligands; however, the ligand in itself may be flexible and therefore cause the

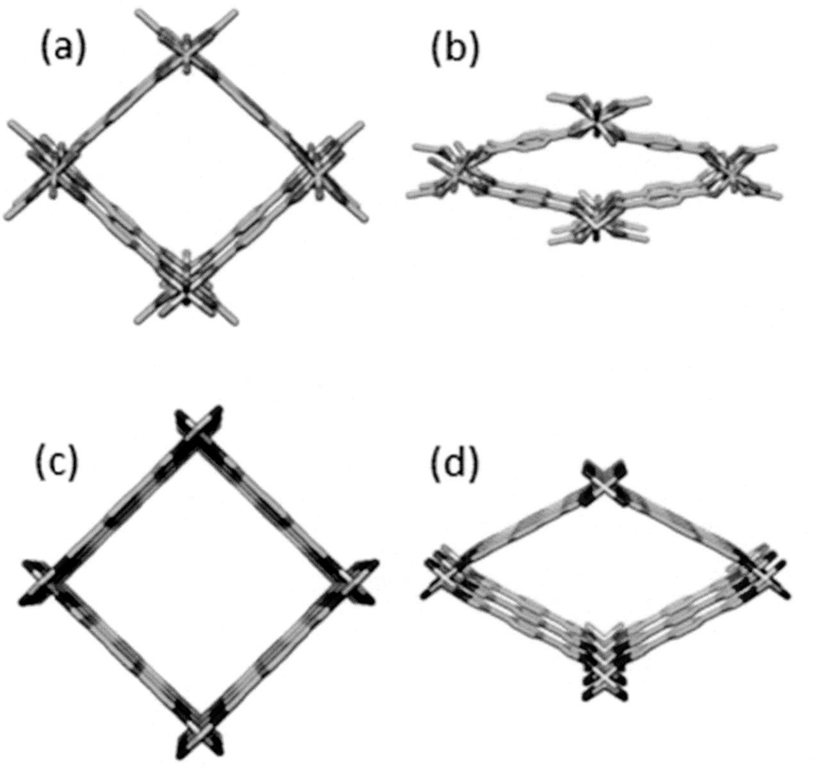

FIGURE 2.26

The open (A and C) and closed (B and D) phases of MIL-53 (CSD code GUSNEN) and [Co(1,4-benzenedipyrazolate)]$_n$ (CSD code COJHIT). *CSD*, Cambridge Structural Database.

coordination network to exhibit flexible behavior. ZIF-8 is a well-studied example of ligand-based flexibility wherein the imidazolate linkers are observed to slightly rotate to allow for adsorption of gas molecules that are larger than expected to enter pores [198]. Kaskel recently reported flexibility in DUT-49 wherein the flexibility of the linker is triggered upon CH_4 adsorption, leading to the pore shrinking, causing what was described as negative adsorption (the amount of gas adsorbed decreased with an increase in pressure) [199,200].

Interpenetrated coordination networks may also show flexible behavior similar to described earlier. However, that the mode of interpenetration could be altered upon inclusion of guest molecules (e.g., offset and centered modes of interpenetration) means that there is an added type of flexibility upon inclusion of guest molecules. Indeed, Kitagawa reported a twofold interpenetrated **pcu** coordination network constructed from the MN_4X_2 MBB wherein the metal center was Ni^{2+}, the pyridyl-type linker ligands were 4,4′-bipyridyl, and the pillaring linker ligand was dicyanamide. Upon exchange of dicyanamide guests for azide molecules, the networks were found to shift with respect to one another, leading to offset interpenetration and an increase in the effective pore size of one of the cavities [201]. Two-dimensional coordination networks offer flexibility similar to that of minerals wherein guest molecules may be intercalated between sheets. Indeed, ELM-11, a **sql** coordination network, exhibits adsorption of CO_2 via a gate-opening event at *ca.* 35 kPa. ELM-11 switches between closed and open phases [202].

Given the requirements for coordination networks to be used in gas storage, flexible coordination networks offer exceptional promise. The targets set by the US Department of Energy mean that the majority of known (especially second generation) coordination networks exhibit a working capacity that is too low for practical applications [203,204]. However, given the host−guest interaction strength and working capacity of most coordination networks, a coordination network could conceivably be crystal engineered wherein the structure "switches" from a closed to open phase at a desired pressure, thereby maximizing working capacity within a specific pressure range. Long was the first to demonstrate the potential advantages of this approach with a flexible coordination network constructed from Co^{2+} or Fe^{2+} and 1,4-benzenedipyrazolate linker ligands (Fig. 2.26C and D) [205]. Our group has recently reported two other examples of switching networks that exhibit high working capacity [206,207]. Hysteresis in flexible coordination networks may hinder their potential application and is a subject of topical interest [208].

2.1.4 Fourth-generation coordination networks

The fourth generation of coordination networks are described as being porous materials whose pores are modifiable and affected by material anisotropy [209].

In essence, fourth-generation coordination networks retain both the qualities of second and third-generation coordination networks wherein the pores of the coordination network are in some way altered by, for example, postsynthetic modification [138,139], defects [210], or solid solutions [206,211,212]. Additionally, a "hard−soft" approach can be realized wherein the coordination network itself remains rigid but the size, shape, and chemistry of the pores adapt to host–guest interactions [213]. This approach can be realized with anions that are not bound to the network and so can move freely within the pore or by additional ligand functional groups that reorient themselves as a response to stimuli (Fig. 2.27). Although a number of the

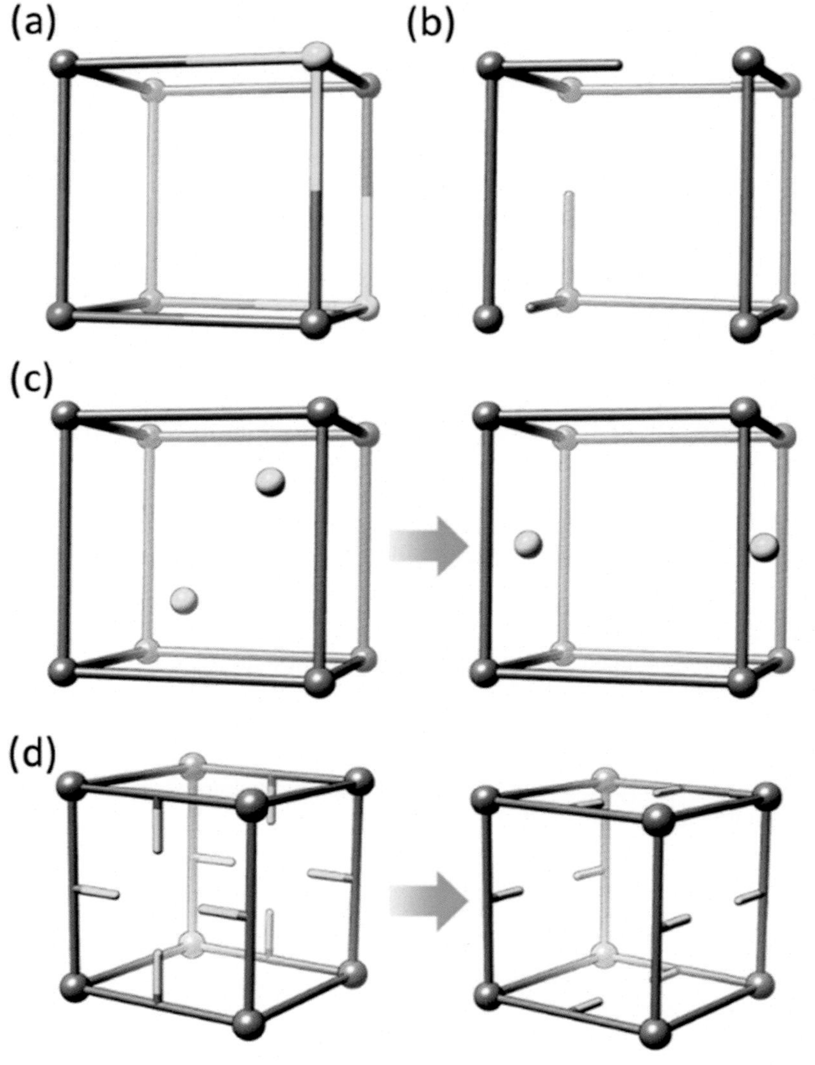

FIGURE 2.27

Fourth-generation coordination networks can (A) be solid solutions; (B) exhibit defects; (C) contain anions or guests within the pore which alter pore shape and chemistry in response to stimuli; (D) contain functional groups which reorient themselves in response to stimuli.

aspects surrounding fourth-generation coordination networks have been studied for several years, control via crystal engineering principles is only recently a subject of interest.

3. Conclusions

If one can control the crystal structure, and the crystal structure is fine-tuneable, then one may have control over the structure's function.

For the crystal engineering of functional coordination networks, it should be expected that all information needed for design is present in the chemistry and symmetry of the underlying molecules/MBBs. Through the use of established design principles, we are at a time when crystal engineering is more than ever focused upon properties and applications rather than design. The design tools available include the following: CSD data mining; elucidation of underlying network topologies; accumulated knowledge of MBBs and linker ligands; better understanding of supramolecular interactions; improved instrumentation for in situ characterization; and advanced modeling that yields better structural characterizations. It is not difficult to envisage that we are undergoing a paradigm shift in that it is a realistic goal to custom-design a new coordination network to solve a particular problem. Furthermore, once a prototypal coordination network has been obtained, it may be fine-tuned (or postsynthetically modified) to improve properties and performance. The crystal engineer should therefore start the design process of solving the problem by understanding and then controlling the needed properties. The phases of crystal engineering could then be redefined as follows:

1. Identify the target property or key problem to be addressed (e.g., selective gas separations, asymmetric catalysis, nonlinear optics etc.).
2. Identify the criteria required by the coordination network to afford the desired properties to address the challenge (e.g., large void space, strong H-bond acceptors, polar space groups, etc.).
3. Identify the appropriate symmetry of the nodes and linkers as well as whether or not interpenetration is desired.
4. Identify the best MBBs and ligands that may afford the symmetry and properties of the desired coordination network.
5. Identify the best method to obtain the coordination network (e.g., layering reactions, solvothermal reactions, template-based reactions, postsynthetic modification, two-step crystal engineering, etc.).
6. Characterise the structure and properties of the coordination network (e.g., atomic structure, thermal stability, porosity etc.).
7. Implement and assess the performance of the coordination network for its intended application.
8. Fine-tune the structure to afford a derivative coordination network with improved properties.

Despite the best laid plans, it is often the case that coordination networks yield unexpected and sometimes exciting results serendipitously. Nevertheless, once these properties are observed and understood, a crystal engineering approach might afford new and even better materials.

Crystal engineering of functional coordination networks covers a broad scope. Indeed, chemistry (synthetic and supramolecular), crystallography, and mathematics all play important roles in crystal engineering. To further understand properties and structure–function relationships, computational studies and in situ experiments are becoming more and more routine and important to further develop the principles of crystal engineering.

References

[1] W.L. Bragg, Proc. Camb. Phil. Soc. 17 (1913) 43–57.
[2] M.v. Laue, Die Interferenzen von Roentgen- und Elektronenstrahlen, Springer Verlag Berlin Heidelburg, Berlin, 1935.
[3] P.P. Ewald, Kristalle und Roentgenstrahlen, Springer Verlag, Berlin, 1923.
[4] G. Breit, I.I. Rabi, Phys. Rev. 46 (1934) 230–231.
[5] E. Ruska, Rev. Mod. Phys. 59 (1987) 627–638.
[6] R. Feynman, There's Plenty of Room at the Bottom, American Physical Society, California, USA, 1959.
[7] A. von Hippel, Science 123 (1956) 315–317.
[8] H.S. Scott, N. Ogiwara, K.-J. Chen, D.G. Madden, T. Pham, K.A. Forrest, B. Space, S. Horike, J.J.P. IV, S. Kitagawa, M.J. Zaworotko, Chem. Sci. 7 (2016) 5470–5476.
[9] R. Pepinsky, Crystal Engineering: A New Concept in Crystallography, American Physical Society, Mexico City, Mexico, 1955.
[10] G.M.J. Schmidt, Pure Appl. Chem. 27 (1971) 647–678.
[11] G.R. Desiraju, Crystal Engineering: The Design of Organic Solids, Elsevier, 1989.
[12] B. Moulton, M.J. Zaworotko, Chem. Rev. 101 (2001) 1629–1658.
[13] J.-M. Lehn, Angew. Chem. Int. Ed. 27 (1988) 89–112.
[14] E.J. Corey, Pure Appl. Chem. 14 (1967) 19–38.
[15] E. Fischer, Ber. Dtsch. Chem. Ges. 27 (1894) 2985–2993.
[16] O. Ermer, J. Am. Chem. Soc. 110 (1988) 3747–3754.
[17] G.R. Desiraju, J.A.R.P. Sarma, J. Chem. Soc., Chem. Commun. (1983) 45–46.
[18] M.C. Etter, G.M. Frankenbach, J. Bernstein, Tetrahedron Lett. 30 (1989) 3617–3620.
[19] G.R. Desiraju, Angew. Chem. Int. Ed. 34 (1995) 2311–2327.
[20] R.D.B. Walsh, M.W. Bradner, S. Fleischman, L.A. Morales, B. Moulton, N. Rodriguez-Hornedo, M.J. Zaworotko, Chem. Commun. (2003) 186–187.
[21] M.C. Etter, Acc. Chem. Res. 23 (1990) 120–126.
[22] J.C. MacDonald, G.M. Whitesides, Chem. Rev. 94 (1994) 2383–2420.
[23] J. Maddox, Nature 335 (1988) 201.
[24] M. Simard, D. Su, J.D. Wuest, J. Am. Chem. Soc. 113 (1991) 4696–4698.
[25] O.M. Yaghi, G. Li, T.L. Groy, J. Chem. Soc., Dalton Trans. (1995) 727–732.
[26] G.B. Gardner, D. Venkataraman, J.S. Moore, S. Lee, Nature 374 (1995) 792.
[27] P.J. Fagan, M.D. Ward, J.C. Calabrese, J. Am. Chem. Soc. 111 (1989) 1698–1719.

[28] T.L. Hennigar, D.C. MacQuarrie, P. Losier, R.D. Rogers, M.J. Zaworotko, Angew. Chem. Int. Ed. Engl. 36 (1997) 972–973.
[29] H. Abourahma, B. Moulton, V. Kravtsov, M.J. Zaworotko, J. Am. Chem. Soc. 124 (2002) 9990–9991.
[30] J.-P. Zhang, X.-C. Huang, X.-M. Chen, Chem. Soc. Rev. 38 (2009) 2385–2396.
[31] H.D. Clarke, K.K. Arora, H. Bass, P. Kavuru, T.T. Ong, T. Pujari, L. Wojtas, M.J. Zaworotko, Cryst. Growth Des. 10 (2010) 2152–2167.
[32] A. Bajpai, H.S. Scott, T. Pham, K.-J. Chen, B. Space, M. Lusi, M.L. Perry, M.J. Zaworotko, IUCrJ 3 (2016) 430–439.
[33] P. Metrangolo, H. Neukirch, T. Pilati, G. Resnati, Acc. Chem. Res. 38 (2005) 386–395.
[34] C.A. Hunter, J.K.M. Sanders, J. Am. Chem. Soc. 112 (1990) 5525–5534.
[35] J.C. Ma, D.A. Dougherty, Chem. Rev. 97 (1997) 1303–1324.
[36] B.L. Schottel, H.T. Chifotides, K.R. Dunbar, Chem. Soc. Rev. 37 (2008) 68–83.
[37] A.F. Wells, Acta Crystallogr. 7 (1954) 535.
[38] A.F. Wells, Acta Crystallogr. 7 (1954) 545.
[39] A.F. Wells, Acta Crystallogr. 7 (1954) 842.
[40] A.F. Wells, Acta Crystallogr. 7 (1954) 849.
[41] M. Sindoro, N. Yanai, A.-Y. Jee, S. Granick, Acc. Chem. Res. 47 (2014) 459–469.
[42] A.M. Spokoyny, D. Kim, A. Sumrein, C.A. Mirkin, Chem. Soc. Rev. 38 (2009) 1218–1227.
[43] J.T.A. Jones, T. Hasell, X. Wu, J. Bacsa, K.E. Jelfs, M. Schmidtmann, S.Y. Chong, D.J. Adams, A. Trewin, F. Schiffman, F. Cora, B. Slater, A. Steiner, G.M. Day, A.I. Cooper, Nature 474 (2011) 367.
[44] A. Pulido, L. Chen, T. Kaczorowski, D. Holden, M.A. Little, S.Y. Chong, B.J. Slater, D.P. McMahon, B. Bonillo, C.J. Stackhouse, A. Stephenson, C.M. Kane, R. Clowes, T. Hasell, A.I. Cooper, G.M. Day, Nature 543 (2017) 657.
[45] F.-X. Coudert, M. Jeffroy, A.H. Fuchs, A. Boutin, C. Mellot-Draznieks, J. Am. Chem. Soc. 130 (2008) 14294–14302.
[46] O. Almarsson, M.L. Peterson, M.J. Zaworotko, Pharm. Pat. Anal. 1 (2012) 313–327.
[47] D. O'Nolan, M.L. Perry, M.J. Zaworotko, Cryst. Growth Des. 16 (2016) 2211–2217.
[48] O.N. Kavanaugh, D.M. Croker, G.M. Walker, M.J. Zaworotko, Drug Discov. Today 24 (2019) 796–804.
[49] S.R. Batten, N.R. Champness, X.-M. Chen, J. Garcia-Martinez, S. Kitagawa, L. Ohrstrom, M. O'Keeffe, M.P. Suh, J. Reedijk, Pure Appl. Chem. 85 (2013) 1715–1724.
[50] Nobel Lectures, Chemistry 1901-1921, Elsevier Publishing Company, Amsterdam, 1966.
[51] J.C. Bailar, Prep. Inorg. React. 1 (1964) 1.
[52] H.G. Voelz, J. Kischkewitz, P. Woditsch, A. Westerhaus, W.-D. Griebler, M.D. Liedekerke, G. Buxbaum, H. Printzen, M. Mansmann, D. Raede, G. Trenczek, V. Wilhelm, S. Schwarz, H. Wienand, J. Adel, G. Adrian, K. Brandt, W.B. Cork, H. Winkeler, W. Mayer, K. Schneider, L. Leitner, H. Katherin, E. Schwab, H. Jakusch, M. Ohlinger, R. Veitch, G. Etzrodt, G. Pfaff, K.-D. Franz, R. Emmert, K. Nitta, R. Besold, H. Gaedcke, Pigments, Inorganic. Ullmann's Encyclopedia of Industrial Chemistry, Wiley-VCH, Weinheim, Germany, 2006.
[53] I. Huskic, I.V. Pekov, S.V. Krivovichev, T. Friscic, Sci. Adv. 2 (2016) e1600621.
[54] H.M. Powell, R.H. Rayner, Nature 163 (1949) 566.

[55] S.-I. Nishikiori, T. Iwamoto, J. Inclusion Phenom. 2 (1984) 341–349.
[56] Y. Kinoshita, I. Matsubara, T. Higuchi, Y. Saito, Bull. Chem. Soc. Jpn. 32 (1959) 1221–1226.
[57] A. Aumüller, P. Erk, G. Klebe, S. Hünig, J.U. von Schütz, H.-P. Werner, Angew. Chem. Int. Ed. in Engl. 25 (1991) 740–741.
[58] B.F. Hoskins, R. Robson, J. Am. Chem. Soc. 111 (1989) 5962–5964.
[59] B.F. Hoskins, R. Robson, J. Am. Chem. Soc. 112 (1990) 1546–1554.
[60] R.W. Gable, B.F. Hoskins, R. Robson, J. Chem. Soc., Chem. Commun. 0 (1990) 1677–1678.
[61] H.-C. Zhou, S. Kitagawa, Chem. Soc. Rev. 43 (2014) 5415–5418.
[62] J.R. Long, O.M. Yaghi, Chem. Soc. Rev. 38 (2009) 1213–1214.
[63] H.-C. Zhou, J.R. Long, O.M. Yaghi, Chem. Rev. 112 (2012) 673–674.
[64] M. Schroeder, Functional Metal-Organic Frameworks: Gas Storage, Separation and Catalysis, Springer, Berlin, 2010.
[65] D. Farrusseng, Metal-Organic Frameworks: Applications from Catalysis to Gas Storage, Wiley-VCH, Weinheim, 2011.
[66] L.R. MacGillivray, Metal-Organic Frameworks: Design and Application, Wiley-VCH, Weinheim, 2010.
[67] S.R. Batten, S.M. Neville, D.R. Turner, Coordination Polymers: Design, Analysis and Application, The Royal Society of Chemistry, Cambridge, 2009.
[68] S.R. Batten, R. Robson, Angew. Chem. Int. Ed. 37 (1998) 1460–1494.
[69] M. Fujita, Y.J. Kwon, S. Washizu, K. Ogura, J. Am. Chem. Soc. 116 (1994) 1151–1152.
[70] S.R. Batten, CrystEngComm 3 (2001) 67–72.
[71] M. Kondo, T. Yoshitomi, K. Seki, H. Matsuzaka, S. Kitagawa, Angew. Chem. Int. Ed. Engl. 36 (1997) 1725–1727.
[72] H. Li, M. Eddaoudi, T.L. Groy, O.M. Yaghi, J. Am. Chem. Soc. 120 (1998) 8571–8572.
[73] S.S.Y. Chui, S.M.-F. Lo, J.P.H. Charmant, A.G. Orpen, I.D. Williams, Science 283 (1999) 1148–1150.
[74] H. Li, M. Eddaoudi, O.M. Yaghi, Nature 402 (1999) 276–279.
[75] O.K. Farha, I. Eryazici, N.C. Jeong, B.G. Hauser, C.E. Wilmer, A.A. Sarjeant, R.Q. Snurr, S.T. Nguyen, A.O. Yazaydin, J.T. Hupp, J. Am. Chem. Soc. 134 (2012) 15016–15021.
[76] I.M. Hoenicke, I. Senkovska, V. Bon, I.A. Baburin, N. Boenisch, S. Raschke, J.D. Evans, S. Kaskel, Angew. Chem. Int. Ed. 57 (2018) 13780–13783.
[77] A.K. Cheetham, C.N.R. Rao, R.K. Feller, Chem. Commun. 0 (2006) 4780–4795.
[78] P. Groth, Chemische Krystallographie, Wilhelm Engelmann, Leipzig, Germany, 1906.
[79] Chem. Eng. News Arch. 36 (1958) 911–995.
[80] C.R. Groom, I.J. Bruno, M.P. Lightfoot, S.C. Ward, Acta Crystallogr. B72 (2016) 171–179.
[81] O. Kennard, in: I. Butterworth (Ed.), The Impact of Electronic Publishing on the Academic Community, Portland Press Ltd, London, 1997, pp. 159–166.
[82] W.D.S. Motherwell, Cryst. Rev. 14 (2008) 97–116.
[83] R.G. Pearson, J. Chem. Educ. 45 (1968) 581.
[84] F.A. Cotton, G. Wilkinson, C.A. Murillo, M. Bochmann, Advanced Inorganic Chemistry, sixth ed., Wiley-Interscience, New York, 1999.
[85] A. Rodger, B.F.G. Johnson, Inorg. Chem. 27 (1988) 3061–3062.

[86] B. Moulton, J. Lu, A. Mondal, M.J. Zaworotko, Chem. Commun. 0 (2001) 863–864.
[87] M. Eddaoudi, J. Kim, J.B. Wachter, H.K. Chae, M. O'Keeffe, O.M. Yaghi, J. Am. Chem. Soc. 123 (2001) 4368–4369.
[88] S.A. Bourne, J. Lu, A. Mondal, B. Moulton, M.J. Zaworotko, Angew. Chem. Int. Ed. 40 (2001) 2111–2113.
[89] B. Moulton, J. Lu, R. Hajndl, S. Hariharan, M.J. Zaworotko, Angew. Chem. Int. Ed. 41 (2002) 2821–2824.
[90] M. Eddaoudi, J. Kim, M. O'Keeffe, O.M. Yaghi, J. Am. Chem. Soc. 124 (2002) 376–377.
[91] X. Lin, I. Telepeni, A.J. Blake, A. Dailly, C.M. Brown, J.M. Simmons, M. Zoppi, G.S. Walker, K.M. Thomas, T.J. Mays, P. Hubberstey, N.R. Champness, M. Schroeder, J. Am. Chem. Soc. 131 (2009) 2159–2171.
[92] D. Zhao, D. Yuan, D. Sun, H.-C. Zhou, J. Am. Chem. Soc. 131 (2009) 9186–9188.
[93] T. Pham, K.A. Forrest, K. MacDonald, B. Space, Cryst. Growth Des. 14 (2014) 5599–5607.
[94] K. Hirozo, S. Yoshihiko, Bull. Chem. Soc. Jpn. 27 (1954) 112–114.
[95] W.H. Bragg, Nature 111 (1923) 532.
[96] M. Eddaoudi, J. Kim, N. Rosi, D. Vodak, J. Wachter, M. O'Keeffe, O.M. Yaghi, Science 295 (2002) 469–472.
[97] O.M. Yaghi, M. O'Keeffe, Nature 423 (2003) 705–714.
[98] R. Weinland, P. Dinkelacker, Ber. Dtsch. Chem. Ges. 42 (1909) 2997–3018.
[99] G. Ferey, C. Serre, C. Mellot-Draznieks, F. Millange, S. Surblé, J. Dutour, I. Margiolaki, Angew. Chem. Int. Ed. 43 (2004) 6296.
[100] P. Horcajada, S. Surblé, C. Serre, D.-Y. Hong, Y.-K. Seo, J.-S. Chang, J.-M. Greneche, I. Margiolaki, G. Ferey, Chem. Commun. 0 (2007) 2820–2822.
[101] G. Ferey, C. Mellot-Draznieks, C. Serre, F. Millange, J. Dutour, S. Surblé, I. Margiolaki, Science 309 (2005) 2040–2042.
[102] F. Nouar, J.F. Eubank, T. Bousquet, L. Wojtas, M.J. Zaworotko, M. Eddaoudi, J. Am. Chem. Soc. 130 (2008) 1833–1835.
[103] Q.-G. Zhai, C.-Z. Lu, S.-M. Chen, X.-J. Xu, W.-B. Yang, Cryst. Growth Des. 6 (2006) 1393–1398.
[104] M.J. Zaworotko, Chem. Soc. Rev. 23 (1994) 283–288.
[105] S.K. Elsaidi, M.H. Mohamed, L. Wojtas, A. Chanthapally, T. Pham, B. Space, J.J. Vittal, M.J. Zaworotko, J. Am. Chem. Soc. 136 (2014) 5072–5077.
[106] K.S. Park, Z. Ni, A.P. Cote, J.Y. Choi, R. Huang, F.J. Uribe-Romo, H.K. Chae, M. O'Keeffe, O.M. Yaghi, Proc. Natl. Acad. Sci. U.S.A. 103 (2006) 10186–10191.
[107] Y. Liu, V.C. Kravtsov, R. Larsen, M. Eddaoudi, Chem. Commun. 0 (2006) 1488–1490.
[108] X.-C. Huang, Y.-Y. Lin, J.-P. Zhang, X.-M. Chen, Angew. Chem. Int. Ed. 118 (2006) 1587–1589.
[109] C. Heering, I. Boldog, V. Vasylyeva, J. Sanchiz, C. Janiak, CrystEngComm 15 (2013) 9757–9768.
[110] V.N. Dorofeeva, S.V. Kolotilov, M.A. Kiskin, R.A. Polunin, Z.V. Dobrokhotova, O. Cador, S. Golhen, L. Ouahab, I.L. Eremenko, V.M. Novotortsev, Chem. Eur. J. 18 (2012) 5006–5012.
[111] R.A. Polunin, S.V. Kolotilov, M.A. Kiskin, O. Cador, S. Golhen, O.V. Shvets, L. Ouahab, Z.V. Dobrokhotova, V.I. Ovcharenko, I.L. Eremenko, V.M. Novotortsev, V.V. Pavlishchuk, Eur. J. Inorg. Chem. 2011 (2011) 4985–4992.

[112] D.N. Dybstev, H. Chun, K. Kim, Angew. Chem. Int. Ed. 43 (2004) 5033–5036.
[113] S.W. Lee, H.-J. Kim, Y.K. Lee, K. Park, J.-H. Son, Y.-U. Kwon, Inorg. Chim. Acta. 353 (2003) 151.
[114] H. Jasuja, K.S. Walton, Dalton Trans. 42 (2013) 15421–15426.
[115] H.S. Quah, W. Chen, M.K. Schreyer, H. Yang, M.W. Wong, W. Ji, J.J. Vittal, Nat. Commun. 6 (2015) 7954.
[116] S. Subramanian, M.J. Zaworotko, Angew. Chem. Int. Ed. Engl. 34 (1995) 2127–2129.
[117] M.H. Mohamed, S.K. Elsaidi, L. Wojtas, T. Pham, K.A. Forrest, B. Tudor, B. Space, M.J. Zaworotko, J. Am. Chem. Soc. 134 (2012) 19556–19559.
[118] T. Aharen, F. Habib, I. Korobkov, T.J. Burchell, R. Guillet-Nicolas, F. Kleiz, M. Murugesu, Dalton Trans. 42 (2013) 7795–7802.
[119] A.W. Augustyniak, M. Fandzloch, M. Domingo, I. Lakomska, J.A.R. Navarro, Chem. Commun. 51 (2015) 14724–14727.
[120] J.J. Perry IV, J.A. Perman, M.J. Zaworotko, Chem. Soc. Rev. 38 (2009) 1400–1417.
[121] B.F. Abrahams, M.G. Haywood, R. Robson, Chem. Commun. 0 (2004).
[122] Y.-X. Tan, Y.-P. He, J. Zhang, Chem. Commun. 47 (2011) 10647–10649.
[123] D.J. Xiao, M.I. Gonzalez, L.E. Darago, K.D. Vogiatzis, E. Haldoupis, L. Gagliardi, J.R. Long, J. Am. Chem. Soc. 138 (2016) 7161–7170.
[124] Z. Guo, D. Yan, H. Wang, D. Tesfgaber, X. Li, Y. Chen, W. Huang, B. Chen, Inorg. Chem. 54 (2015) 200–204.
[125] K. Sumida, S. Horike, S.S. Kaye, Z.R. Herm, W.L. Queen, C.M. Brown, F. Grandjean, G.J. Long, A. Dailly, J.R. Long, Chem. Sci. 1 (2010) 184–191.
[126] Y.-B. Zhang, H.-L. Zhou, R.-B. Lin, C. Zhang, J.-B. Lin, J.-P. Zhang, X.-M. Chen, Nat. Commun. 3 (2012) 642.
[127] J.H. Cavka, S. Jakobsen, U. Olsbye, N. Guillou, C. Lamberti, S. Bordiga, K.P. Lillerud, J. Am. Chem. Soc. 130 (2008) 13850–13851.
[128] J.E. Mondloch, M.J. Katz, W.C. Isley III, P. Ghosh, P. Liao, W. Bury, G.W. Wagner, M.G. Hall, J.B. DeCoste, G.W. Peterson, R.Q. Snurr, C.J. Cramer, J.T. Hupp, O.K. Farha, Nat. Mater. 14 (2015) 512–516.
[129] D. Feng, Z.-Y. Gu, J.-R. Li, H.-L. Jiang, Z. Wei, H.-C. Zhou, Angew. Chem. Int. Ed. 51 (2012) 10307–10310.
[130] V. Bon, I. Senkovska, I.A. Baburin, S. Kaskel, Cryst. Growth Des. 13 (2013) 1231–1237.
[131] D. Feng, W.-C. Chung, Z. Wei, Z.-Y. Gu, H.-L. Jiang, Y.-P. Chen, D.J. Darensbourg, H.-C. Zhou, J. Am. Chem. Soc. 135 (2013) 17105–17110.
[132] H. Furukawa, F. Gandara, Y.-B. Zhang, J. Jiang, W.L. Queen, M.R. Hudson, O.M. Yaghi, J. Am. Chem. Soc. 136 (2014) 4369–4381.
[133] J.-R. Li, D.J. Timmons, H.-C. Zhou, J. Am. Chem. Soc. 131 (2009) 6368–6369.
[134] G. Ferey, M. Latroche, C. Serre, F. Millange, T. Loiseau, A. Percheron-Guegan, Chem. Commun. 0 (2003) 2976–2977.
[135] P.D.C. Dietzel, B. Panella, M. Hirscher, R. Blom, H. Fjellvag, Chem. Commun. 0 (2006) 959–961.
[136] A. Schoedel, L. Wojtas, S.P. Kelley, R.D. Rogers, M. Eddaoudi, M.J. Zaworotko, Angew. Chem. Int. Ed. 50 (2011) 11421–11424.
[137] A.P. Cote, A.I. Benin, N.W. Ockwig, M. O'Keeffe, A.J. Matzger, O.M. Yaghi, Science 310 (2005) 1166–1170.
[138] Z. Wang, S.M. Cohen, Chem. Soc. Rev. 38 (2009) 1315–1329.
[139] K.K. Tanabe, S.M. Cohen, Chem. Soc. Rev. 40 (2011) 498–519.

[140] S.M. Cohen, Chem. Rev. 112 (2012) 970–1000.
[141] N.W. Ockwig, O. Delgado-Friedrichs, M. O'Keeffe, O.M. Yaghi, Acc. Chem. Res. 38 (2005) 176–182.
[142] O. Delgado-Friedrichs, M. O'Keeffe, O.M. Yaghi, Acta Crystallogr. A62 (2006) 350–355.
[143] O. Delgado-Friedrichs, M. O'Keeffe, O.M. Yaghi, Acta Crystallogr. A63 (2007) 344–347.
[144] O. Delgado-Friedrichs, M. O'Keeffe, O.M. Yaghi, Phys. Chem. Chem. Phys. 9 (2007) 1035–1043.
[145] M. O'Keeffe, Acta Crystallogr. A64 (2008) 425–429.
[146] V.A. Blatov, M. O'Keeffe, D.M. Proserpio, CrystEngComm 12 (2010) 44–48.
[147] V.A. Blatov, A.P. Shevchenko, D.M. Proserpio, Cryst. Growth Des. 14 (2014) 3576–3586.
[148] V.A. Blatov, L. Carlucci, G.I. Ciani, D.M. Proserpio, CrystEngComm 6 (2004) 377–395.
[149] L. Carlucci, G. Ciani, D.M. Proserpio, Coord. Chem. Rev. 246 (2003) 247–289.
[150] M. O'Keeffe, M.A. Peskov, S.J. Ramsden, O.M. Yaghi, Acc. Chem. Res. 41 (2008) 1782–1789.
[151] A. Karmakar, A. Paul, A.J.L. Pombeiro, CrystEngComm 19 (2017) 4666–4695.
[152] T.A. Makal, A.A. Yakovenko, H.-C. Zhou, J. Phys. Chem. Lett. 2 (2011) 1682–1689.
[153] G. Yuan, C. Zhu, Y. Liu, W. Xuan, Y. Cui, J. Am. Chem. Soc. 131 (2009) 10452–10460.
[154] E.V. Alexandrov, V.A. Blatov, D.M. Proserpio, CrystEngComm 19 (2017).
[155] P. Nugent, Y. Belmabkhout, S.D. Burd, A.J. Cairns, R. Luebke, K. Forrest, T. Pham, S. Ma, B. Space, L. Wojtas, M. Eddaoudi, M.J. Zaworotko, Nature (2013) 495.
[156] K. Biradha, A. Mondal, B. Moulton, M.J. Zaworotko, J. Chem. Soc., Dalton Trans. 0 (2000) 3837–3844.
[157] S. Yang, X. Lin, W. Lewis, M. Suyetin, E. Bichoutskaia, J.E. Parker, C.C. Tang, D.R. Allan, P.J. Riskallah, P. Hubbertey, N.R. Champness, K.M. Thomas, A.J. Blake, M. Schroeder, Nat. Mater. 11 (2012) 710–716.
[158] A. Ferguson, L. Liu, S.J. Tapperwijn, D. Perl, F.-X. Coudert, S.V. Cleuvenbergen, T. Verbiest, M.A. van der Veen, S.G. Telfer, Nat. Chem. 8 (2016) 250–257.
[159] G. Verma, S. Kumar, T. Pham, Z. Niu, L. Wojtas, J.A. Perman, Y.-S. Chen, S. Ma, Cryst. Growth Des. 17 (2017) 2711–2717.
[160] K.S. Chichak, S.J. Cantrill, A.R. Pease, S.-H. Chiu, G.W.V. Cave, J.L. Atwood, J.F. Stoddart, Science 304 (2004) 1308–1312.
[161] J. Crassous, Chem. Soc. Rev. 38 (2009) 830–845.
[162] S.-Y. Zhang, D. Li, D. Guo, H. Zhang, W. Shi, P. Cheng, L. Wojtas, M.J. Zaworotko, J. Am. Chem. Soc. 137 (2015) 15406–15409.
[163] A. Mackay, Glass Phys. Chem. 4 (2017) 196–206.
[164] S. Kitagawa, R. Kitaura, S.-I. Noro, Angew. Chem. Int. Ed. 43 (2004) 2334–2375.
[165] S. Horike, S. Shimomura, S. Kitagawa, Nat. Chem. 1 (2009) 695–704.
[166] N.C. Burtch, H. Jasuja, K.S. Walton, Chem. Rev. 114 (2014) 10575–10612.
[167] K. Leus, T. Bogaerts, J.D. Decker, H. Depauw, K. Hendrickx, H. Vrielinck, V.V. Speybroeck, P.V.D. Voort, Microporous Mesoporous Mater. 225 (2016) 110–116.
[168] A.J. Howarth, Y. Liu, P. Li, Z. Li, T.C. Wang, J.T. Hupp, O.K. Farha, Nat. Rev. Mater. 1 (2016) 15018.
[169] O.R. Evans, W. Lin, Acc. Chem. Res. 35 (2002) 511–522.

[170] O.R. Evans, W. Lin, Chem. Mater. 13 (2001) 2705–2712.

[171] G.J. Halder, C.J. Kepert, B. Moubaraki, K.S. Murray, J.D. Cashion, Science 298 (2002) 1762–1765.

[172] S. Horike, D. Umeyama, M. Inukai, T. Itakura, S. Kitagawa, J. Am. Chem. Soc. 134 (2012) 7612–7615.

[173] M.D. Allendorf, C.A. Bauer, R.K. Bhakta, R.J.T. Houk, Chem. Soc. Rev. 38 (2009) 1330–1352.

[174] Y. Cui, Y. Yue, G. Qian, B. Chen, Chem. Rev. 112 (2012) 1126–1162.

[175] D. O'Nolan, A. Kumar, M.J. Zaworotko, J. Am. Chem. Soc. 139 (2017) 8508–8513.

[176] P.M. Schoenecker, C.G. Carson, H. Jasuja, C.J.J. Fleming, K.S. Walton, Ind. Eng. Chem. Res. 51 (2012) 6513–6519.

[177] Department of Energy, Fuel Cell Technologies Office, Hydrogen Storage, 2018. https://energy.gov/eere/fuelcells/hydrogen-storage.

[178] N.L. Rosi, J. Eckert, M. Eddaoudi, D.T. Vodak, J. Kim, M. O'Keeffe, O.M. Yaghi, Science 300 (2003) 1127–1129.

[179] Department of Energy, ARPA-E, MOVE (Methane Opportunities for Vehicular Energy), 2018. https://arpa-e.energy.gov/?q=arpa-e-programs/move.

[180] Y. He, W. Zhou, G. Qian, B. Chen, Chem. Soc. Rev. 43 (2014) 5657–5678.

[181] Y. Peng, V. Krungleviciute, I. Eryazici, J.T. Hupp, O.K. Farha, T. Yildirim, J. Am. Chem. Soc. 135 (2013) 11887–11894.

[182] O.K. Farha, A.O. Yazaydin, I. Eryazici, C.D. Malliakas, B.G. Hauser, M.G. Kanatzidis, S.T. Nguyen, R.Q. Snurr, J.T. Hupp, Nat. Chem. 2 (2010) 944–948.

[183] E.D. Bloch, W.L. Queen, R. Krishna, J.M. Zadrozny, C.M. Brown, J.R. Long, Science 335 (2012) 1606–1610.

[184] T. Pham, K.A. Forrest, R. Banerjee, G. Orcajo, J. Eckert, B. Space, J. Phys. Chem. C 119 (2015) 1078–1090.

[185] A. Kumar, D.G. Madden, M. Lusi, K.-J. Chen, E.A. Daniels, T. Curtin, J.J. Perry IV, M.J. Zaworotko, Angew. Chem. Int. Ed. 54 (2015) 14372–14377.

[186] T.M. McDonald, W.R. Lee, J.A. Mason, B.M. Wiers, C.S. Hong, J.R. Long, J. Am. Chem. Soc. 134 (2012) 7056–7065.

[187] Z. Qiao, N. Wang, J. Jiang, J. Zhou, Chem. Commun. 52 (2015) 974–977.

[188] Y.-S. Wei, M. Zhang, P.-Q. Liao, R.-B. Lin, T.-Y. Li, G. Shao, J.-P. Zhang, X.-M. Chen, Nat. Commun. 6 (2015) 8348.

[189] J.-S. Seo, D. Whang, H. Lee, S.I. Jun, J. Oh, Y.J. Jeon, K. Kim, Nature 404 (2000) 982–986.

[190] A.M. Schultz, O.K. Farha, J.T. Hupp, S.T. Nguyen, J. Am. Chem. Soc. 131 (2009) 4204–4205.

[191] M.H. Alkordi, Y. Liu, R.W. Larsen, J.F. Eubank, M. Eddaoudi, J. Am. Chem. Soc. 130 (2008) 12639–12641.

[192] Y. Inokuma, T. Arai, M. Fujita, Nat. Chem. 2 (2010) 780–783.

[193] M. Hoshino, A. Khutia, H. Xing, Y. Inokuma, M. Fujita, IUCrJ 3 (2016) 139–151.

[194] Y. Matsuda, T. Mitsuhashi, S. Lee, M. Hoshino, T. Mori, M. Okada, H. Zhang, F. Hayashi, M. Fujita, I. Abe, Angew. Chem. Int. Ed. 55 (2016) 5785–5788.

[195] K. Barthelet, J. Marrot, D. Riou, G. Ferey, Angew. Chem. Int. Ed. 41 (2002) 281–284.

[196] T. Loiseau, C. Serre, C. Huguenard, G. Fink, F. Taulelle, M. Henry, T. Bataille, G. Ferey, Chem. Eur. J. 10 (2004) 1373–1382.

[197] C. Mellot-Draznieks, C. Serre, S. Surblé, N. Audebrand, G. Ferey, J. Am. Chem. Soc. 127 (2005) 16273–16278.

[198] D. Fairen-Jimenez, S.A. Moggach, M.T. Wharmby, P.A. Wright, S. Parsons, T. Duren, J. Am. Chem. Soc. 133 (2011) 8900–8902.
[199] S. Krause, V. Bon, I. Senkovska, U. Stoeck, D. Wallacher, D.M. Toebbens, S. Zander, R.S. Pillai, G. Maurin, F.-X. Coudert, S. Kaskel, Nature 532 (2016) 348–352.
[200] J.D. Evans, L. Bocquet, F.-X. Coudert, Inside Chem. 1 (2016) 873–886.
[201] T. Kumar-Maji, R. Matsuda, S. Kitagawa, Nat. Mater. 6 (2007) 142–148.
[202] A. Kondo, H. Noguchi, S. Ohnishi, H. Kajiro, A. Tohdoh, Y. Hattori, W.-C. Xu, H. Tanaka, H. Kanoh, K. Kaneko, Nano Lett. 6 (2006) 2581–2584.
[203] D.A. Gómez-Gualdrón, Y.J. Colón, X. Zhang, T.C. Wang, Y.-S. Chen, J.T. Hupp, T. Yildrim, O.K. Farha, J. Zhang, R.Q. Snurr, Energy Environ. Sci. 9 (2016) 3279–3289.
[204] C.M. Simon, J. Kim, D.A. Gomez-Gualdron, J.S. Camp, Y.G. Chung, R.L. Martin, R. Mercado, M.W. Deem, D. Gunter, M. Haranczyk, D.S. Sholl, R.Q. Snurr, B. Smit, Energy Environ. Sci. 8 (2015) 1190–1199.
[205] J.A. Mason, J. Oktawiec, M.K. Taylor, M.R. Hudson, J. Rodriguez, J.E. Bachman, M.I. Gonzalez, A. Cervellino, A. Guagliardi, C.M. Brown, P.L. Llewellyn, N. Masciocchi, J.R. Long, Nature 527 (2015) 357–361.
[206] Q.Y. Yang, P. Lama, S. Sen, M. Lusi, K.J. Chen, W.Y. Gao, M. Shivanna, T. Pham, N. Hosono, S. Kusaka, J.J. Perry IV, S. Ma, B. Space, L.J. Barbour, S. Kitagawa, M.J. Zaworotko, Angew. Chem. Int. Ed. 57 (2018) 5684–5689.
[207] A.X. Zhu, Q.Y. Yang, A. Kumar, C. Crowley, S. Mukherjee, K.J. Chen, S.Q. Wang, D. O'Nolan, M. Shivanna, M.J. Zaworotko, J. Am. Chem. Soc. 140 (2018) 15572–15576.
[208] P. Lama, H. Aggarwal, C.X. Bezuidenhout, L.J. Barbour, Angew. Chem. Int. Ed. 55 (2016) 13271–13275.
[209] S.-Y. Zhang, S. Jensen, K. Tan, L. Wojtas, M. Roveto, J. Cure, T. Thonhauser, Y.J. Chabal, M.J. Zaworotko, J. Am. Chem. Soc. 140 (2018) 12545–12552.
[210] T.D. Bennett, A.K. Cheetham, A.H. Fuchs, F.X. Coudert, Nat. Chem. 9 (2016) 11.
[211] P. Deria, J.E. Mondloch, O. Karagiaridi, W. Bury, J.T. Hupp, O.K. Farha, Chem. Soc. Rev. 43 (2014) 5896–5912.
[212] M. Lestari, M. Lusi, A. O'Leary, D. O'Nolan, M.J. Zaworotko, CrystEngComm 20 (2018) 5940–5944.
[213] S.-Y. Zhang, L. Wojtas, M.J. Zaworotko, J. Am. Chem. Soc. 137 (2015) 12045–12049.

CHAPTER

3

From mechanical properties to mechanochemistry of molecular crystals: role of nanointendation and AFM indentation techniques in crystal engineering

Kashyap Kumar Sarmah[1,2], Ranjit Thakuria[1], C. Malla Reddy[3]

[1]*Department of Chemistry, Gauhati University, Guwahati, Assam, India;* [2]*Department of Chemistry, Behali Degree College, Borgang, Biswanath, Assam, India;* [3]*Department of Chemical Science, Indian Institute of Science Education and Research (IISER), Mohanpur, Kolkata, India*

1. Introduction

Mechanical property is a fundamental property of materials. Study of mechanical property in molecular solids has gained much attention in the recent years, due to their applications in pharmaceuticals, flexible electronics, waveguides, piezo-/ferro-electrics, soft robotics, mechanical actuators, plastic electrolytes, and biomedical devices [1–3]. Mechanical behavior of molecular solids is significantly different from that of the extended solids due to the much weaker nature of their noncovalent interactions such as hydrogen bonds, halogen bonds, van der Waals interactions, and so on. These weak interactions can make the ordered molecular solids soft, mechanically flexible, structurally anisotropic, and labile with high surface reactivity. For instance, Desiraju and Reddy reported the exceptional plasticity in a series of organic crystals, which led to further exploration of various classes of such soft crystals. Reddy and Ghosh reported the exceptional elastic flexibility in caffeine cocrystals. Naumov's group studied the thermosalient and photosalient behaviors, as well as various other mechanical effects in organic crystals. All these fascinating examples demonstrate the enormous importance of intermolecular interactions and their directionality in crystal structures. Hence, crystal engineering provides a great opportunity to tune the properties of organic solids, creating new opportunities in various branches of materials science. Hence, attempts have been made to enhance the understanding of structure–mechanical property correlation using various experimental and computational techniques.

Hot Topics in Crystal Engineering. https://doi.org/10.1016/B978-0-12-818192-8.00005-6

Instrumented nanoindentation and atomic force microscopy (AFM) experimental techniques have gained prominence in recent years for not only probing the mechanical anisotropy but also for understanding the role of molecular (size and shape) and supramolecular factors (nature and directionality of noncovalent interactions, and arrangement of constituents) in molecular crystals. A deeper understanding of mechanical properties of molecular crystals is critical for assessing the stability of different pharmaceutical solid forms as well as various mechanochemical processes such as mechanical milling, grinding, micronization, compaction, granulation, flowability, and so on. Mechanically durable and adaptable materials are also critical for future smart materials for foldable and wearable miniature devices. Hence, it is important to gain molecular-level insights into the structure and mechanical properties of organic materials. Here in this chapter, we cover recent progress on utilization of the experimental techniques, nanoindentation and AFM, for understanding the mechanical properties of organic crystals. There have been excellent reviews in recent years on the specialized topics within a broader perspective of mechanical properties of molecular materials [3–8]. The work covered here aims to provide insights into the mechanochemistry of organic crystals and is presented from the own perspective of the authors. The literature covered here is limited; purely the choice of the authors is by no means exhaustive.

One of the major topics in mechanochemistry is the reactivity of solids. Mechanical behavior of molecular crystals plays a critical role in determining the outcome in mechanochemical processes. While the studies correlating mechanical properties with the reactivity of solids have been a topic of considerable interest for long time, systematic studies are very limited in the literature. It has been shown that high molecular mobility is the basis for the observation of solid-state reactivity as well as the facile mechanical deformation. Kaupp et al. in their seminal nanoscratching work using AFM of organic molecular solids have demonstrated the correlation between mechanical pressure inducing molecular movements and the solid-state reactivity [9]. These nanoscratching experiments give enough evidence to the existence of surface molecular mobility that results high solid-state reactivity of molecular solids. On the other hand, 3D interlocked structures do not undergo solid-state reactions easily. Based on the experimental results, Kaupp proposed a three-stage model to describe mechanochemical transformation of organic molecules as activation of solid reactants, their reaction, and crystallization of the product [10–12]. In this three-step process, mass transfer can proceed either via a gas, a liquid, a solid phase, or any combination of them as shown in Fig. 3.1.

Despite its extensive application, little is known about the molecular level mechanism of mechanochemical reactions. Methodologies have been developed only recently to monitor the progress of the mechanochemical reactions by introduction of ex situ and in situ techniques (synchrotron X-ray power diffraction and Raman spectroscopy), for example, to analyze the coground materials at intermediate intervals [13–18]. Apart from chemical reactions, molecular mobility may also induce polymorphic phase transformation. For example, barbituric acid

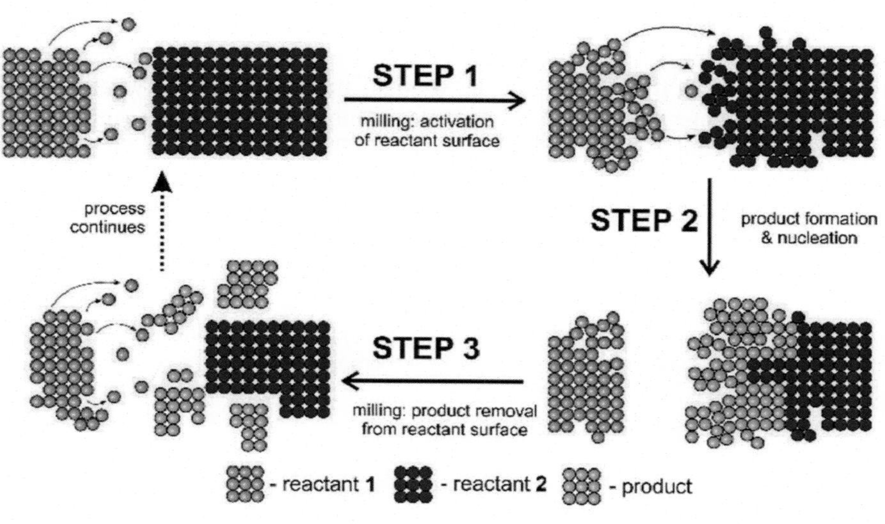

FIGURE 3.1

Schematic representation of three-step mechanism for mechanochemical reaction upon milling or grinding of solids.

Reprinted with permission from Chem. Commun. 52 (2016) 7760–7781. Copyright 2016 Royal Society of Chemistry.

upon neat ball milling undergoes a polymorphic phase transformation from the tri(keto) form to the tri(hydroxyl) tautomer [19]. Similarly, polymorphic phase transformation was also observed for chlorpropamide upon mechanical grinding, as reported by Boldyreva's group [20]. Along with mechanical pressure, amount of material in the jar [21] and polarity of the liquid used during milling (liquid-assisted grinding, LAG) may also result change in crystal packing, for example, anthranilic acid [22], caffeine–anthranilic acid cocrystal [23], caffeine–glutaric acid cocrystal [24], caffeine–citric acid cocrystal [25–27], ethenzamide–gentisic acid cocrystal [28], pyrazinamide–pimelic acid cocrystal [29], isoniazid–DL tartaric acid [30], and phenazine–mesaconic acid system [31]. The difference in mechanical pressure induced during ball milling may sometimes result different products and reaction kinetics in the same milling vessel as reported by Boldyreva and coworkers [32,33]. In a very recent report, Germann et al. observed a very unusual phenomenon, where changes to the choice of milling media (i.e., number and material of milling balls) and/or the choice of milling assembly (i.e., jar material) can be used to control polymorphism of mechanochemical cocrystallization. Considering cocrystallization of nicotinamide and adipic acid by milling under stainless steel (ss) versus poly(methyl methacrylate) (PMMA) grinding jar and/or ø 7 mm ss versus ø 10 mm zirconia balls, they could synthesize two polymorphic cocrystals of the same. Difference in weight of the milling jars and balls that lead to different energy input could be the probable reason for the polymorphic outcome in closely related enantiotropic system [34]. Apart from molecular movement, the presence of crystal defects also induces change in mechanical properties of molecular solids. Recently, Naumov et al. showed that even mechanical bending can induce phase transformation in organic crystals [35]. Reddy et al. showed that exceptionally soft plastic crystals (orientationally disordered crystals or rotator phases) transform partially to another phase upon mechanical thinning, compression, or elongation of single crystals [36]. Takamizawa and coworkers showed ferroelasticity in molecular crystals, in which the mechanical stress interconverts two polymorphic forms of a compound [37–39]. These examples indicate that the mechanically soft molecular crystals may undergo phase transformation under mechanochemical conditions.

Mechanical properties of organic crystals can be influenced by other factors as well. For instance, Ramos et al. using indentation measurements showed how the onset of plasticity of an RDX single crystal changes with washing and mechanical cleavage during sample preparation [40]. In another report, Liu et al. showed that ~0.01%–0.1% doping of uric acid crystal with dye additives decreases elastic modulus by a factor of ~25%–50% [41].

Identifying possible metastable polymorphs that do not crystallize under ambient conditions has been done efficiently in recent years by crystal structure prediction. Aspirin is one of such examples, where prior to discovery of the metastable form II, Price and coworkers predicted the possibility of existence of additional polymorphic phase based on gas-phase conformers and ab initio calculations [42]. In a later stage, form II was found to be mechanically unstable and has been shown to convert to form I under mechanical pressure. In another report, Beyer et al. predicted the crystal structure of another metastable form of paracetamol with better tabletting properties than the existing stable polymorphic form [43]. With progression of the field, a large number of case studies have been reported where better correlation of mechanical property to crystal packing has been established. For example, indomethacin exists as α and γ forms with variable compressibility, tabletability, and compactibility profile based on their difference in molecular packing. The presence of slip planes in γ form offered it increased compressibility and deformation behavior, whereas α form showed better compactibility due to higher crystal packing density [44]. Stevens and coworkers reported enhancement in tabletability of *p*-aminobenzoic acid with incorporation of alkyl chains for a series of its benzoate ester based on molecular modification [45]. The very first example of a cocrystal system that showed better powder compaction properties was caffeine–methyl gallate cocrystal reported by Sun and Hou [46]. The presence of molecularly flat slip planes in the cocrystal is the primary reason for having ~8-fold tablet tensile strength at 350 MPa compared with pure caffeine. In the following sections, we will discuss about the various mechanical properties exhibited by molecular organic solids and their correlation with the structural packing.

2. Mechanical properties of organic crystals

Every material shows initial elastic response followed by plastic or brittle behavior. A typical stress–strain curve for materials is shown in Fig. 3.2. Reddy and Desiraju categorized molecular crystals into two broad classes, brittle or compliant (shearing and plastic), based on their gross mechanical response. Deformation in complaint crystals can be further classified into reversible (elastic) or irreversible (plastic) [3]. Reversible or elastic crystals can recover their original shape upon removal of the applied force. Plastic crystals acquire permanent deformation (irreversible) beyond the elastic limit. Molecular crystals that show large elastic response (typically with strains exceeding ~2%) are termed as elastic crystals, while those that quickly move into plastic region with negligible elasticity (stresses <1%–2%) are called plastically deformable crystals. Depending on the number of slip or weak interaction planes, molecular crystals may show one-, two-, or three-dimensional (1D, 2D, or 3D, respectively) plastic deformation. Crystals that show 3D plasticity generally belong to orientationally disordered crystals, e.g., those formed by nearly globular molecular systems, known as plastic crystals [36]. The name "*plastic crystals*" comes from their typical exceptional plasticity, which is often comparable with atomic plastic metals (e.g., Cu, Ag, or Au) [47,48]. Criteria for observing 3D plasticity

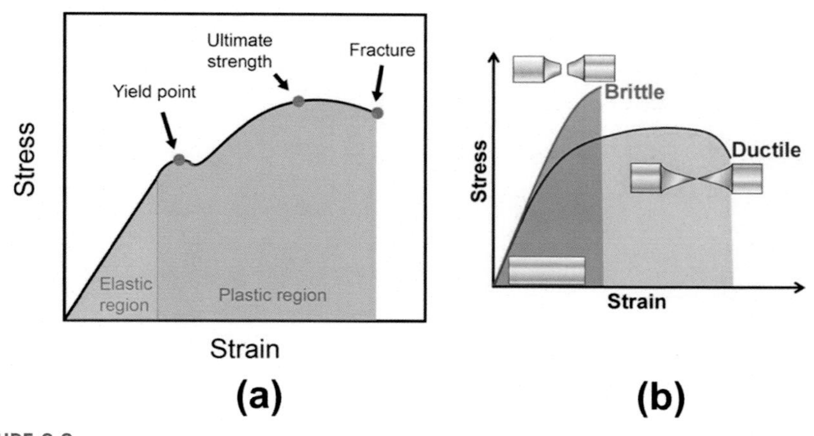

FIGURE 3.2

(A) Typical stress–strain curve of a material showing the elastic and plastic regions of deformation. (B) Stress–strain curve of a brittle material. Notice the much steeper slope compared with ductile materials.

(A) Reprinted with permission from Angew. Chem. Int. Ed. 58 (2019) 10052–10060. Copyright 2019 John Wiley and Sons. (B) Reprinted with permission from S. Varughese, M.S.R.N. Kiran, U. Ramamurty, G.R. Desiraju, Nanoindentation in crystal engineering: quantifying mechanical properties of molecular crystals. Angew. Chem. Int. Ed. 52 (2013) 2701–2712. Copyright 2013 John Wiley and Sons.

are not yet fully understood in molecular crystals. Ferroelasticity and superelasticity are special cases of reversible plastic deformation due to phase or gross domain transformation under applied stress. Brittle crystals quickly fracture under mechanical stress after an initial small but steep elastic response.

Molecular-level mechanism in various classes of compliant organic molecular crystals can be understood based on the movement of molecules in response to the external stress (Fig. 3.3). The response is reversible and/or irreversible depending on the energy related to the intermolecular interactions that define the characteristics of the molecular solids.

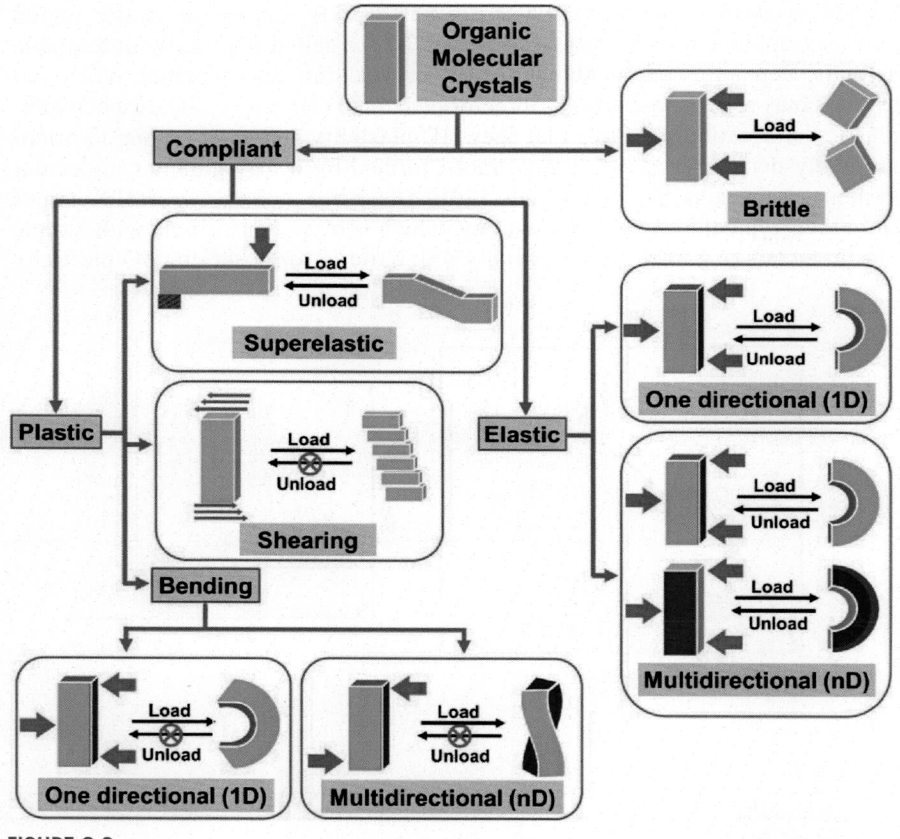

FIGURE 3.3

Classification of soft organic crystals.

Reprinted with permission from S. Saha, M.K. Mishra, C.M. Reddy, G.R. Desiraju, From molecules to interactions to crystal engineering: mechanical properties of organic solids. Acc. Chem. Res. 51 (11) (2018) 2957–2967. Copyright 2018 American Chemical Society.

2.1 Brittle crystals

Most of the organic molecular crystals belong to this category. They break into pieces on application of a mechanical stress in any of the three dimensions. The reason behind this is the internal arrangement of molecules. Structural analysis of most of the brittle crystals reveals that they have isotropic intermolecular interactions in all three dimensions. They typically possess strong hydrogen bond networks or weak but specific interactions with zigzag- or herringbone-type crystal packing. In these crystals, shear stress is greater than fracture stress, and hence, fracture ensues. Due to the isotropic interactions (and absence of facile slip planes), these crystals do not display compliant nature. Desiraju and coworkers carried out a comprehensive study on about 60 examples of organic molecular crystals, out of which 43 systems ($\sim 71\%$) were found to be brittle [4]. Some of the most significant examples considered in the study include anti-TB drug isoniazid and δ-polymorph of pyrazinamide, form I of venlafaxine hydrochloride, benzoic acid, trimesic acid, naphthalene, nicotinamide, etc. Molecular packing of a few brittle crystals along with their optical images is shown in Fig. 3.4.

2.2 Plastic compliant crystals

Plastic materials when subjected to bending, compressive, tensile, torsion, or shear stresses (>yield strength) may undergo bending, compression (or buckling), elongation, twisting, or shearing (via slip or twin) deformation, respectively. Since most plastic compliant organic crystals are anisotropic, all the types of deformations may not be possible in a single material. Crystal structure determines the type of deformation in a material.

2.2.1 Plastic shearing in layered structures

Shearing is generally observed in materials with layered structures. Shearing may occur upon application of a stress parallel to layers, if the fracture stress of a material is greater than its yield shear stress. Plastic shearing was systematically investigated by Desiraju and coworkers in a series of halogenated benzene derivatives [49]. This phenomenon is mostly observed for layered structures with small d-spacing and relatively weak interlayer interactions. 1,3,5-trichloro-2,4,6-triiodobenzene (135C246I) is the best considered example available in the literature with layered packing of molecules in a nearly hexagonal arrangement. Short I···I interactions between I_3 clusters, and loosely packed Cl···Cl between Cl_3 clusters constitute the 2D layers of molecules in 135C246I crystals. Weak, nonspecific interlayer interactions allow sliding of layers one over the other or shear deformation when a pristine crystal is pushed parallel to the layers using a shear force on the crystal. Application of the mechanical force orthogonal to these layers results in fracture through its cleavage planes rather than smooth shear deformation.

Similar to 135C246I, other isostructural halo- and/or methyl derivatives, viz., 135B246I-triclinic, 135I246M, 135B246M, and 135C246M (M = methyl), and their solid solutions also undergo shear deformation. Compounds without layered

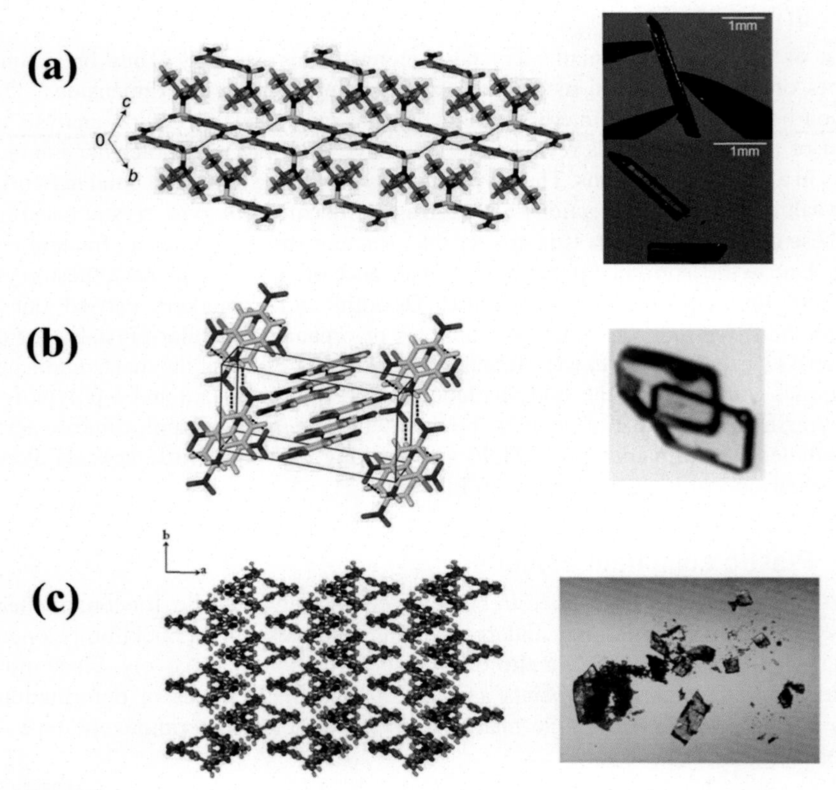

FIGURE 3.4

Molecular packing of a few brittle-type single crystals of (A) probenecid amide. (B) Form III of 6-chloro-2,4-dinitroaniline and (C) 1:1 cocrystal of caffeine·2-fluoro-5-nitrobenzoic acid.

(A) Reprinted with permission from P. Gupta, S.A. Rather, B.K. Saha, T. Panda, D.P. Karothu, N.K. Nath, Mechanical flexibility of molecular crystals achieved by exchanging hydrogen bonding synthons. Cryst. Growth Des. 20 (5) (2020) 2847–2852. Copyright 2020 American Chemical Society; (B) Reprinted with permission from Chem. Commun. (2005) 2439–2441; (C) Reprinted with permission from S. Ghosh, C. Malla Reddy, Cocrystals of caffeine with substituted nitroanilines and nitrobenzoic acids: structure–mechanical property and thermal studies. CrystEngComm 14 (2012) 2444–2453. Copyright 2005, 2012 Royal Society of Chemistry.

structure generally do not show such shearing. Although the flat layer structure is a desirable condition, it is not a sufficient condition. For instance, a large number of layered crystals in literature do not show shear deformation, for example, δ-polymorph of pyrazinamide, 4-hydroxybenzoic acid-monohydrate, 4-nitroacetophenone, and so on [4]. In such crystals, the interlayer interactions are found to be strong, making the sliding of the layers difficult (shear yield stress is greater than fracture stress). Apart from the halogenated series, other examples of organic molecular crystals

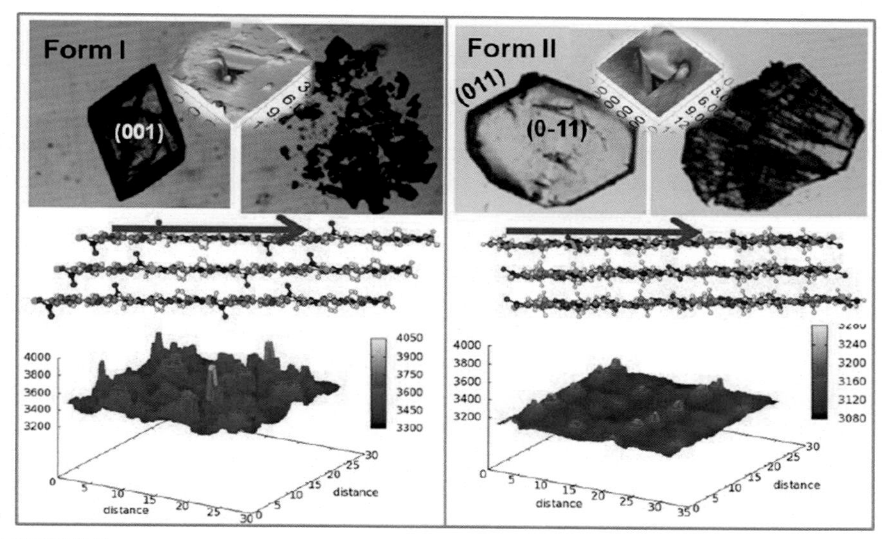

FIGURE 3.5

Brittle and shearing nature of polymorphic form I and form II of caffeine·4-chloro-3-nitrobenzoic acid cocrystal along with their molecular packing and potential energy surfaces (kcal/mol).

Reprinted with permission from S. Ghosh, A. Mondal, M.S.R.N. Kiran, U. Ramamurty, C.M. Reddy, The role of weak interactions in the phase transition and distinct mechanical behavior of two structurally similar caffeine co-crystal polymorphs studied by nanoindentation. Cryst. Growth Des. 13 (10) (2013) 4435–4441. Copyright 2013 American Chemical Society.

showing shear deformation include form I of trimorphic 6-chloro-2,4-dinitroaniline (CDNA) molecule [50], 2D layered cocrystal of CDNA with vanillin isomers [51], 2D layered cocrystals of caffeine with 2,4-dinitrobenzoic acid, 2-chloro-5-nitroaniline and 2-iodo-4-nitroaniline [52], theophylline multicomponent solids with picolinamide, 3,4-dichlorobenzoic acid, 4-chloro-3-nitrobenzoic acid and 4-fluoro-3-nitrobenzoic acid [53], and many more [54,55]. Reddy and coworkers prepared two polymorphic cocrystals of caffeine·4-chloro-3-nitrobenzoic acid (CAF·CNB) (form I and form II) with comparable 2D layered packing. Despite their similar layered structure, form I is brittle, whereas form II shows shearing of molecular layers under mechanical stress (Fig. 3.5) [56]. This is attributed to the smooth potential energy surface of the layers in form II as compared with the rough topology of the same in form I (Fig. 3.5).

2.2.2 Plastic bending

Mechanical plastic bending in molecular crystals was first observed and studied systematically by Desiraju, Reddy and coworkers [4,57]. When a crystal bends, the two parallel opposite faces become nonplanar with irreversible change in shape even on release of the mechanical stress (plastic bending). Crystal structure analysis of hexachlorobenzene (HCB) and many other organic crystals allowed them to propose a general bending model for molecular crystals. According to the Desiraju–Reddy's plastic bending model, the presence of facile slip or weak interaction planes, generally parallel to the length of acicular crystals, and anisotropic distribution of interaction energies in crystal structure facilitate plastic bending [3,57]. It has been observed that the relatively strong and weak interaction patterns are in nearly orthogonal directions in such crystals.

The mechanism of plastic bending can be understood from the crystal structure of 2-(methylthio)nicotinic acid (MTN) [57]. In the MTN crystals, nearly planar dimer consisting two MTN molecules is formed via centrosymmetric carboxylic acid synthon involving O–H···O hydrogen bonds. The dimers stack along the needle direction [100] to form molecular columns, which in other two directions, [101] and [010] are connected only by weak C–H···O and Me···Me interactions, respectively (Fig. 3.6). The weakly interacting Me···Me interactions among the columns act as facile slip planes, allowing slippage of molecules parallel to the columns during bending (Fig. 3.6C and D). A similar molecular packing as well as mechanical property has been observed for other plastically bendable crystals, including HCB crystals. The bending deformation leads to strain in the crystals and formation of defects while the long-range molecular motions have been shown to occur via slip and/or twinning deformation [36,58].

Following these early reports of Desiraju and coworkers, a large number of plastically bendable crystals have been reported in literature [3]. The crystals that show plastic bending on more than one face have also been shown to undergo twisting on application of torsional stress around the crystal length [36,59].

2.2.3 Role of water in plasticity

An example of twistable pharmaceutical crystal, namely caffeine hydrate, was reported by Sun and coworkers [60]. The water molecules form catemeric O–H···O hydrogen-bonded network along the crystal length, which is also the stacking direction. The caffeine hydrate crystal has a two-dimensional hydrogen-bonded network, supported by C–H···O and O–H···O hydrogen bonds with identical crystal packing on two faces. This facilitates two-face bending in the crystals due to slippage of molecules along the needle direction, i.e., parallel to water chain. Here the water molecules play a key role in facilitating the two-face bending and twisting. The presence of multidirectional plasticity (2D plasticity) results exceptional compressibility property or tabletability (tensile strength of 5 MPa) of caffeine hydrate, which is superior compared with other 1D plastic crystals. In a much earlier report, Sun and Grant already discussed the ability of water molecules to act as a "molecular lubricant" in the hydrated *p*-hydroxybenzoic acid crystals [61]. Incorporation of water molecule in the crystal lattice of *p*-hydroxybenzoic acid acts as a space filler, increasing the separation of the molecular layers and easier slip between layers in the hydrated form as compared with the anhydrous crystal structure under the same compaction pressure. A similar structure–property correlation has been observed for uric acid dihydrate that has higher plasticity and compaction property

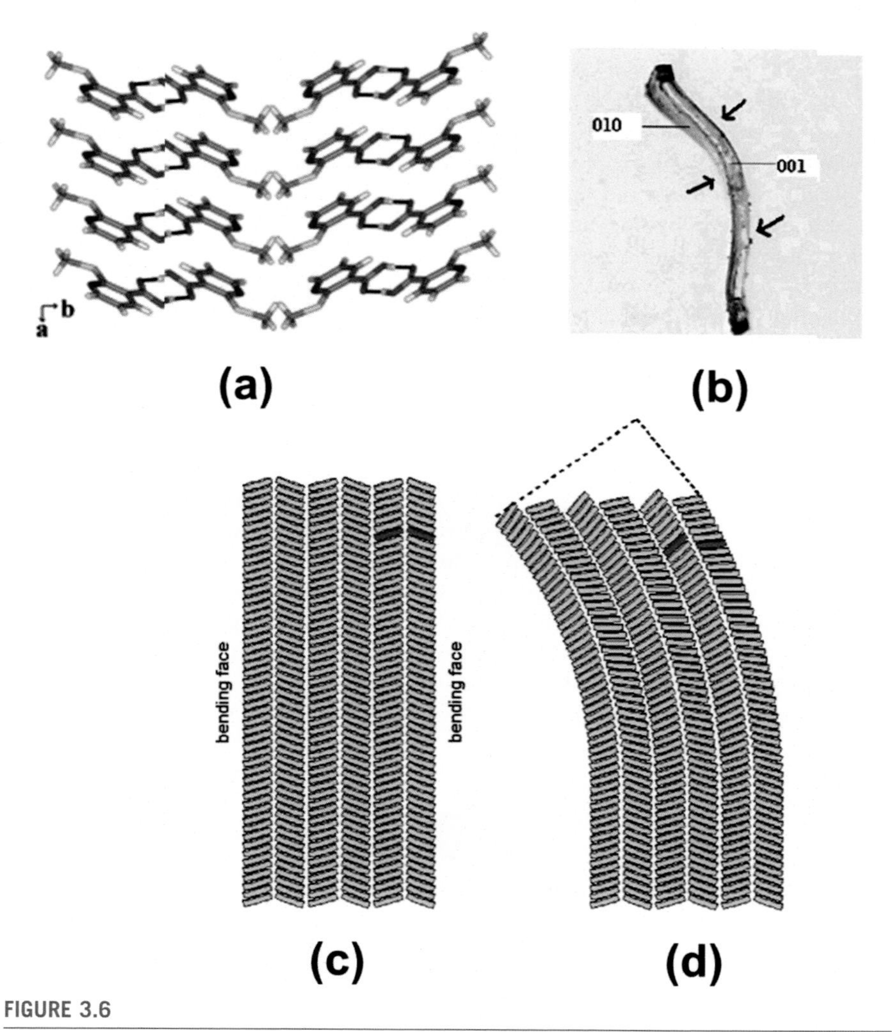

FIGURE 3.6

(A) Crystal packing of MTN along crystal length. (B) Snapshot of crystal bent on (010) face. (C) and (D) A model of Desiraju—Reddy showing plastic deformation. *MTN*, 2-(methylthio)nicotinic acid.

Reprinted with permission from C.M. Reddy, R.C. Gundakaram, S. Basavoju, M.T. Kirchner, K.A. Padmanabhan, G.R. Desiraju, Structural basis for bending of organic crystals. Chem. Commun. (2005) 3945–3947. Copyright 2005 Royal Society of Chemistry.

compared with the anhydrous form [62]. Panda et al. showed that inclusion of water molecule in the crystal structure of carbohydrates (derivatives of galactose, glucose, and mannose) facilitates slip during plastic deformation [63]. Although majority of the plastic crystals contain weak interaction-based slip planes in anisotropic crystal structure, the hydrates with strong hydrogen bonding interactions show that it is not mandatory to have such conventional slip planes and anisotropy for achieving plasticity in molecular crystals. The hydrate crystals also suggest that the presence of low energy barrier, irrespective of the absolute strength of the interactions across the plane, is important for observing the slip or molecular motions in crystals. In theophylline anhydrate and monohydrate crystal structure, Chang and Sun noted that although theophylline anhydrate has orthogonal distribution of strong and weak interactions with no mechanical interlocking, its plasticity is lower than that of theophylline monohydrate [64]. Despite interdigitated structure in theophylline monohydrate, under mechanical pressure, theophylline dimers slip parallel to the 2D sheet of water chains (Fig. 3.7), in spite of strong hydrogen bonding interactions among molecules. Pregabalin and gabapentin are additional examples, discussed by Khandavilli et al. where the hydrated forms have better plasticity and tabletability as compared with the anhydrous form [65]. In hydrates, the one-dimensional water chains can facilitate slippage due to the ability of nearly globular water molecules to reorganize their bonds on slippage in plastic deformation. These examples further suggest that the barrier for molecular movements is more important than the absolute strength of interactions for slippage to take place.

In a recent report on the dimorphic 4-chlorobenzonitrile, Rather and Saha reported plastic deformation in one of the polymorphs with isotropic interactions [66]. The centrosymmetric ($P2_1/c$) form exhibits elastic bending (Form-E), whereas the noncentrosymmetric (Pc) form exhibits plastic bending (Form-P). Energy frameworks analysis of plastic form showed the presence of nearly isotropic interactions

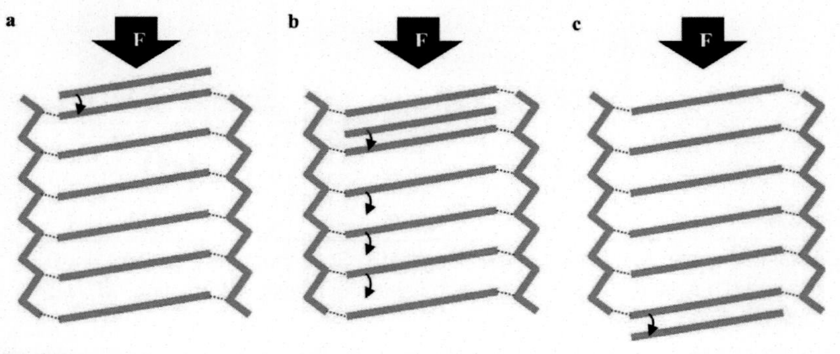

FIGURE 3.7

Schematic representation of crystal deformation of theophylline monohydrate: (A) Initial state when subjected to stress. (B) When the hydrogen bonds (blue dotted line) of theophylline dimers were overcome under external stress, the dimers are consecutively displaced and moved forward along water chains. (C) Theophylline dimer is moved to the other end of the crystal at the end of the deformation process.

Reprinted with permission from S.-Y. Chang, C.C. Sun, superior plasticity and tabletability of theophylline monohydrate. Mol. Pharmaceutics 14 (2017) 2047–2055. Copyright 2017 American Chemical Society.

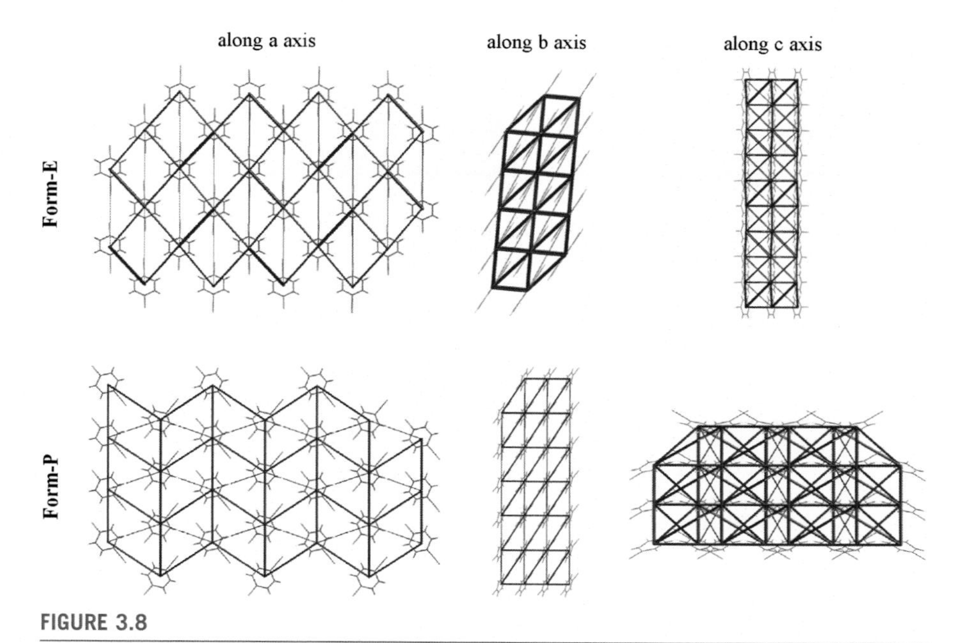

FIGURE 3.8

Energy frameworks showing total interaction energy (*blue*) for Form-E (elastic) and Form-P (plastic) along the three crystallographic axes; the interactions are isotropic in nature. The energy scale factor is 25, and the energy threshold is 5 kJ/mol.

in its crystal structure (Fig. 3.8). Based on expansion−contraction bending mechanism, they explained the unusual mechanical property of this isotropic molecular solid.

As the major thermal expansion axis was along the direction of the axis of crystal bending, they postulated that the rearrangement of the molecules at the concave side of the bent crystal may be correlated to the structural features that are obtained at low temperature. Similarly, the rearrangement of the molecules at the convex side may correspond to some of the structural features that are obtained at high temperature. This rearrangement of the molecules is possible by offset decrease or increase by sliding of the molecules over each other in the bent region as shown in Fig. 3.9. These thermal studies indeed capture to some extent the molecular rotations and change in the intermolecular distances (compression on inner arc and expansion on outer arc) that are observed in mechanically bent crystals. However, the anisotropic structural changes in mechanically bent crystals are more complex and depend on the system [67]. Dimethyl sulfone is another peculiar example that showed similar quasi-isotropic crystal packing. Spackman and coworkers explained the plastic bending behavior in this example based on H···H dihydrogen interactions and differences in electrostatic complementarity between molecular layers [68]. It is to be noted that the topic on the mechanism of plastic and elastic crystals is still

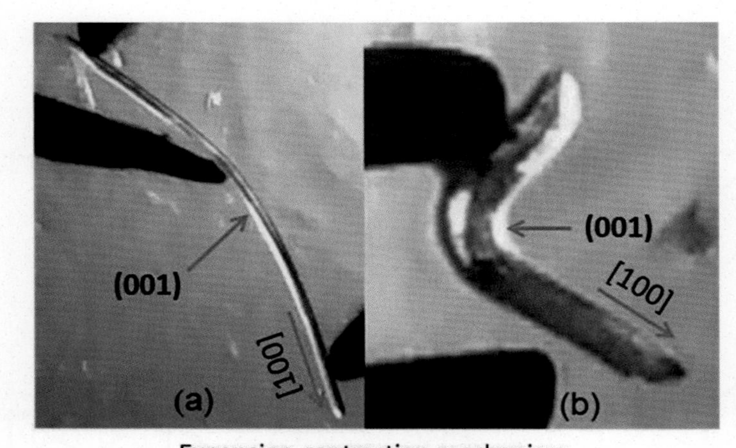

FIGURE 3.9

(A) Elastic and (B) plastic bending of Form-E and Form-P single crystals, respectively. (C) Schematic representation of the expansion—contraction mechanism that occurs during the elastic bending of Form-E. It shows how the length of the "a" axis decreases or increases due to decrease (concave side) or increase (convex side) in the offset of the $\pi \cdots \pi$ stacking but still maintains close packing.

Reprinted with permission from S.A. Rather, B.K. Saha, Thermal expansion study as a tool to understand the bending mechanism in a crystal. Cryst. Growth Des. 18 (5) (2018) 2712–2716. Copyright 2018 American Chemical Society.

evolving, and the role of dislocations and the kinetics of solid-state dynamics is yet to be understood.

Based on crystal engineering approach, it is also possible to design plastically deformable crystals. Reddy and coworkers proposed shape synthon approach to introduce active slip planes via incorporation of selected noninterfering functional groups, such as van der Waals (e.g., —t-Bu, —Me, —OMe, —Cl/Br, etc.), π-stacking, and hydrogen bonding groups, which was implemented on three different classes of

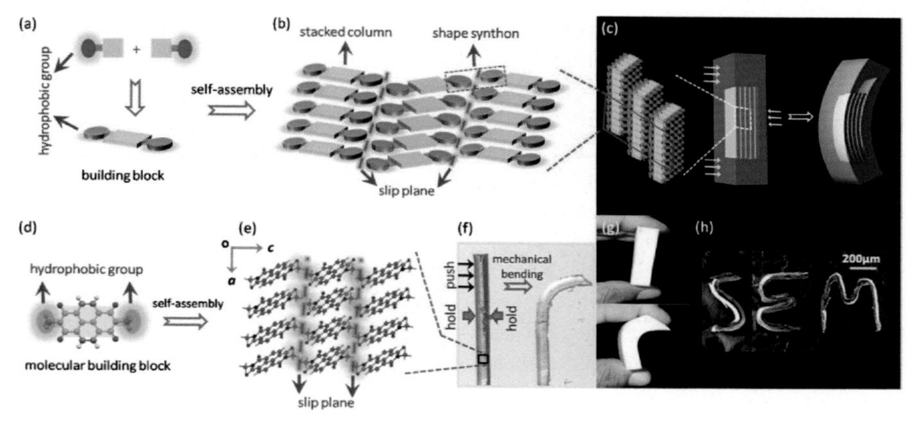

FIGURE 3.10

Illustration of the supramolecular shape synthon-based, slip plane model for achieving plastic flexibility in single crystals.

Reprinted with permission from G.R. Krishna, R. Devarapalli, G. Lal, C.M. Reddy, Mechanically flexible organic crystals achieved by introducing weak interactions in structure: supramolecular shape synthons. J. Am. Chem. Soc. 138 (2016) 13561–13567. Copyright 2016 American Chemical Society.

compounds [69]. The supramolecular shape synthon-based, slip plane model for achieving plastic flexibility in single crystals is illustrated in Fig. 3.10.

Other examples that show plastic flexibility include cocrystal of 6-chloro-2,4-dinitroaniline with Schiff base product of *ortho*-vanillin and ethylene diamine [51]; 1D plastic, 1,4-dibromobenzene; hand-twistable 2D plastic crystals of 4-bromophenyl-4′-nitrobenzoate [59] and 4-pyridinyl-4-nitrobenzoate hydrate [70]; cocrystal/salt of probenecid with coformers having two acceptor sites, viz., bipyridine and piperazine [71,72]; cocrystal of flufenamic acid with nicotinamide [73]; cocrystal of vitamin D3 with 3α-cholesterol [74]; and molecular salts of pregabalin with oxalic acid and salicylic acid [75]. Nath and coworkers used a different approach where simple modification of supramolecular synthon can also lead to transfer of plastic bending property from the parent probenecid molecule to probenecid hydrazide, whereas the structural analog probenecid amide resulted brittle crystals due to the formation of interdigitated molecular packing of *N*-propyl chain attached at the opposite ends of the molecule [76]. Introduction of metal coordination also directs slip plane formation that may result plastic deformation of molecular solid. Nath et al. considered a series of molecular analogs of 4-*n*-alkoxybenzoic acid and their metal salts for this study [77].

2.3 Elastic bending

In 2012, Ghosh and Reddy reported [78] an exceptional elastic bending deformation in a caffeine cocrystal solvate, which led to the exploration of many other elastic functional crystals with implications to flexible devices and adaptive materials–based technologies. The cocrystal, with one molecule of CAF, one molecule of CNB, and a disordered methanol molecule, crystallizes in orthorhombic polar space group, *Fdd*2. CAF and CNB form dimers through a supramolecular synthon involving a strong O–H···N and supportive C–H···O hydrogen bonds. Adjacent CAF dimers are connected via weak C–H···O interactions to form comb-like 1D tapes. The 1D tapes stack over each other with flanking CNB molecules on either side of the sheet resulting a double-sided comb-like sheet. These sheets pack in a zipper-like manner and close pack via van der Waal's interactions. The open channels, formed between the 2D sheets, are occupied by disordered methanol molecule that may act as local buffer zones between the comb-like 2D sheet. The crystal packing is modular enough to facilitate elastic bending of the needle-shaped single crystal perpendicular to the needle direction (Fig. 3.11).

Ghosh and Reddy attributed the exceptional elasticity of the CAF-CNB methanol solvate to the absence of slip planes or existence of interlocked packing with high proportion of dispersive interactions in crystal structure. Moreover, they pointed out that the morphological features in the crystal before and after the elastic bending are different from that of the plastically bent crystals. In case of plastic crystals, the angles between the needle end face (shown in inset in Fig. 3.12A) and bending faces (θ) change markedly to preserve the original length (l_O) in the inner (l_{in}) and outer (l_{out}) arcs of the crystal (i.e., $l_O = l_{out} = l_{in}$), whereas for elastic bending, no such change in interfacial angle (i.e., $\theta = \theta_O$) is observed (Fig. 3.12B). As a consequence, the $l_{out} > l_O > l_{in}$ in the elastically bent crystal, i.e., stretching and compression of outer and inner arc induces strain energy (E_S) (Fig. 3.12). To confirm the mechanical stress to be truly elastic in nature, they mounted the bent crystal on a single-crystal X-ray diffractometer (SCXRD) and investigated the diffraction spots. Along with sharp spots, in certain regions, they observed elongated spots that signify localized movement of molecules away from their idealized lattice points. Once the crystal relaxed, it produces sharp spots that signify reversible elastic nature of the single crystal. Further spatially resolved studies by Dey et al. revealed the molecular-level changes in the elastically bent caffeine cocrystals [79].

Corrugated crystal packing with elastic bending was reported by Desiraju and coworkers for several single component [80–82] as well as multicomponent systems [83–85]; Hayashi's group studied a series of fluorescent π-conjugated molecular crystals [86–90]. While the studies on elastic crystals by Desiraju and several other groups pointed out to the molecular rotation based bending model, the experimental evidence was provided by Worthy et al. using spatially resolved synchrotron X-ray diffraction studies on the elastically bent single crystals of $Cu(acac)_2$ [91]. Several other groups explored elastic deformation in organic crystals as well as coordination complexes [92–99].

Wang et al. reported the mechanical property of celecoxib, as a first example of elastically bendable single-component pharmaceutical crystal. Structural analysis showed interlocked molecular packing of celecoxib dimers with isotropic hydrogen bond network and without the presence of slip planes for exceptional elastic

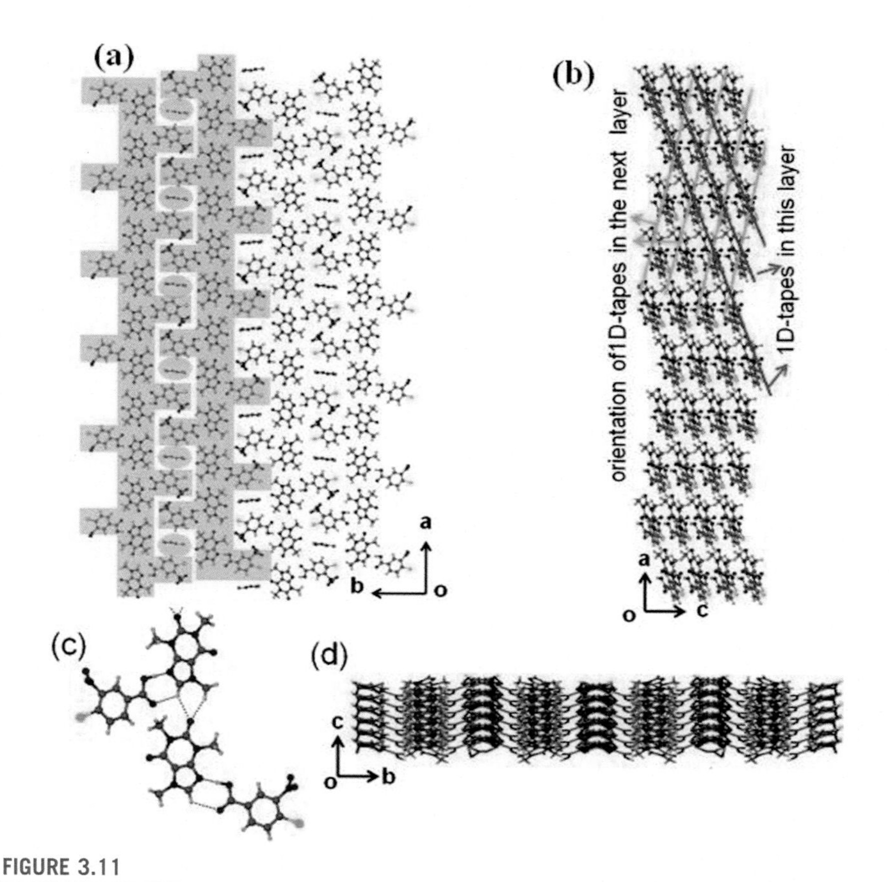

FIGURE 3.11

(A) Interlocked 2D comblike sheets with channels occupied by disordered methanol molecule; (B) the 2D layers in (010) plane; (C) CAF-CNB dimers connected through O—H···N and C—H···O interactions and CAF molecules form 1D tapes; views perpendicular to (D) (100). *CAF*, caffeine; *CNB*, 4-chloro-3-nitrobenzoic acid.

Adapted with permission from S. Ghosh, C.M. Reddy, Elastic and bendable caffeine cocrystals: implications for the design of flexible organic materials. Angew. Chem. Int. Ed. 124 (2012) 10465–10469. Copyright 2012 John Wiley and Sons.

flexibility and stiffness [100]. Moreover, inhomogeneous spatial separation of molecules in the bent region of the single crystal was further characterized by in situ micro-Raman spectroscopy (Fig. 3.13).

Based on the aforementioned discussion, the characteristic molecular interactions present in brittle and compliant crystals can be summarized as shown in Table 3.1. Table 3.1 is based on observed molecular interactions in majority of the case studies reported in literature [3]. It is to be noted that the exceptions are always encountered during investigation of mechanical properties.

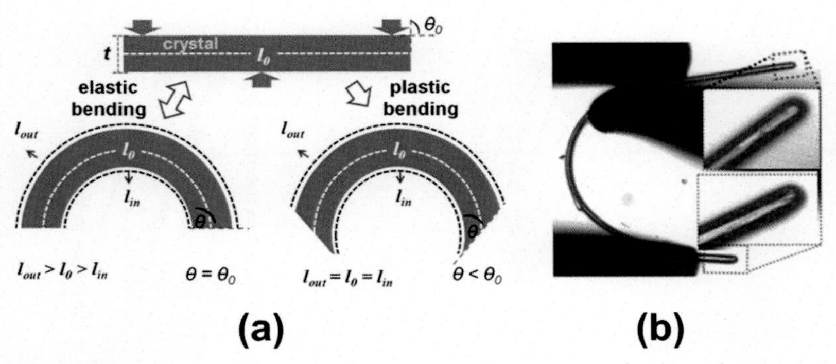

FIGURE 3.12

(A) Schematic representation of the single crystal before and after elastic and plastic bending and its respective morphological differences; (B) elastic bending of CAF-CNB methanol-solvated cocrystal. *CAF*, caffeine; *CNB*, 4-chloro-3-nitrobenzoic acid.

Adapted with permission from S. Ghosh, C.M. Reddy, Elastic and bendable caffeine cocrystals: implications for the design of flexible organic materials. Angew. Chem. Int. Ed. 124 (2012) 10465–10469. Copyright 2012 John Wiley and Sons.

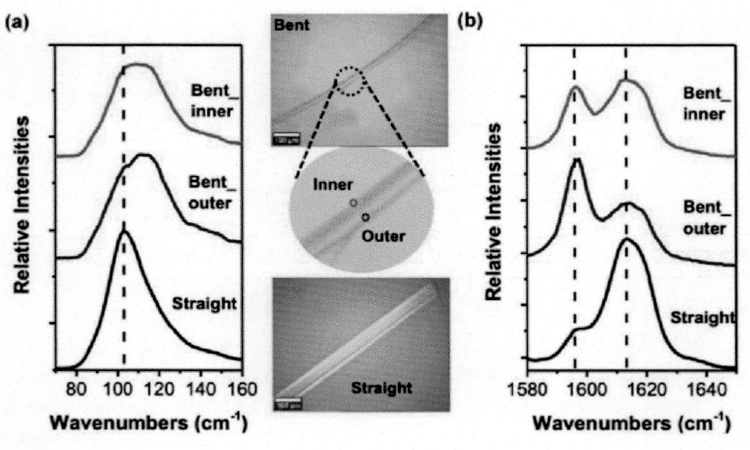

FIGURE 3.13

Raman spectra corresponding to the bent and straight crystals of celecoxib with *red and blue circles* indicating the point of data collection (inner arc and outer arc, respectively). (A) Lattice vibration and (B) aromatic symmetric in-plane bending δ(C—H) modes of inner and outer arcs of a bent celecoxib single crystal, as well as in the straight (stress-free) state.

Reprinted with permission from K. Wang, M.K. Mishra, C.C. Sun, Exceptionally elastic single-component pharmaceutical crystals. Chem. Mater. 31 (2019) 1794–1799. Copyright 2019 American Chemical Society.

Table 3.1 Structure—property correlation of different mechanical responses of organic molecular crystals.

Type	Characteristics
Plastic shearing	**1.** Anisotropic planar sheet structure generally with two strong directions and one weak direction. **2.** *Intra*layer interactions stronger than *inter*layer. **3.** Mechanism involves layer movement perpendicular to the stacking direction.
Plastic bending	**1.** Crystal structure generally with columnar packing. **2.** Presence of facile slip plane(s) among columns with low stacking fault energy. **3.** Irreversible migration or slippage of molecules along columns without need for any significant change in distances of molecules between and within columns.
Elastic bending	**1.** Nonrigid structure with rugous molecular layers. **2.** Near-isotropy along orthogonal directions. **3.** Buffering region, typically consisting of soft interactions, that accommodate elastic strain without permanent slippage of molecules. **4.** Reversible movement of parallel domains with change in the intermolecular distances/orientation from outer through inner arcs.
Brittle	**1.** Lack of significant anisotropic layering or the presence of rigid packing. **2.** Lack of facile slip lanes. **3.** Lack of buffering or compressible zones.

Adapted with slight modifications with permission from S. Saha, M.K. Mishra, C.M. Reddy, G.R. Desiraju, From molecules to interactions to crystal engineering: mechanical properties of organic solids. Acc. Chem. Res. 51 (11) (2018) 2957–2967. Copyright 2018 American Chemical Society.

In a recent report, Saha and Desiraju studied a polymorphic compound, 4-bromophenyl 4-bromobenzoate (BPBB) that shows distinct mechanical behavior due to subtle changes in molecular interactions. Form I is elastic where molecules interact via π-stacking along [010], Br···Br along [001], and C—H···O, C—H···Br along [100], resulting isotropic interlocked packing in three mutually perpendicular directions. In form II, the presence of C—H···π interactions and orientation of the phenyl rings along different directions prohibits the coordinated structural changes required for elastic bending, making it brittle in nature. A closely related structural analogue, identified to be a solid- solution between BPBB (site occupancy factor, s.o.f. 25%) and 4-bromophenyl 4-nitrobenzoate (s.o.f. 75%; BPNB) does have Br···Br interactions perpendicular to the bendable faces and experience maximum deformation strain; however, weaker Br···Br interactions of this solid-solution (108 kJ/mol) compared with form I (140 kJ/mol) are not enough to restore the bent crystal to its initial stage, leaving a permanent deformation. Hence, the solid-solution crystal shows 2D plastic bending upon application of a mechanical stress [101, 102, 103]. In a recent report, Masunov et al. reexamined these three solids using a novel methodology for computational prediction of the mechanical properties of single crystals. The methodology was based on constrained optimization using dispersion-corrected density functional theory and can be termed as virtual tensile test. The change in cell volume and equilibrium lattice energy with respect to simulated stretching of crystal structure along the three directions provides insights into the mechanical properties of the resultant molecular crystal [104]. These studies showed that the elastic form I withstands a large amount of stretch with a gradual increase in energy, while the energy in the brittle form II quickly raises for a small amount of stretch, leading to abrupt change in basic structure. In case of plastic form, the intermolecular interactions switch from one molecule to other allowing slip and stabilizing the energy, thus leading to large plastic deformation. These results, which are consistent with the experimental observations made by Desiraju and coworkers, showed how one could use the computational tools to classify the molecular crystals into elastic, brittle, and plastic in a reliable manner. Apart from mechanical force–induced actuating behavior, Naumov and several other groups investigated mechanically responsive molecular crystals that are not included here and discussed elsewhere [6–8,35,99,105–116].

3. Superelasticity and ferroelasticity in organic crystals

Takamizawa and coworkers reported an unusual behavior in organic molecular crystals, known as superelasticity or pseudo elasticity that involves reversible plastic deformation [117]. Although this phenomenon is mostly observed in case of inorganic solids such as alloys and ceramics, several organic molecular solids have been shown to display superelasticity due to the presence of unique structure and intermolecular interactions. This reversible deformation is not because of bond stretching or introduction of discrete defects in the crystal lattice; rather due to the single-crystal-to-single-crystal (SCSC) phase or gross domain transformation under applied stress. During this SCSC solid transformation, interactions between neighboring molecules in both phases are preserved, and the elastic strain that aids the spontaneous shape recovery is conserved once the applied stress has been removed. A few of them even exhibit thermal shape recovery, i.e., regain the initial crystal shape by a thermal phase transition similar to the well-known shape memory alloys. All the reported organosuperelastic crystals have well-separated slip planes and high degree of structural freedom of molecular components. This structural freedom can simultaneously contribute to easy and fast diffusionless transformability with the structural regularity of components. Their packing in the interface by depressing the undesired deformation by slip, dislocation, and so on, which can lower the energy of the structure at the interface, results this unique property. The superelastic deformation of terephthalamide (first example of organosuperelastic crystal) [117] and 3,5-difluorobenzoic acid [37] (35DFBA) exposed to external force and corresponding crystal packing on the twinning interfaces are shown in Fig. 3.14.

Apart from superelasticity, Takamizawa reported another phenomenon known as ferroelasticity in which organic crystals exhibit a spontaneous strain on application

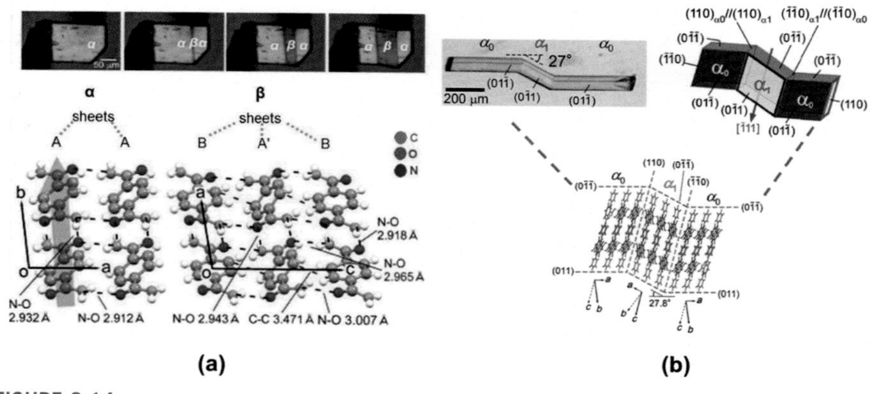

FIGURE 3.14

(A) Snapshot of a single crystal of terephthalamide under applied stress that shows superelasticity and the respective crystal packing in the two phases (α and β). (B) Snapshot of a superelastic single crystal of 35DFBA, face indexes of 35DFBA bent along [$\bar{1}$10], and the respective crystal packing on twinning interfaces.

(A) Adapted with permission from S. Takamizawa, Y. Miyamoto, Superelastic organic crystals. Angew. Chem. Int. Ed. 53 (2014) 6970–6973. (B) Adapted with permission from S. Takamizawa, Y. Takasaki, Superelastic shape recovery of mechanically twinned 3,5-difluorobenzoic acid crystals. Angew. Chem. Int. Ed. 54 (2015) 4815–4817. Copyright 2014, 2015 John Wiley and Sons.

of mechanical stress [38]. For this particular case, when the shear stress is applied, the crystal undergoes deformation with the generation of twin domains. As soon as the stress is removed, both the domains exist without change in volume and position. Next, switching the direction of the applied stress restores the initial shape of the crystal with elimination of the twinning interface. The mechanism of ferroelasticity involves phase or gross domain transformation followed by partial rotation at the twin interface resulting restoration of both the domain and initial shape of the crystal (5-chloro-2-nitroaniline) (Fig. 3.15A). The difference between the superelastic and ferroelastic behavior of an organic crystal can be shown by the schematic model given in Fig. 3.15B.

The molecular structures of organosuperelastic and ferroelastic materials reported in the literature [37–39,117–133] are listed in Table 3.2.

Superplasticity is observed in certain classes of crystalline materials at high homologous temperature where the material is uniformly deformed over 600% strain under tensile stress without fracture or necking. In a very recent study, Takamizawa and coworkers reported superplasticity in a simple organic compound: *N,N*-dimethyl-4-nitroaniline [134]. In general, organic crystals are brittle in nature; however, depending on internal arrangement of molecules, the crystals may show plastic or elastic deformation. According to classical mechanics, a material initially undergoes elastic (reversible) deformation followed by plastic (irreversible) deformation, unless a fracture is generated beyond the fracture (breaking) point. But in

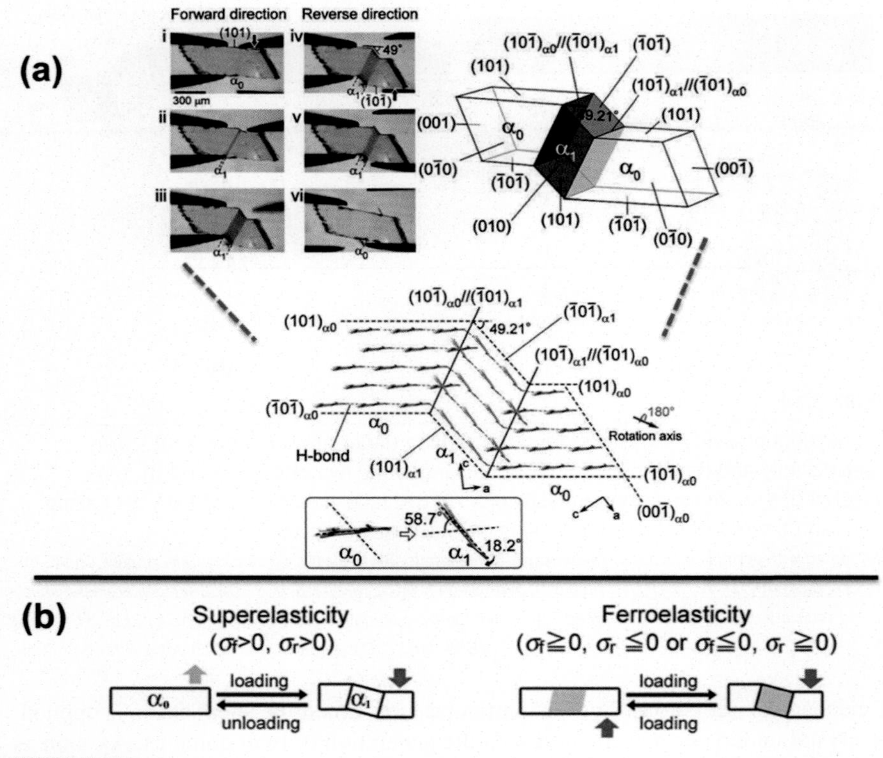

FIGURE 3.15

(A) Snapshots of the twinning deformation of 5-chloro-2-nitroaniline induced by shear stress using tweezers under a polarizing microscope: (i–iii) forward direction, (iv–vi) reverse direction; crystal face indices of the representative twinned crystal showing the presence of two packing domains and the corresponding molecular packing of mother (α_0) and daughter (α_1) phase. (B) Schematic representation of a superelastic and ferroelastic crystal showing the presence of α_0 and α_1 packing domain under shear stress for forward (σ_f) and reverse (σ_r) deformation.

(A) Adapted with permission from S.H. Mir, Y. Takasaki, E.R. Engel, S. Takamizawa, Ferroelasticity in an organic crystal: a macroscopic and molecular level study. Angew. Chem. Int. Ed. 56 (2017) 15882–15885. Copyright 2017 John Wiley and Sons. (B) Adapted with permission from T. Sasaki, S. Sakamoto, S. Takamizawa, Flash shape-memorization processing and inversion of a polar direction in a chiral organosuperelastic crystal of 1,3,5-tricyanobenzene. Cryst. Growth Des. 20 (7) (2020) 4621–4626. Copyright 2020 American Chemical Society.

superplastic state, materials can deform past their usual breaking point with excessive deformation ranging 500%–600% in tensile test. Due to the presence of slip planes and anisotropic interactions, *N*,*N*-dimethyl-4-nitroaniline can undergo two types of plastic deformations, viz., superplastic deformation with 500% strain and superelastic deformation from different crystal face without any loss of crystallinity

Table 3.2 Molecular structures of organosuperelastic/ferroelastic crystals.

Superelastic Crystals		Ferroelastic Crystals
	Saturated Fatty acid (C15–C21)[126]	

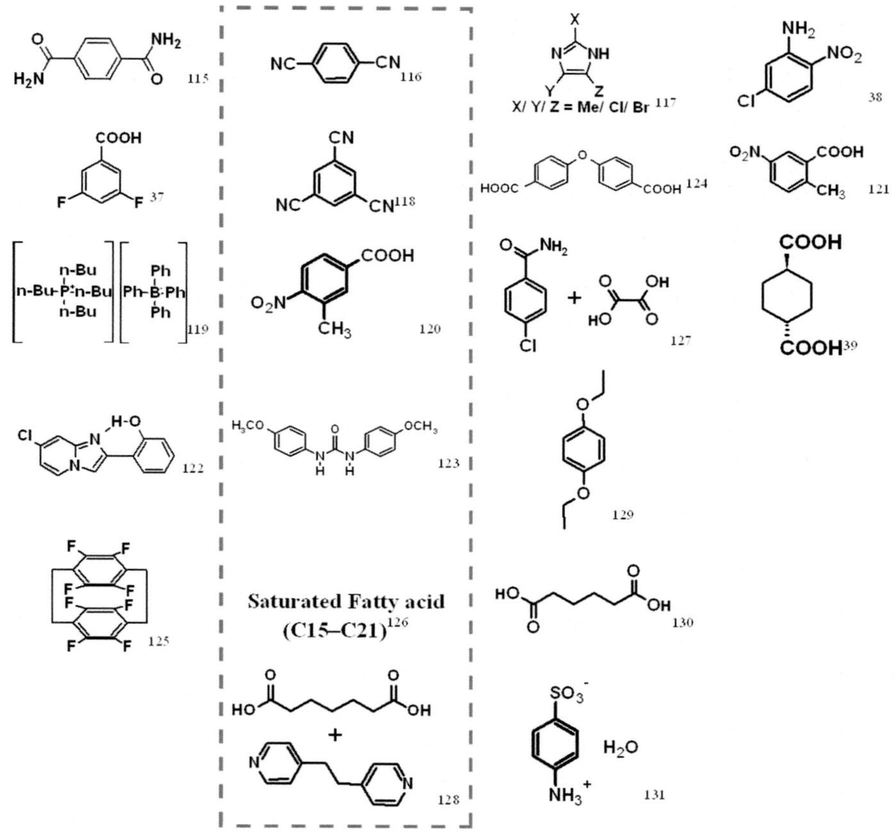

Compounds presented inside the red block show both the superelastic and ferroelastic behaviors.

in a SCSC manner. The mechanism of superplasticity and superelasticity of *N,N*-dimethyl-4-nitroaniline is explained based on slip and twinning deformations, respectively, as shown in Fig. 3.16.

4. Characterization techniques

Quantification of mechanical properties of molecular crystals can be done using experimental [135–138] and computational techniques [139]. Among the

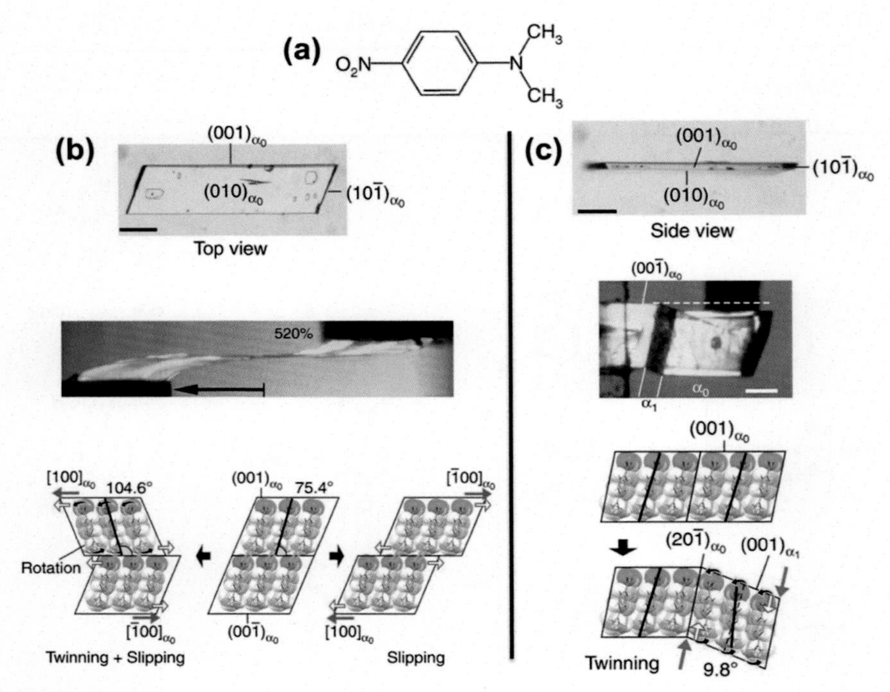

FIGURE 3.16

(A) Molecular structure of *N,N*-dimethyl-4-nitroaniline. Crystal micrographs with face indices before deformation, (B) top view and (C) side view. (B) Deformed shape of crystal in superplastic deformation (middle panel), and slip mechanism (bottom panel), showing shearing in the opposite (bottom left) and same (bottom right) directions as the molecular tilting direction in the single crystals. (C) Bending shape (middle panel), microstructure around an interface between twins (middle), and orientation change (bottom) in superelastic deformation.

Adapted with permission from S. Takamizawa, Y. Takasaki, T. Sasaki, N. Ozaki, Superplasticity in an organic crystal. Nat. Commun. 9 (2018) 3984. Copyright 2018 Springer Nature.

experimental techniques, nanoindentation is very popular while AFM indentation is also used for molecular crystals. There are other efficient experimental techniques such as Brillouin light scattering [140] but have not been fully explored yet due to the limited availability of the technique to nonexperts. Computational tools are currently widely available and become very reliable with recent developments [58,139,141]. The computational tools such as energy frameworks analysis software (Crystal Explorer) help in analyzing the pairwise intermolecular interaction energies for understanding the distribution of interactions in the three-dimensional crystal structures [142]. Here, we focus mainly on the nanoindentation and AFM indentation techniques.

4.1 Nanoindentation

Nanoindentation is a technique to quantitatively measure mechanical property, viz., elastic modulus or Young's modulus (E), hardness (H), fracture toughness (K_C), and brittleness index (BI) of small volumes of materials [143]. This method is extensively used for the mechanical property measurement of metallic and inorganic solids. As metals and other inorganic solids are held together by strong metallic bonds, macroindentation can be conveniently used to determine their mechanical behavior; however, due to the availability of only small sample sizes and presence of weak noncovalent interactions, such as hydrogen bonding, halogen bonding, van der Waals interactions, and π-stacking interactions, the macro- or microindentation techniques are not effective to measure mechanical strength of organic crystals. With respect to that, when the applied load P and penetration depth h are in the range of μN and nm respectively, it is referred to as nanoindentation.

Hardness, H is the resistance offered by the material to plastic deformation, and elastic modulus E is a measure of the resistance to elastic deformation that enables the material to return back to its original shape once the applied stress is withdrawn. Fracture toughness K_c is the quantitative measure of the resistance of a material to cracking and brittleness index, which is the ratio of H/K_c and gives a relative measure of the resistance to fracture. From the understanding of compliant crystals, plasticity in a crystal is generally achieved through irreversible glide of slip, twinning, and kinking motion of molecular layers, whereas elasticity is achieved from the ability of the material to store elastic strain from the resistance from hydrogen bonds, halogen interactions, mechanical interlocking, and so on, as they are of restorative character. Quantitative measurement of mechanical property is of great importance in pharmaceutical industry, as micronization, granulation, milling, tableting, and so on are the mechanical processes frequently used for each and every pharmaceutical material during formulation. For example, voriconazole (VOR) is a soft material, and it is very difficult to mill [144]; nanocrystals of paracetamol form better tablets than their respective polymorphs [145]; by preparing cocrystal of paracetamol and caffeine, powder compaction property of the parent drug could be profoundly improved [146], and mechanical milling of sulfathiazole induces partial transformation from form III to form I [147].

Molecular crystals are held together by periodic arrangement of molecules connected to each other using noncovalent interactions. Due to the difference in molecular interactions in three dimensions, mechanical property may vary in different directions and result anisotropic reactivity toward external stimuli, which is well explored in literature. In general, equivalent or isotropic specific interactions in three principal directions make the molecular crystals brittle in nature; isotropic interactions with predominantly dispersive nature, however, make molecular crystals compliant (plastic or elastic). The response of molecular crystals in indentation experiments depends on their crystal structure and nature of intermolecular interactions with respect to the indentation direction. As shown in Fig. 3.17, indentation

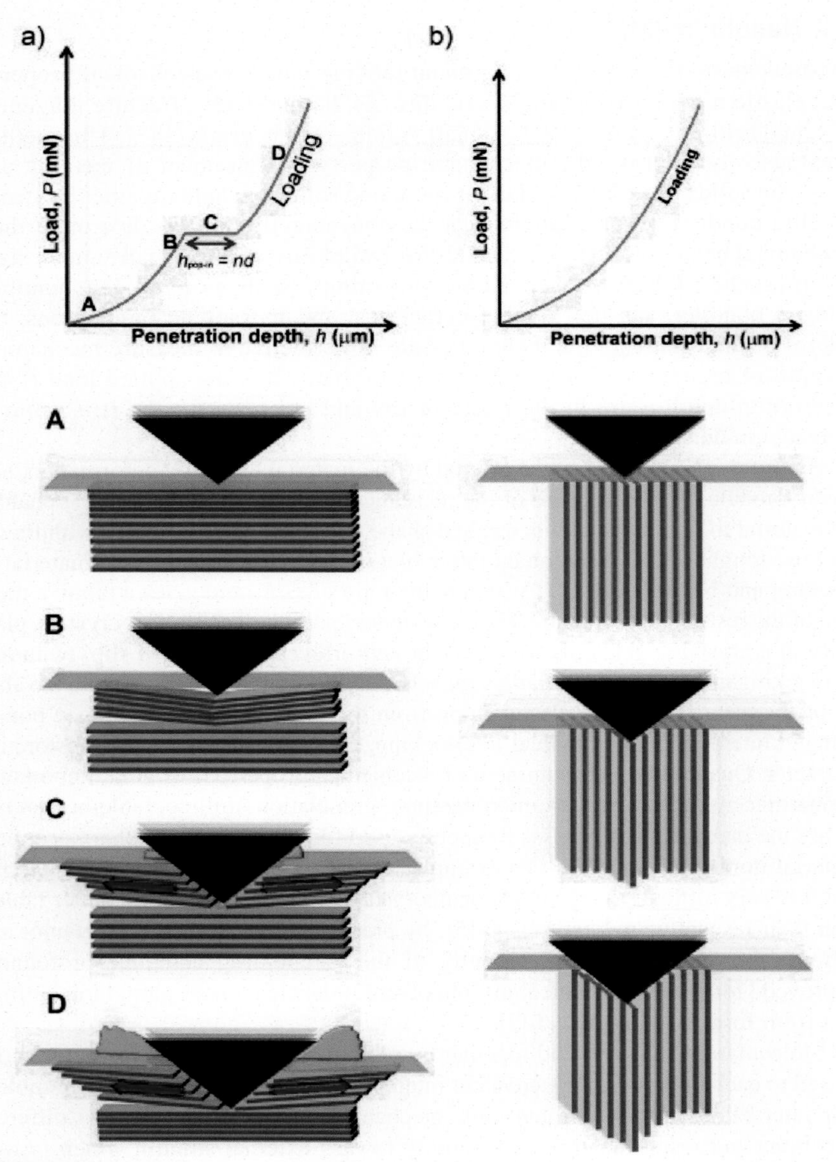

FIGURE 3.17

Schematic representation of (A) the possible mechanism of pop-in and pile-up during indenting normal to slip direction in molecular crystal; (B) indenting along slip direction results smooth P–h curve. The bottom panels of (A) and (B) show the response of layers at different stages of indentation.

Adapted with permission from S. Varughese, M.S.R.N. Kiran, U. Ramamurty, G.R. Desiraju, Nanoindentation in crystal engineering: quantifying mechanical properties of molecular crystals. Angew. Chem. Int. Ed. 52 (2013) 2701–2712. Copyright 2013 John Wiley and Sons.

normal to active slip planes results pop-in and pile-up, whereas indentation along the slip plane shows smooth $P-h$ curves. Generally, the softer crystal faces show higher maximum depth of penetration (h_{max}) as compared with the harder materials faces, for the same amount of applied maximum indentation load (P_{max}).

Ramamurty and Jang described various factors that affect the quality of nanoindentation data and the precautions that one needs to take while studying molecular crystals [148]. The sample size needed for nanoindentation is at least 0.1 mm^2 × 0.1 mm^2 in cross section and a few hundreds of nanometer thick. The safer thickness limit of the sample being probed is a minimum of 10 × h_{max} (in case of hard materials) or 20 × h_{max} (in case of compliant materials). Following these thickness limits ensures that the obtained response is purely from the sample and not from the substrate. One may measure submicron size crystals by dispersing in a matrix with superior mechanical properties. Sample surface is also an extremely important factor, as the indentation is done normal to the surface and uneven surfaces can lead to underestimation or overestimation of the measured values. However, since the area required to perform an indent is very small (~5–20 μm), finding smooth regions even on an extremely rough surface is not difficult. In a very recent report, Cruz-Cabeza and coworkers [149] discussed various factors that can influence the measured value of H and E of a molecular crystal. Considering polymorphic form I of aspirin as a model system, they investigated the importance of growing a good-quality and appropriate-size single crystal with face indexing (experimental as well as BFDH morphology prediction), preparation of smooth crystal surface (cleaving, washing with solvents/antisolvents) prior to indentation measurements, tip calibration using standard calibration material having the mechanical properties comparable with that of the organic crystals, effect of solvent occlusion in single crystals, etc. To obtain high-quality indentation data, it is always recommended to carry out the measurements over a wide range of depth penetration (h_{max}). At low penetration depth, the measured mechanical properties are influenced by the depth of indentation. Above a minimum indentation depth, when a "plateau" is reached, the data becomes independent of h_{max}. Indentation measurements should be carried out above this critical value of depth of penetration (Fig. 3.18).

Saccharin is one among the first examples studied in detail, by Desiraju and coworkers, using nanoindentation technique [143]. Saccharin is an artificial sweetener and used as a coformer in the pharmaceutical industry. It crystalizes in monoclinic space group $P2_1/c$ with (100) and (011) as major faces. The attachment energy (energy released on the attachment of a growth slice to a growing crystal face) are −15.858 and −27.790 kJ/mol for the (100) and (011) faces, respectively. The molecules form a centrosymmetric N–H···O dimer, with stack down [100] axis and oblique angle to (100) plane. The molecular stacks are arranged as bilayers with respect to (100) consisting of weak interactions between them (Fig. 3.19A). In (011), stacked dimers make a criss-cross arrangement with the vector of two adjacent dimers, which is along the indentation direction on (011) (Fig. 3.19B). Therefore, the loading part of the $P-h$ curve obtained on (011) is smooth, whereas due to the oblique angle between the indenter and slip planes, pop-ins are observed

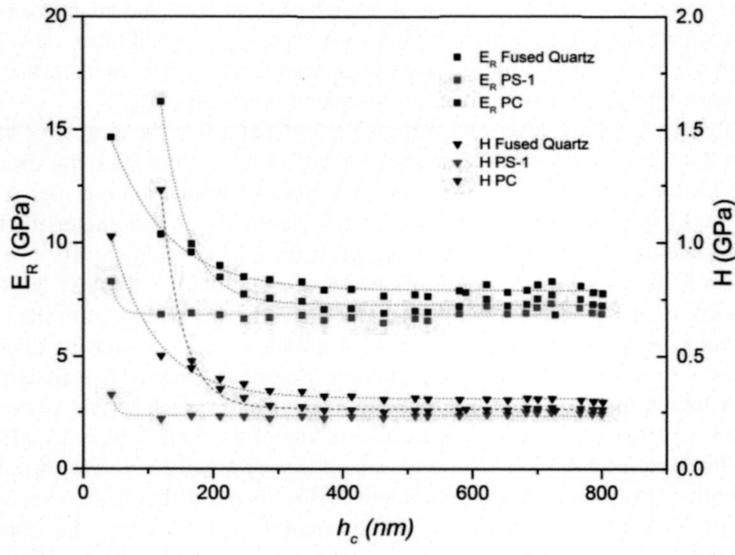

FIGURE 3.18

Series of measurements carried out on a unique single crystal of aspirin for increasing depths of contact h_C. E_R and H were extracted using three calibrations, made on the usual fused quartz, on the low modulus photopolymer PS-1 sheet sample, and on polycarbonate. Dashed lines were added as a visual guide only.

Reprinted with permission from B.P.A. Gabriele, C.J. Williams, M.E. Lauer, B. Derby, A.J. Cruz-Cabeza, Nanoindentation of molecular crystals: lessons learned from aspirin. Cryst. Growth Des. 20 (9) (2020) 5956–5966. Copyright 2020 American Chemical Society.

corresponding to (100) plane (Fig. 3.19C). Moreover, the discrete displacement bursts associated with the pop-ins are multiples of 18 nm, which is an integral multiple of 0.9 nm, the distance between the bilayers (Fig. 3.19D). This observation clearly explains the mechanism of pop-ins and pile-up in molecular crystals. Moreover, the residual depths upon unloading in P–h curve signify the plastic deformation of saccharin during indentation.

Nanoindentation could be a very effective technique to characterize intergrowth polymorphs (a single crystal containing different domains of polymorphic forms). Aspirin is a pharmaceutical molecule that exhibits two polymorphic modifications. Due to structural similarity with comparable energies, the two polymorphic forms coexist together as intergrowth polymorphs in the single crystal of form II. However, form I crystal could be prepared in pure form. Although crystal structures are determined for both the polymorphic forms, intergrowth in form II is challenging to characterize based on SCXRD analysis. In both forms I and II, the carboxylic acid groups form centrosymmetric O—H···O dimers that are arranged as two-dimensional layers parallel to (100) plane. The structural difference between the two forms exists in the

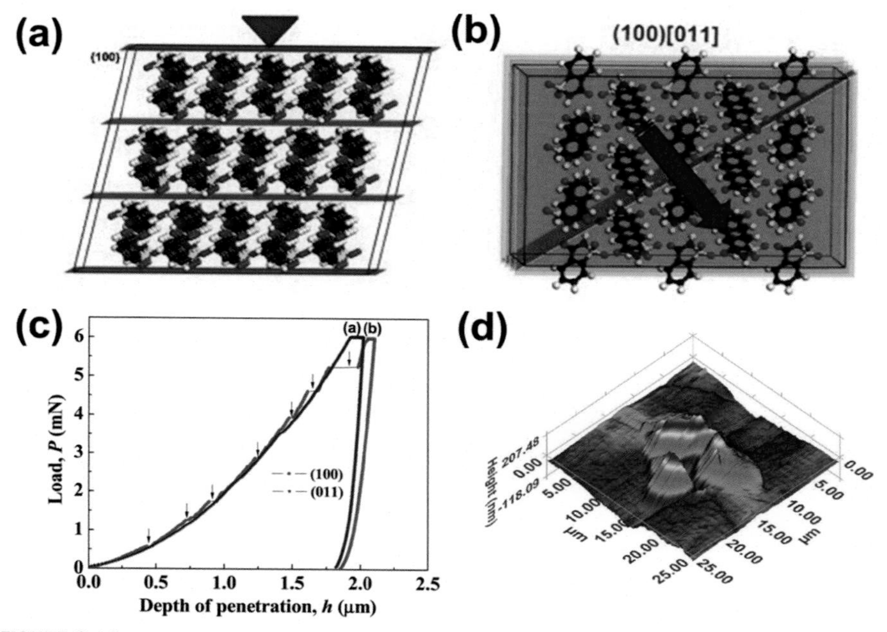

FIGURE 3.19

Structure–mechanical property correlation in saccharin crystals. (A) and (B) Crystal packing showing the active slip planes. The *blue triangle* in (A) and *blue arrow* in (B) represent the direction of indentation; (C) *P–h* curves; (D) AFM image showing an indentation impression with anisotropic pile-up on (100) plane.

Reprinted with permission from M.S.R.N. Kiran, S. Varughese, C.M. Reddy, U. Ramamurty, G.R. Desiraju, Mechanical anisotropy in crystalline saccharin: nanoindentation studies. Cryst. Growth Des. 10 (10) (2010) 4650–4655. Copyright 2010 American Chemical Society.

intermolecular interactions across the slip planes and position of the symmetry element, i.e., relative shift of adjacent layers parallel to one of the crystallographic axes (specifically 1/2*c*). Nanoindentation is carried out along the structurally equivalent crystal faces of the two polymorphs, i.e., along the *c*-axis or on (001) face of form I, and $(10\bar{2})$ of form II. *P–h* curve showed appearance of pop-ins for form I which is due to the application of stress normal to the sip planes (Fig. 3.20A). Based on the values of nanohardness and elastic modulus, form I was found to be stiffer and harder as compared with form II. The softer nature of form II along the potential shearing direction results solid-state transformation of form II → form I under mechanical stress. In a very recent report, Mishra and coworkers [150] investigated the in situ stress-induced polymorphic transformation of aspirin form II → form I using a spatially synchronized Raman integrated nanoindentation setup. Storage stiffness as a function of indentation load was used to identify the two polymorphic forms. On $\{10\bar{2}\}$ face of form II crystal above 4 mN load, polymorphic phase transformation

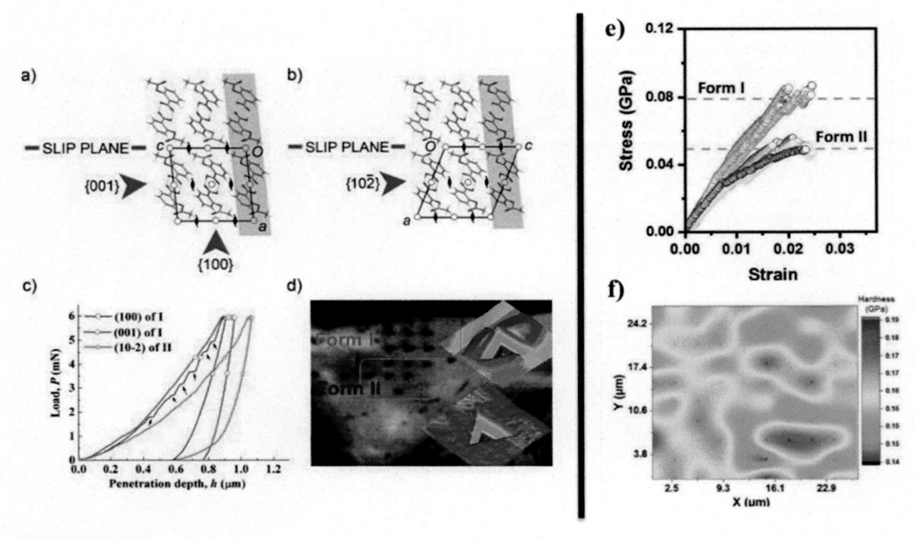

FIGURE 3.20

Crystal packing showing a representation of the slip planes in aspirin (A) polymorph I and (B) polymorph II. The *gray slab* represents the layer of molecules that are involved in stabilizing interactions across the slip plane. (C) Representative *P—h* curves of aspirin polymorphs. (D) Optical micrograph and AFM images show domain coexistence in polymorph II crystals. (E) Indentation stress—strain curves generated from spherical nanoindentation data. Aspirin form II had lower yield strength and showed more plastic behavior than form (I). (F) Hardness map recorded from the {10$\bar{2}$} face of form II crystals showing the coexistence of forms I and II microdomains. *AFM*, atomic force microscopy.

(A-D) Adapted with permission from S. Varughese, M.S.R.N. Kiran, U. Ramamurty, G.R. Desiraju, Nanoindentation in crystal engineering: quantifying mechanical properties of molecular crystals. Angew. Chem. Int. Ed. 52 (2013) 2701—2712. Copyright 2013 John Wiley and Sons. (E and F). Adapted with permission from P. Manimunda, S.A. Syed Asif, M.K. Mishra, Probing stress induced phase transformation in aspirin polymorphs using Raman spectroscopy enabled nanoindentation. Chem. Commun. 55 (2019) 9200—9203. Copyright 2019 Royal Society of Chemistry.

was observed. Hardness map recorded from the {10$\bar{2}$} face showed the existence of form I and form II domains, reconfirming the existence of intergrowth polymorphism (Fig. 3.20B). Apart from aspirin, felodipine is another example that shows intergrowth polymorphism. Mishra et al. carried out nanoindentation on {100} face of form II that showed different values of *H* and *E* corresponding to form I domain as the intergrowth polymorph in its microstructure [151].

An important example in the context of nanoindentation study is the pentamorphic omeprazole. Omeprazole is a blockbuster antiulcer drug that shows tautomeric polymorphism with positional disorder of a methoxy group. Form I is the pure 6-methoxy tautomer, whereas all other polymorphs (forms II—V) consist of varying proportion of the 5-methoxy (T_1) and 6-methoxy (T_2) tautomers resulting a solid-

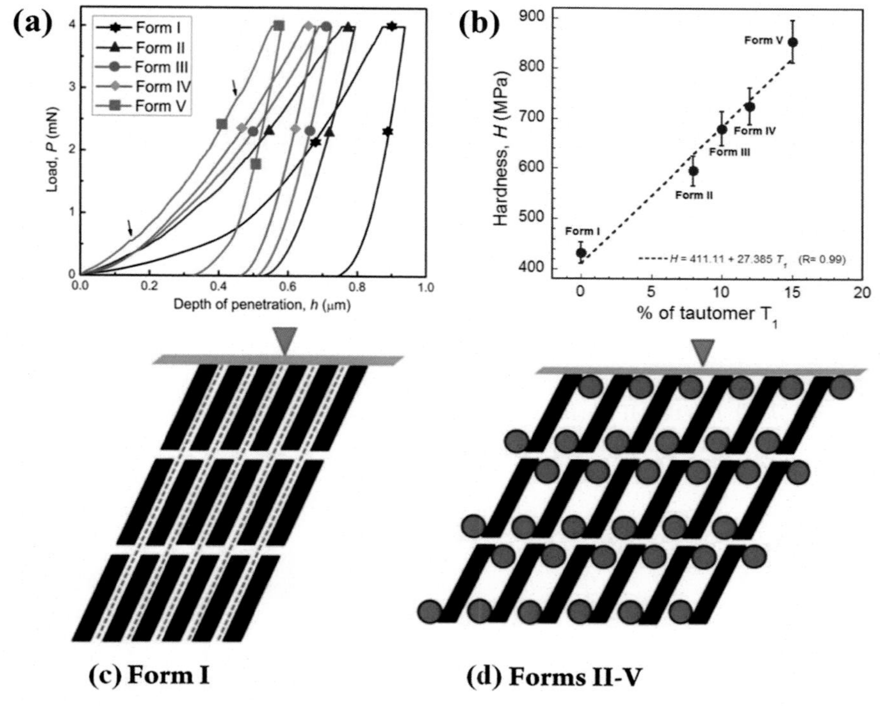

FIGURE 3.21

(A) Representative *P–h* curves of the five forms of omeprazole. (B) Correlation between *H* and proportion of the 5-methoxy tautomer, T_1 in omeprazole polymorphs. (C) Schematic representation of the crystal packing of omeprazole form I and (D) forms II–V. The dimers are represented as *solid parallelograms*. The indentation direction is shown as a *solid triangle*. *Red dotted lines* represent slip planes in form I. Methoxy groups are shown as *red circles* in forms II–V.

Reprinted with permission from M.K. Mishra, U. Ramamurty, G.R. Desiraju, Solid solution hardening of molecular crystals: tautomeric polymorphs of omeprazole. J. Am. Chem. Soc. 137 (5) (2015) 1794–1797. Copyright 2015 American Chemical Society.

solution mixture. Mishra et al. carried out nanoindentation measurements on the major face {100} of tautomeric polymorphs of omeprazole to correlate material hardness (*H*) with the solid-solution composition [152]. Pure 6-methoxy tautomer, form I, has *H* value of ~432 MPa; on the other hand, increase in the occupancy of the 5-methoxy tautomer from 8% to 15% results increase in the value of *H* from ~596 to ~855 MPa (Fig. 3.21A and B). As *H* of an organic crystal depends on the relative ease with which molecular layers irreversibly slide past each other, the increase in hardness can be attributed to the increase in the positional occupancy of the 5-methoxy group that causes a higher friction for shear sliding of omeprazole dimer across the slip planes (Fig. 3.21C and D).

Another highly polymorphic molecule ROY (5-methyl-2-[(2-nitrophenyl) amino]-3-thiophenecarbonitrile) showed distinct bending behaviors of its two polymorphic forms, viz., the yellow needle (YN) and orange needle (ON) [154]. Crystal structure analysis and interaction energy topology showed the key difference between the YN and ON, which lies in the separation distance between the adjacent stacks of aromatic rings of ROY in the molecular columns, 3.322 Å in elastic YN and 3.489 Å in the brittle ON. The smaller separation between the aromatic rings with several weak interactions, i.e., C—H···O, O—H···N, and C—H···π in the buffering region gives the YN a higher ability to accommodate reversible elastic flexibility, whereas in case of ON, the larger separation between neighboring aromatic rings made them to reach the elastic limit soon, as the strain develops during bending and the corresponding brittle behavior of ON crystal. Nanoindentation measurements on major crystal facets showed higher *E* of YN (9.98 ± 0.28 GPa) compared with ON (7.39 ± 0.09 GPa) (Fig. 3.22).

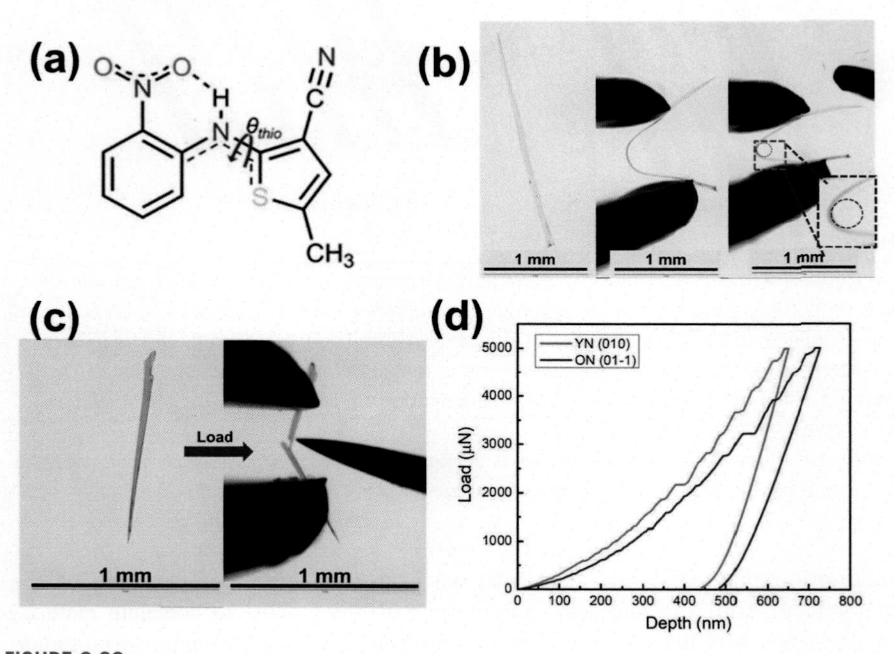

FIGURE 3.22

(A) Molecular structure of ROY. (B) Elastic bending of YN crystal. (C) Brittle fracture of an ON crystal during three-point bending. (D) Nanoindentation characterization of YN (*red curve*) and ON (*blue curve*) crystals, *P–h* curves.

Reprinted with permission from M.K. Mishra, C.C. Sun, Conformation directed interaction anisotropy leading to distinct bending behaviors of two ROY polymorphs. Cryst. Growth Des. 20 (7) (2020) 4764–4769. Copyright 2020 American Chemical Society.

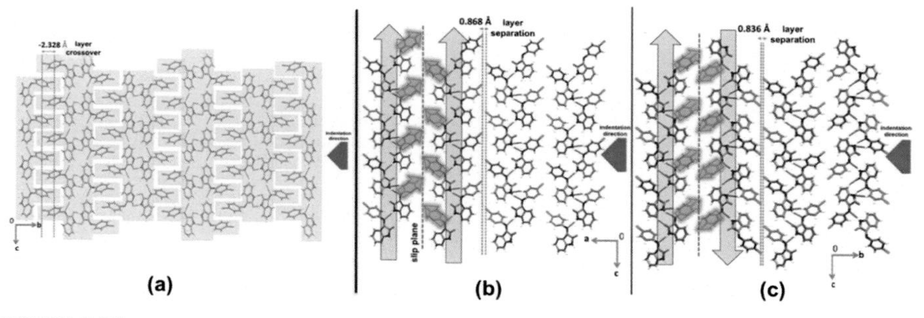

FIGURE 3.23

Crystal packing arrangement of (A) form I, (B) form II, and (C) form III of 3-((4-chlorophenyl)imino)indolin-2-one. *Red arrows* indicate indentation direction on major crystal faces (010) of form I, (100) of form II, and (001) of form III, respectively.

Reprinted with permission from K.B. Raju, S. Ranjan, V.S. Vishnu, M. Bhattacharya, B. Bhattacharya, A.K. Mukhopadhyay, C.M. Reddy, Rationalizing distinct mechanical properties of three polymorphs of a drug adduct by nanoindentation and energy frameworks analysis: role of slip layer topology and weak interactions. Cryst. Growth Des. 18 (7) (2018) 3927–3937. Copyright 2018 American Chemical Society.

3-((4-Chlorophenyl)imino)indolin-2-one is another example that yields three polymorphic forms during crystallization with distinct mechanical behavior [154]. Form I has 1D molecular tapes along [001] direction involving centrosymmetric N—H···O dimer synthon further connected via C—H···N hydrogen bond. These adjacent tapes are further connected via weak C—H···Cl interactions forming a corrugated sheet. These 1D tapes are mechanically interlocked in (010) by zipper-like arrangement that made the crystal structure 3D interlocked type with brittle behavior under mechanical stress. Forms II and III have very similar 1D molecular tapes; in form II, these 1D tapes are parallel to each other, whereas in form III, they are antiparallel to each other (Fig. 3.23). This slight difference in the topology of the layers in crystal packing made a huge difference to the mechanical behavior of the two polymorphic systems; form II is plastic in nature, whereas form III shows elastic behavior under mechanical stress.

Nanoindentation study on the major crystal faces of the three polymorphic systems showed substantial difference in mechanical properties (Fig. 3.24). From the P–h curves, it was observed that form II has highest depth of penetration (~1367 nm) compared with forms I and III (~860 and 790 nm, respectively). In other words, the plastic nature of form II nicely correlates with its highest observed h_{max} value. The pop-ins observed in case of form I are attributed to the elastic compression of the interlocked layers followed by a sudden slip upon reaching a critical load. The nanohardness of forms I and III are comparable but significantly higher than that of form II. Lower value of nanohardness clearly explains the intrinsic plastic behavior of form II, whereas higher value of both the E and stiffness of form III quantitatively confirms its superior elastic nature as compared with forms I and II.

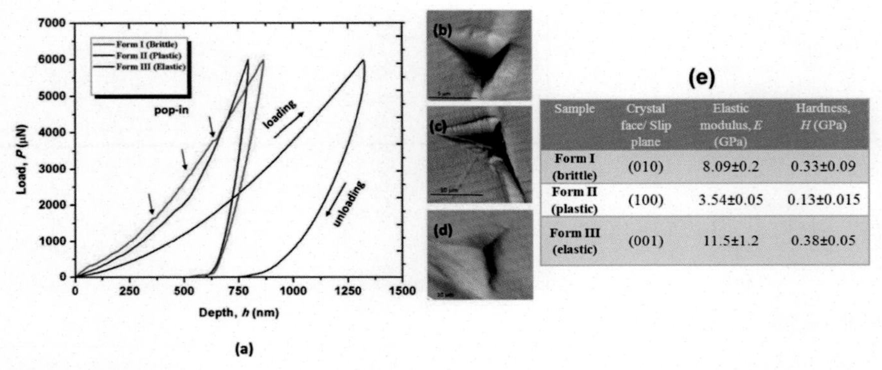

Sample	Crystal face/ Slip plane	Elastic modulus, *E* (GPa)	Hardness, *H* (GPa)
Form I (brittle)	(010)	8.09±0.2	0.33±0.09
Form II (plastic)	(100)	3.54±0.05	0.13±0.015
Form III (elastic)	(001)	11.5±1.2	0.38±0.05

FIGURE 3.24

(A) Typical *P*–*h* curves of forms I, II, and III of 3-((4-chlorophenyl)imino)indolin-2-one obtained from nanoindentation experiments. Scanning probe microscopy images showing the nanoindentation impressions of (B) form I (brittle), (C) form II (plastic), and (D) form III (elastic) crystals along with *E* and *H* parameters in the tabular form (right).

Reprinted with permission from K.B. Raju, S. Ranjan, V.S. Vishnu, M. Bhattacharya, B. Bhattacharya, A.K. Mukhopadhyay, C.M. Reddy, Rationalizing distinct mechanical properties of three polymorphs of a drug adduct by nanoindentation and energy frameworks analysis: role of slip layer topology and weak interactions. Cryst. Growth Des. 18 (7) (2018) 3927–3937. Copyright 2018 American Chemical Society.

Chinnasamy et al. synthesized a few structural analogs of halogenated Schiff base molecules, out of which two are elastically bendable [155]. One of them was dimorphic, i.e., one form was brittle, while the other one was elastically bendable. Nanoindentation measurements showed ~70%–80% elastic load recovery; loading–unloading patterns were consistent with their mechanical properties and structural packing.

Desiraju and coworkers reported the plasticity in HCB, a classic example that can be bent up to 360 degrees without fracture when external local pressure is applied on the (001) face. These crystals break readily when impacted on the (100) face. Naumov, Reddy and coworkers carried out an extensive investigation on the mechanism of plastic bending of HCB based on microscopic, spectroscopic, diffraction analysis, and nanoindentation on both pristine and mechanically bent crystals [156]. Scanning electron microscopy and AFM analysis of the bent crystals showed sliding of layers on top of each other along the (001) slip planes, leading to segregation of molecular layers into sheets. The sliding of layers is aided by restorable Cl···Cl interactions. The structural perturbation is further confirmed using spatially resolved techniques with micrometer (μ)-scale beam sizes, namely μ-IR and μ-Raman spectroscopy, μ-X-ray diffraction analyses, and nanoindentation by probing in different regions of the bent crystal. The nanoindentation study revealed that the *E* and *H* gradually decrease from the straight to the bent portion of the crystal, which also supports that the bent portion of the crystal is softer in nature due to

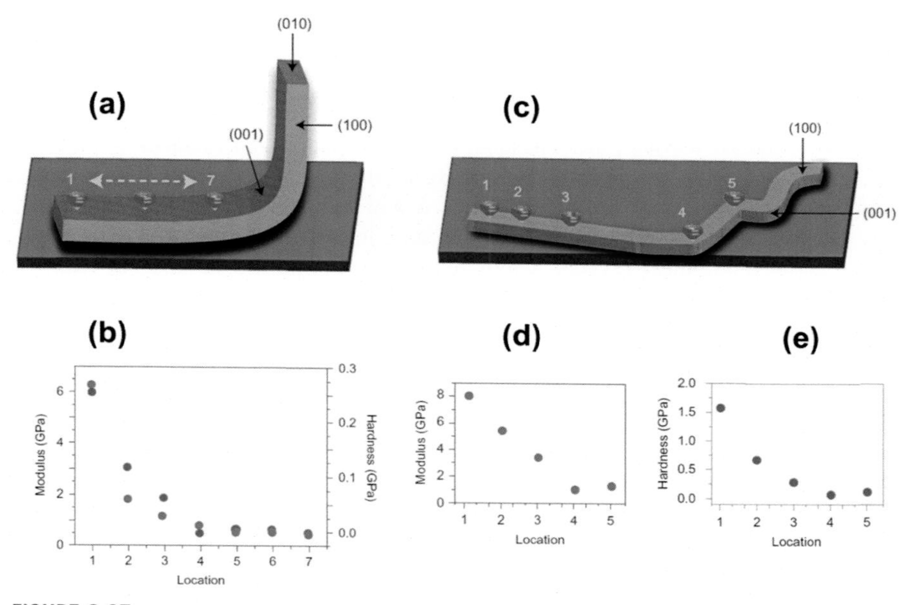

FIGURE 3.25

Nanoindentation at fixed load (2 mN) of two hexachlorobenzene (HCB) crystals bent normal to the (001) face. (A) Schematic of the locations where a bent crystal with a single kink was indented with indenter axis ⊥(001). (B) Modulus (*red filled circles*, left axis) and hardness (*blue filled circles*, right axis) obtained for each point in (A). (C) Schematic of locations where a multiply bent crystal was indented with indenter axis ⊥(100). Plots of modulus, *E* (D), and hardness, *H* (E), obtained for locations 1–5 of the crystal in (C). The results in (B), (D), and (E) indicate a decrease in the *E* and *H* on moving from the straight part to the kink, in line with the conclusion that the bent region is softer than the unbent portion of the crystal.

Adapted with permission from M.K. Panda, S. Ghosh, N. Yasuda, T. Moriwaki, G.D. Mukherjee, C.M. Reddy, P. Naumov, Spatially resolved analysis of short-range structure perturbations in a plastically bent molecular crystal. Nat. Chem. 7 (2015) 65–72. Copyright 2015 Springer Nature.

the structural strain (Fig. 3.25). This was further complemented by slight depression in the melting point in bent crystals.

As an extension to this work, Naumov and coworkers investigated the effect of softening and depression of melting point in the bent region of a plastically deformed single crystal of 1,4-dibromobenzene. The lowering of melting point of the bent section of about 0.3–0.4 K is attributed to the introduction of defects on mechanical deformation. Also, the bent section is mechanically softer and has lower hardness and Young's modulus as compared with the straight sections. The softness of the bent section is due to increased density of dislocations, generation of multiple domains, and crystal imperfections. The softening of the bent crystal prior to melting was analyzed by spatially resolved nanoscale thermal analysis using AFM [157]. Hence, these studies suggest that the structural strain stored in the bent crystals is responsible for the depression of melting point.

Reddy and coworkers studied two isostructural globular molecular solids, BH_3NMe_3 and BF_3NMe_3 [36]. These solids show exceptional metal-like ductility and malleability (isotropic plastic deformation) at room temperature. The crystals can be bent, twisted, stretched, thinned, rolled, and compressed with large deformation strains over 500%. The detailed structure–mechanical property correlation studies showed that these crystals belong to the class of "plastic crystals" or "rotator phases" with reorientational motion of molecules at room temperature in solid state. Plastic crystals, which have been studied since the 1920s, possess long-range order but lack short-range order (orientationally disordered). Structural analysis and comparison with another closely related crystalline solid of globular molecule, BH_3NHMe_2, with strong dihydrogen bonding formed by −NH and −BH groups, reveals that molecular globularity is a desirable condition, but not a sufficient condition for forming plastic crystals. Due to the presence of the strong dihydrogen bonds in BH_3NHMe_2, it has only one facile slip plane and hence shows only 1D plastic deformation. On the other hand, the presence of a large number of slip planes separated by dispersive interactions, among molecular columns in the high-symmetry crystals, facilitates exceptional ductility in the crystals of BH_3NMe_3 and BF_3NMe_3. Although nanoindentation experiments could not be carried out for BF_3NMe_3 crystal due to exceptionally small-size crystals, the measurement on 3D plastic BH_3NMe_3 crystal showed that these crystals are extremely soft compared with the regular ordered crystals. The three times larger residual displacement on indentation in BH_3NMe_3 crystals ($H = \sim 10$ MPa; $E = \sim 0.2$ GPa) as compared with the 1D plastic BH_3NHMe_2 crystals ($H = \sim 115$ MPa; $E = \sim 2.2$ GPa) suggests the softer nature of the former (Fig. 3.26). However, all these crystals are softer compared with most other molecular crystals.

Nanoindentation technique has also been used to discriminate polymorphic cocrystal systems. Reddy and coworkers synthesized two polymorphic cocrystals of caffeine with 4-chloro-3-nitrobenzoic acid and investigated their mechanical behavior [56]. Both the polymorphic forms have comparable 2D layer structure; however, form I is brittle in nature, whereas form II exhibits plastic shear deformation. Nanoindentation measurements showed form I to be relatively hard compared with form II that nicely correlates with their internal structure. In another report, Thakuria et al. carried out indentation measurements on caffeine–glutaric acid cocrystal polymorphs (forms I and II) and discussed their structure–mechanical property correlation [158]. Vangala and coworkers studied this system further and discussed the possible tabletability order of the polymorphs [159].

With the incorporation of cocrystals in the design of pharmaceutical solids with improved physicochemical property, the nanoindentation technique became a fundamental tool to quantify mechanical properties of pharmaceutical solids. Indentation parameters, viz., H, E, K_c, and BI give better comparison for the synthesized multicomponent pharmaceutical solids with that of the single-component drug molecule.

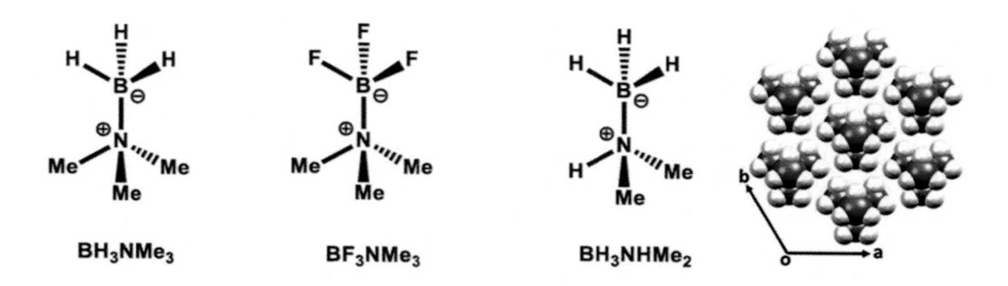

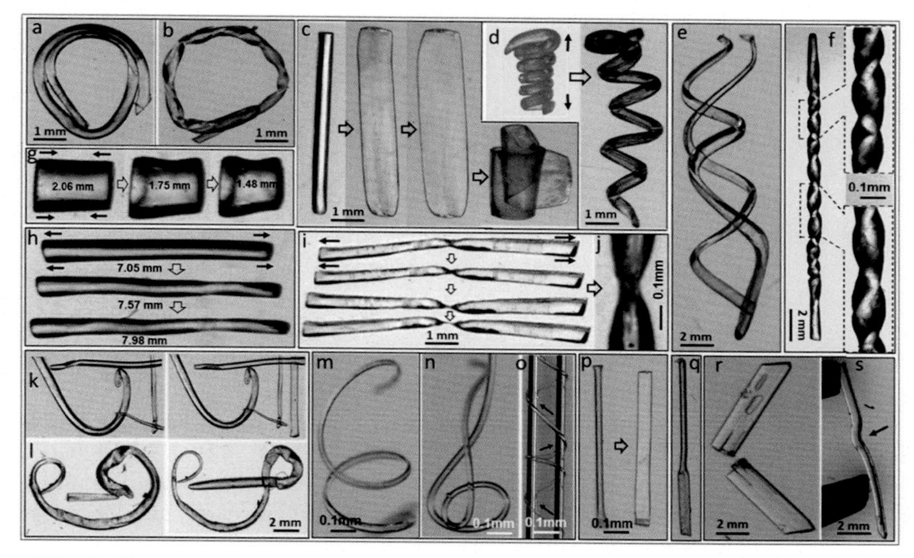

FIGURE 3.26

Molecular structure and crystal packing of the aminoborane compounds (top). Different crystal shapes achieved by mechanical stress on (A–L) BH_3NMe_3, (M–Q) BF_3NMe_3, and (R,S) BH_3NHMe_2.

Reprinted with permission from A. Mondal, B. Bhattacharya, S. Das, S. Bhunia, R. Chowdhury, S. Dey, C.M. Reddy, Metal-like ductility in organic plastic crystals: role of molecular shape and dihydrogen bonding interactions in aminoboranes. Angew. Chem. 132 (2020) 11064–11073. https://doi.org/10.1002/anie.202001060. Copyright 2020 John Wiley and Sons.

Cocrystallization of paracetamol with coformers such as theophylline, naphthalene, oxalic acid, phenazines, and so on greatly enhanced tableting behavior of paracetamol [146]. Sanphui et al. examined the mechanical property of VOR that is very soft in nature and difficult to mill [160]. It results cocrystals with fumaric acid (VOR-FUM), 4-hydroxybenzoic acid (VOR-PHB), and 4-aminobenzoic acid (VOR-PAB); molecular salts with hydrochloric acid (VOR-HCl); and two variable stoichiometric salts with oxalic acid (VOR-OXA1 and VOR-OXA2). Nanoindentation measurements showed high value of H and E for VOR-HCl compared with pure

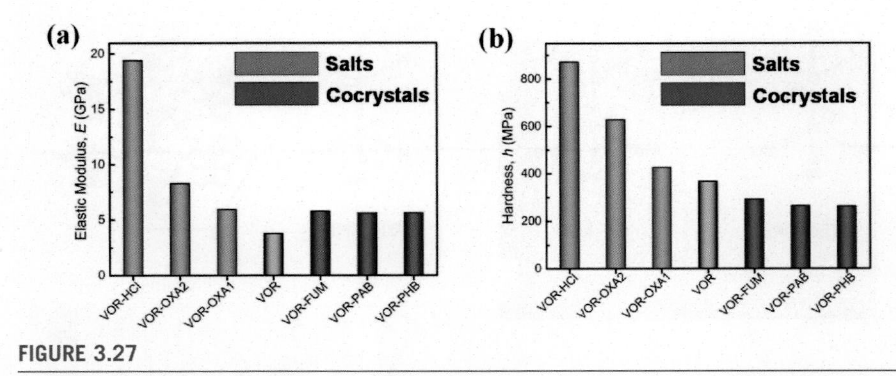

FIGURE 3.27

Variation in (A) *E* and (B) *H* in VOR and its salts and cocrystals. *VOR*, voriconazole.

Adapted with permission from Curr. Opin. Solid State Mater. Sci. 20 (2016) 361–370. Copyright 2016 Elsevier.

drug molecule. Strong ionic interaction and other prominent hydrogen bonds present in between the slip planes create high resistance toward shearing. On the other hand, VOR-OXA1 and VOR-OXA2 have intermediate values of *H* and *E*, whereas the other three cocrystals VOR-FUM, VOR-PHB, and VOR-PAB have similar value of *E* (Fig. 3.27) and are much softer compared with VOR.

Sun and coworkers in two different reports investigated mechanical properties of isoniazid (INZ) with homologous dicarboxylic acid ($HOOC-(CH_2)_n-COOH$, $n = 1-6$ and 8) and hydroxybenzoic acid coformers [161,162]. Mondal et al. measured nanomechanical properties of two polymorphs of 5-fluoroisatin, picolinic acid cocrystal, and a DMSO solvate [163]. Nanoindentation reveals that DMSO solvate is significantly harder and stiffer than picolinic acid cocrystal due to relatively stronger host–guest interlayer interactions. Relatively weak interlayer stabilization in case of 5-fluoroisatin–picolinic acid cocrystal facilitates easy layer movement for smooth plastic deformation. This could lead to better tabletability of 5-fluoroisatin cocrystal compared with the 5-fluoroisatin DMSO solvate. However, cocrystal design may not always result better mechanical property. For instance, 1:1 cocrystal of piroxicam and saccharin, with corrugated layers formed by piroxicam–saccharin dimer, shows resistance to sliding of layers unlike the starting material, under external stress resulting deterioration in the tableting performance [164].

4.2 Atomic force microscopy indentation

AFM is a type of scanning probe microscopy that is used for imaging and measuring features at nanoscale level [165,166]. There are two AFM operating modes: (1) force-distance mode also known as nonimaging mode and (2) imaging mode. Among them, the imaging mode is generally used to investigate surface features and phase identification of nanomaterials. AFM force–distance curve is generated by bringing the cantilever and the sample surface into contact and applying a defined

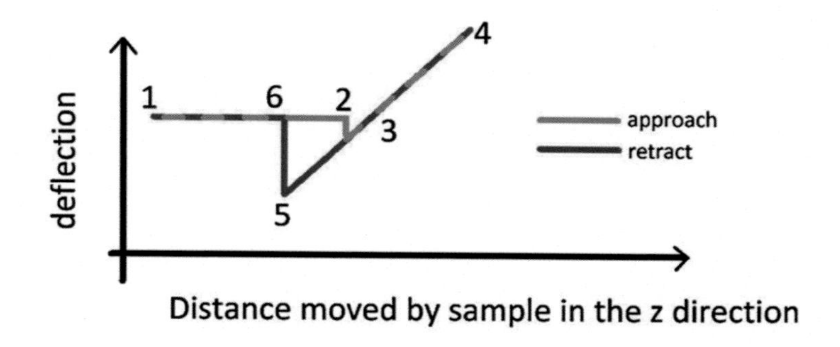

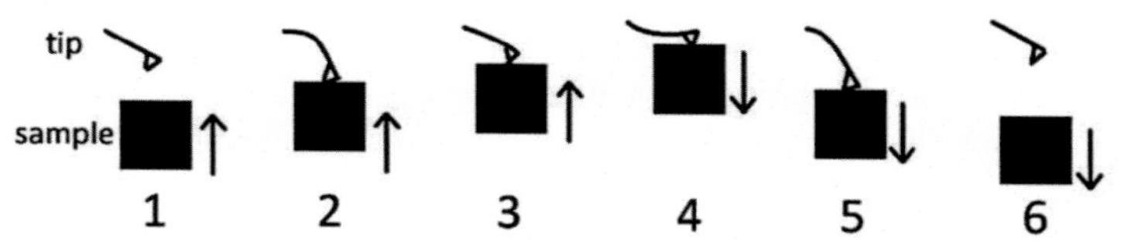

FIGURE 3.28

Typical force–distance curve showing a complete ramping cycle.

Reprinted with permission from E.H.H. Chow, D.-K. Bučar, W. Jones, New opportunities in crystal engineering–the role of atomic force microscopy in studies of molecular crystals. Chem. Commun. 48 (2012) 9210–9226.

force, and subsequently separating them. In this process, the cantilever displacement is plotted against the vertical probe displacement (in the z direction). A typical force–distance curve is shown in Fig. 3.28 with a complete ramping cycle of trace and retrace. The restoring force F_{res} (the force required to restore the cantilever to its equilibrium position) can be measured using Hook's law.

$$F_{res} = -k_{cant}dz$$

where F_{res} is the restoring force, k_{cant} is the spring constant of the cantilever, and dz is the corresponding cantilever displacement.

Based on the nonimaging mode or force–distance mode, AFM can also be used to perform nanoindentation measurement, as the instrument can record the dependence of cantilever deflection on piezo movement. The high sensitivity of the cantilever tip with respect to piezo movement results artifacts and large standard deviation. Series of careful calibrations and preliminary determinations are needed to rely on the AFM nanoindentation measurements compared with depth sensing instruments. Moreover, for the nanoindentation measurement, the stiffness of the cantilever tip should be very high. AFM nanoindentation technique is ideal for very soft materials, including biological samples and soft organic materials, but it has limited use for hard materials when compared with regular nanoindentation technique. Therefore, only a few reports are available on AFM nanoindentation for soft organic crystals or hybrid materials such as metal–organic frameworks

[167]. However, systematic studies outlining advantages and disadvantages of using AFM nanoindentation technique for molecular crystals are scarce.

Karunatilaka et al. used AFM nanoindentation technique to investigate the change in mechanical property of macro- and nanosized organic cocrystals in a SCSC transformation [168]. For the nanomechanical study, they considered a cocrystal system of 5-cyanoresorcinol and trans-1,2-bis(4-pyridyl)ethylene that undergoes intermolecular [2 + 2] photodimerization reaction. Recording force–displacement curves at approximately 100 crystal positions on macro- and nanosized single crystals of the unreacted and reacted crystals, they found that the crystals of millimeter dimensions become 40% softer, while crystals of nanoscale dimensions become 40% harder following photoreaction. To rely on the value of the local stiffness or Young's modulus, they carried out approximately 100 measurements and used a Gaussian fit as shown in Fig. 3.29. This study demonstrated the role of particle size on the photochemical reactions as monitored by measuring mechanical properties. The hardness can be improved just by changing the size of the reactive particles without altering its internal structure.

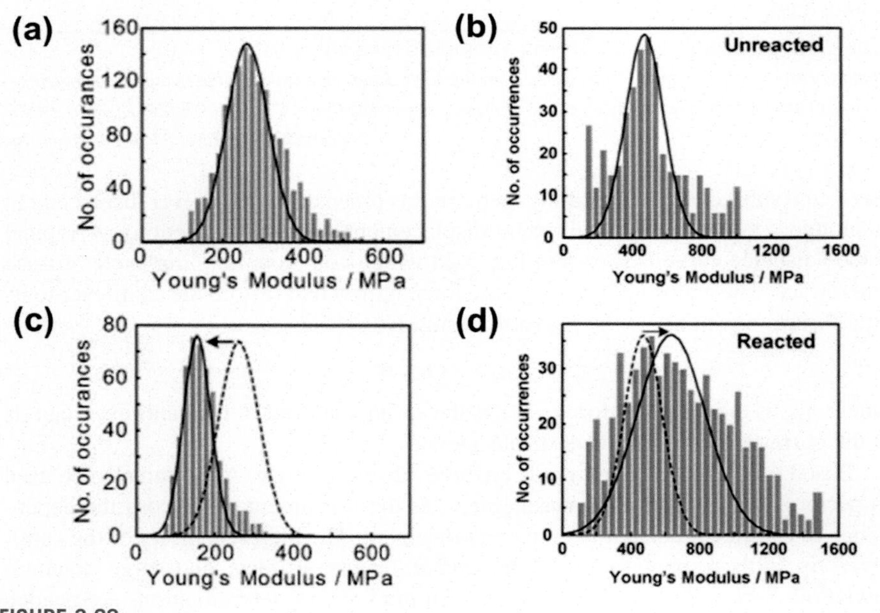

FIGURE 3.29

Young's modulus distribution of the (011) face of millimeter-sized cocrystal before (A) and after (C) photoreaction, showing 42% decrease in stiffness; on the other hand, nanometer-sized cocrystals before (B) and after (D) photoreaction showed a 40% increase in stiffness value.

Reprinted with permission from C. Karunatilaka, D.K. Bučar, L.R. Ditzler, T. Friščić, D.C. Swenson, L.R. MacGillivray, A.V. Tivanski, Softening and hardening of macro and nano sized organic cocrystals in a single crystal transformation. Angew. Chem. 123 (2011) 8801–8805. Copyright 2011 John Wiley and Sons.

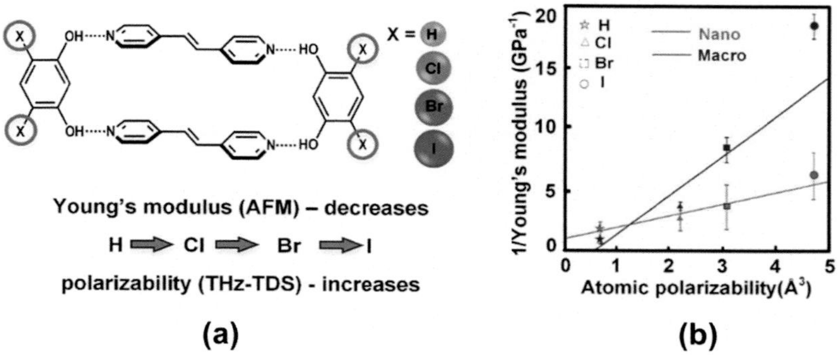

FIGURE 3.30

(A) Correlation between the polarizability of synthesized cocrystal and their mechanical property; (B) inverse Young's modulus versus atomic polarizability for 1–4.

Reprinted with permission from T.P. Rupasinghe, K.M. Hutchins, B.S. Bandaranayake, S. Ghorai, C. Karunatilake, D.-K. Bučar, D.C. Swenson, M.A. Arnold, L.R. MacGillivray, A.V. Tivanski, Mechanical properties of a series of macro- and nanodimensional organic cocrystals correlate with atomic polarizability. J. Am. Chem. Soc. 137 (40) (2015) 12768–12771. Copyright 2015 American Chemical Society.

As an extension to this work, Tivanski and coworkers synthesized a series of cocrystals of resorcinol (res) or 4,6-di-X-res (X = Cl, Br, I) and trans-1,2-bis(4-pyridyl) ethylene (4,4′-bpe). They considered these cocrystal systems to investigate any plausible correlation between Young's modulus, as determined by AFM nanoindentation and atomic polarizability. Prepared macro- and nanosized cocrystals of (res)·(4,4′-bpe) **1**, (4,6-di-Cl-res)·(4,4′-bpe) **2**, (4,6-di-Br-res)·(4,4′-bpe) **3**, and (4,6-di-I-res)·(4,4′-bpe) **4** were considered for AFM nanoindentation study that showed two major features—first, an inverse correlation of Young's modulus with atomic polarizability (H = 0.67, Cl = 2.18, Br = 3.05, I = 4.7 Å^3) with respect to the nature of res substituents, and second, a size-dependent increase in stiffness of nanosized X-substituted cocrystals compared with the macrosized solids (Fig. 3.30) [169]. Based on the experimental result, they concluded that nanomechanical property can be tuned by changing atomic polarizability as well as particle size of solids.

In 2011, Jing et al. developed a novel approach to identify active slip planes of soft organic crystals using AFM nanoindentation technique [170]. Slip occurs along a certain specific crystallographic planes in organic crystals, as they have limited number of slip planes compared with atomic solids due to steric interference imposed by various functional groups; indentation is one of the processes to activate these slip planes that could be observed at the crystal surface as pile-ups. They used succinic acid as the model system and carried out face indexing to identify the major crystal faces exposed to the surface to perform the indentation experiments (Fig. 3.31).

From the indentation impression, the trace angle Θ_{trace} (the angle between the observed trace and reference direction) was measured. The miller indices of the

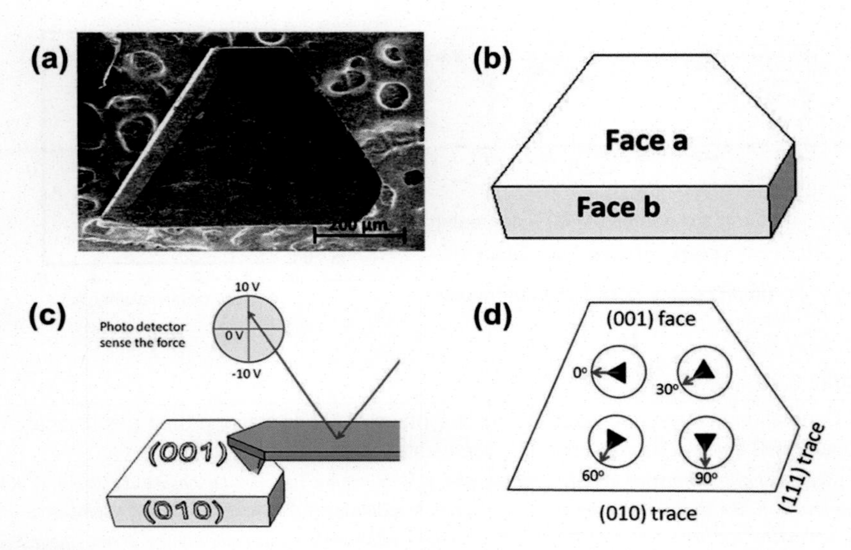

FIGURE 3.31

(A) Scanning electron microscopy image of succinic acid crystal with two unknown faces designated as (B) "face a" and "face b"; (C) schematic of AFM nanoindentation on (001) crystal face of succinic acid; (D) sample rotated by 0, 30, 60, and 90 degrees to study the slip planes.

Reprinted with permission from Y. Jing, Y. Zhang, J. Blendell, M. Koslowski, M.T. Carvajal, Nanoindentation method to study slip planes in molecular crystals in a systematic manner. Cryst. Growth Des. 11 (12) (2011) 5260–5267. Copyright 2011 American Chemical Society.

slip planes were determined by matching the observed Θ_{trace} with the calculated value based on attachment energy calculation. Rotation of the sample at specific angles 0, 30, 60, and 90 degrees, different combinations of slip planes were observed in different sample rotations (Fig. 3.32). Based on the measurements from (001) and (010) crystal faces, major slip planes were identified as (001) and (111). This study also demonstrates that different crystal planes follow different plastic deformation mechanism in response to mechanical loading and anisotropic nature of slip properties of organic single crystals.

Masterson and Cao discussed the difficulties and challenges involved in using the AFM nanoindentation technique compared with depth-sensing instrument [138]. They have stated that *although some of the case studies were successful to measure particle hardness, some differed largely in experimental measurements and result analysis, making them difficult to compare the data*. To justify the statement, they considered a few pharmaceutical solids, viz., sucrose, lactose, ascorbic acid, and ibuprofen for AFM nanoindentation analysis. They fixed the set point at 0.9 V with trigger threshold varying from 0.1 to 1.2 V and threshold steps from 0.1 to 0.3 V. They observed a linear increase in peak load with respect to trigger threshold

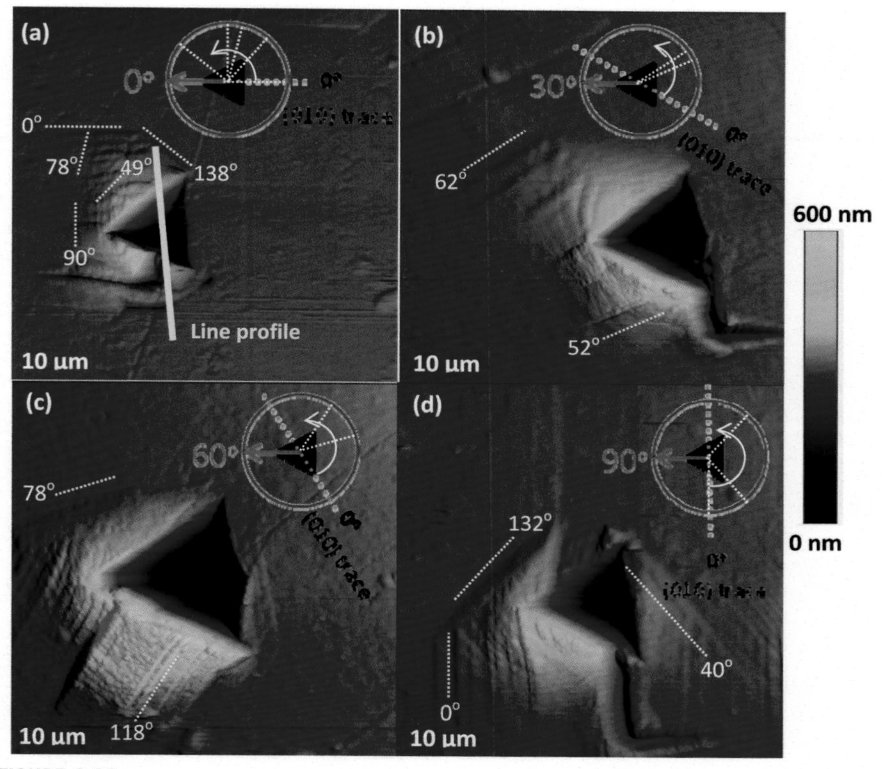

FIGURE 3.32

AFM topography of the succinic acid crystal with indentation on (001) face, with the (010) trace being the reference shown with *blue circle*.

Reprinted with permission from Y. Jing, Y. Zhang, J. Blendell, M. Koslowski, M.T. Carvajal, Nanoindentation method to study slip planes in molecular crystals in a systematic manner. Cryst. Growth Des. 11 (12) (2011) 5260–5267. Copyright 2011 American Chemical Society.

with a relative standard deviation of <2% with a fitting coefficient (R^2) of 1. Nanoindentation results showed that data variation and effect of indentation size or peak load affect the particle hardness; moreover, for lactose sample, no visible indents could be found (Fig. 3.33). The possible reason for lack of visible indents may be the rough surface or diamond tip used for the analysis. Generally, for better resolution, silicon nitride or silicon tips are used during AFM imaging; however, for nanoindentation purpose, diamond tip should be used to get the desired trigger threshold. In that process, sometimes the resolution of the images is compromised. Also, when the particle hardness values are compared for different materials, they should be compared at the same peak load or the same range of the peak load. Moreover, to validate the results, a large number of indentation measurements need to be carried out as reported by Karunatilaka et al. [168]. Therefore, using AFM nanoindentation,

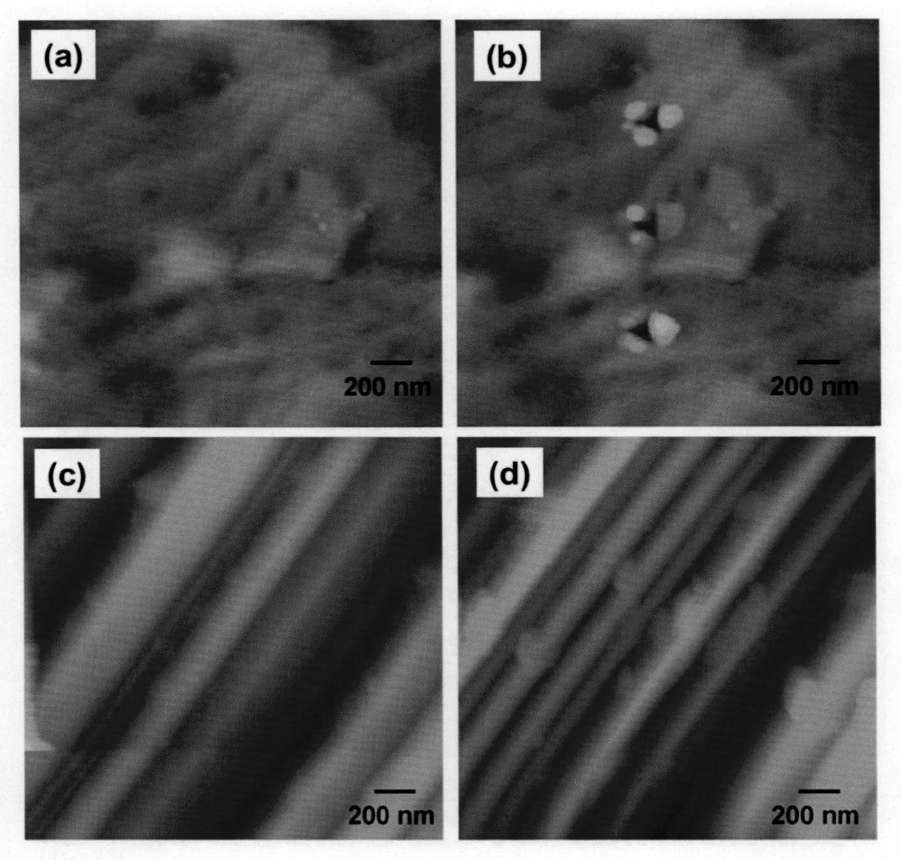

FIGURE 3.33

AFM height image of sucrose (A) before and (B) after indentation at a trigger threshold of 0.6 V; lactose (C) before and (D) after indentation at a trigger threshold of 0.6, 0.9, and 1.2 V. *AFM*, atomic force microscopy.

Reprinted with permission from V.M. Masterson, X. Cao, Evaluating particle hardness of pharmaceutical solids using AFM nanoindentation. Int. J. Pharm. 362 (2008) 163–171. Copyright 2008 Elsevier.

the obtained value of particle hardness may not be accurate; however, depth-sensing nanoindentation measurement can be useful to compare the relative hardness of the different samples.

Jones and coworkers also discussed the difficulties and challenges involved with force–distance measurement and phase imaging mode considering caffeine–glutaric acid polymorphic cocrystal as a model system. In the first report, they discussed the utility of AFM height image to distinguish the two polymorphs of caffeine–glutaric acid [171]. Intermittent contact-mode AFM (IC-AFM) images showed layered structure for both the polymorphic forms I and II. Height images of the surface of forms I

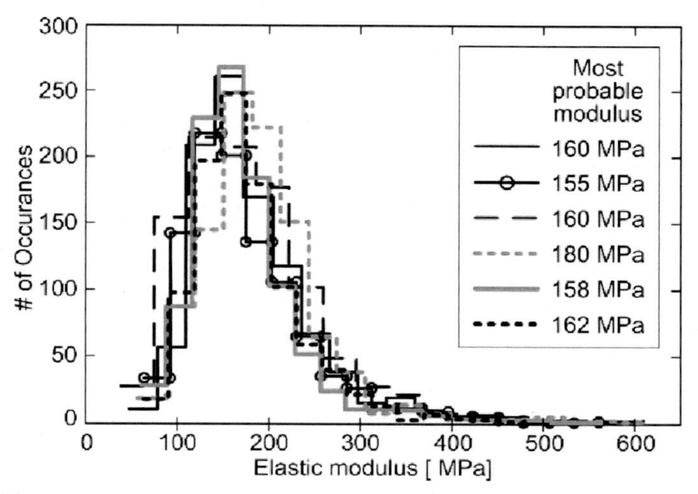

FIGURE 3.34

Elastic modulus distributions collected on four insulin crystals by three AFM tips. *AFM*, atomic force microscopy.

Reprinted with permission from S. Guo, B.B. Akhremitchev, Investigation of mechanical properties of insulin crystals by atomic force microscopy. Langmuir 24 (2008) 880–887. Copyright 2008 American Chemical Society.

and II showed steps with different height corresponding to individual supramolecular layers. Based on the differences of step heights, they could identify the two polymorphic forms as well as their phase transformation under high humidity conditions. As an extension to this work, they also carried out force–distance measurements using contact-mode AFM [158]. Although force–distance measurements are sometimes quite useful to identify polymorphic forms, it is not always effective when the two polymorphs have nearly identical structure. On the other hand, it is one of the most convenient techniques to measure mechanical properties of soft biological samples with a high ($\sim$10 nm) spatial resolution such as proteins. Guo and Akhremitchev used AFM nanoindentation to measure the elasticity of insulin crystals using IC-AFM mode [138]. The measured distribution of elastic modulus of several insulin crystals with different AFM probes gives similar asymmetric shape with similar most probable modulus value. Based on the model, they measured the elastic modulus of insulin crystal and found it to be around 164 $\pm$ 10 MPa along the *c*-axis (Fig. 3.34).

5. Conclusion

Mechanochemistry is a broader term covering both chemical and physical phenomena of solids under stress. While traditionally the mechanochemical reactions have been considered to involve breaking and making of covalent connections, recent appreciation of the solid-state phase transformations under mechanical stress has led to the consideration of breaking and making of supramolecular or noncovalent interactions also as part of the mechanochemistry [172–176]. Since the mechanically compliant crystals also involve the reorganization of supramolecular interactions, including phase transformations, we consider that the mechanical effects, such as plastic, superelastic, ferroelastic, superplastic, elastic, and brittle deformation in crystalline solids, also come under the broad subject of mechanochemistry.

Although the stress-induced mechanochemical reactions of organic, inorganic, and hybrid materials have been studied for long time in materials chemistry, the topic has always remained attractive due to its ease of operation, while the underlying complex molecular level mechanisms have kept the chemists ever intrigued. Availability of various new in situ experimental tools to probe the molecular level mechanism has provided new opportunities to study the chemical and physical transformations in mechanochemical reactions. While various studies have also been carried out to understand the structure–mechanical property correlation in brittle and compliant molecular crystals, the systematic studies evaluating the dependence of mechanical properties of organic solids on the mechanochemical outcome are scarce. It is known that the mechanically soft and unstable solid forms react fast or transform to more stable forms, for instance, the polymorphs of aspirin, but studies revealing how these distinct mechanical behaviors of solid forms influence their cocrystallization and chemical reactions are scarce. Does it mean that the mechanically unstable forms cannot be obtained by mechanochemistry? There are no systematic studies to understand the role of mechanical hardness and elastic modulus of molecular crystals on the solid-state phase transformations and supramolecular reactivity or noncovalent chemistry. The dependence of mechanical properties of crystalline solids on various mechanochemical processes such as milling, granulation, compaction, and flow properties could be evaluated systematically. Generally, the stacking fault energy is low in soft or plastic materials as compared with brittle or elastic materials. Hence the energetics in mechanochemical processes of the solids change with the solid form or its structure, for instance, in polymorphs. Hence a question arises as to how the energetics would impact the mechanochemical outcome when different solid forms are taken as the starting materials for performing a solid-state reaction. Hence, we feel that there is a need to consider the mechanical properties of the materials for understanding the outcome of mechanochemical processes.

References

[1] C. Majidi, Soft-matter engineering for soft robotics, Adv. Mater. Technol. 4 (2019) 1800477.

[2] Z. Shen, F. Chen, X. Zhu, K.-T. Yong, G. Gu, Stimuli-responsive functional materials for soft robotics, J. Mater. Chem. B (2020), https://doi.org/10.1039/D1030TB01585G.

[3] S. Saha, M.K. Mishra, C.M. Reddy, G.R. Desiraju, From molecules to interactions to crystal engineering: mechanical properties of organic solids, Acc. Chem. Res. 51 (2018) 2957–2967.

[4] C.M. Reddy, K.A. Padmanabhan, G.R. Desiraju, Structure–Property correlations in bending and brittle organic crystals, Cryst. Growth Des. 6 (2006) 2720–2731.

[5] C.M. Reddy, G. Rama Krishna, S. Ghosh, Mechanical properties of molecular crystals—applications to crystal engineering, CrystEngComm 12 (2010) 2296–2314.

[6] P. Naumov, S. Chizhik, M.K. Panda, N.K. Nath, E. Boldyreva, Mechanically responsive molecular crystals, Chem. Rev. 115 (2015) 12440–12490.

[7] P. Commins, I.T. Desta, D.P. Karothu, M.K. Panda, P. Naumov, Crystals on the move: mechanical effects in dynamic solids, Chem. Commun. 52 (2016) 13941–13954.

[8] E. Ahmed, D.P. Karothu, P. Naumov, Crystal adaptronics: mechanically reconfigurable elastic and superelastic molecular crystals, Angew. Chem. Int. Ed. 57 (2018) 8837–8846.

[9] G. Kaupp, M.R. Naimi-Jamal, Mechanically induced molecular migrations in molecular crystals, CrystEngComm 7 (2005) 402–410.

[10] G. Kaupp, Solid-state molecular syntheses: complete reactions without auxiliaries based on the new solid-state mechanism, CrystEngComm 5 (2003) 117–133.

[11] G. Kaupp, J. Schmeyers, U.D. Hangen, Anisotropic molecular movements in organic crystals by mechanical stress, J. Phys. Org. Chem. 15 (2002) 307–313.

[12] G. Kaupp, Waste-free large-scale syntheses without auxiliaries for sustainable production omitting purifying workup, CrystEngComm 8 (2006) 794–804.

[13] I.A. Tumanov, A.F. Achkasov, E.V. Boldyreva, V.V. Boldyrev, About the possibilities to detect intermediate stages in mechanochemical synthesis of molecular complexes, Russ. J. Phys. Chem. A 86 (2012) 1014–1017.

[14] I.A. Tumanov, A.F. Achkasov, E.V. Boldyreva, V.V. Boldyrev, Following the products of mechanochemical synthesis step by step, CrystEngComm 13 (2011) 2213–2216.

[15] K. Lien Nguyen, T. Friščić, G.M. Day, L.F. Gladden, W. Jones, Terahertz time-domain spectroscopy and the quantitative monitoring of mechanochemical cocrystal formation, Nat. Mater. 6 (2007) 206–209.

[16] T. Friščić, I. Halasz, P.J. Beldon, A.M. Belenguer, F. Adams, S.A.J. Kimber, V. Honkimäki, R.E. Dinnebier, Real-time and in situ monitoring of mechanochemical milling reactions, Nat. Chem. 5 (2013) 66–73.

[17] D. Gracin, V. Štrukil, T. Friščić, I. Halasz, K. Užarević, Laboratory real-time and in situ monitoring of mechanochemical milling reactions by Raman spectroscopy, Angew. Chem. Int. Ed. 53 (2014) 6193–6197.

[18] L. Batzdorf, F. Fischer, M. Wilke, K.J. Wenzel, F. Emmerling, Direct in situ investigation of milling reactions using combined X-ray diffraction and Raman spectroscopy, Angew. Chem. Int. Ed. 54 (2015) 1799–1802.

[19] M.R. Chierotti, R. Gobetto, L. Pellegrino, L. Milone, P. Venturello, Mechanically induced phase change in barbituric acid, Cryst. Growth Des. 8 (2008) 1454–1457.

[20] T.N. Drebushchak, A.A. Ogienko, E.V. Boldyreva, 'Hedvall effect' in cryogrinding of molecular crystals. A case study of a polymorphic transition in chlorpropamide, CrystEngComm 13 (2011) 4405–4410.

[21] A.V. Trask, J. van de Streek, W.D.S. Motherwell, W. Jones, Achieving polymorphic and stoichiometric diversity in cocrystal formation: importance of solid-state grinding, powder X-ray structure determination, and seeding, Cryst. Growth Des. 5 (2005) 2233–2241.

[22] A.V. Trask, N. Shan, W.D.S. Motherwell, W. Jones, S. Feng, R.B.H. Tan, K.J. Carpenter, Selective polymorph transformation via solvent-drop grinding, Chem. Commun. (2005) 880–882.
[23] F. Fischer, G. Scholz, S. Benemann, K. Rademann, F. Emmerling, Evaluation of the formation pathways of cocrystal polymorphs in liquid-assisted syntheses, CrystEngComm 16 (2014) 8272–8278.
[24] A.V. Trask, W.D.S. Motherwell, W. Jones, Solvent-drop grinding: green polymorph control of cocrystallisation, Chem. Commun. (2004) 890–891.
[25] S. Karki, T. Friščić, W. Jones, W.D.S. Motherwell, Screening for pharmaceutical cocrystal hydrates via neat and liquid-assisted grinding, Mol. Pharmaceutics 4 (2007) 347–354.
[26] D. Hasa, M. Marosa, D.-K. Bučar, M.K. Corpinot, D. Amin, B. Patel, W. Jones, Mechanochemical formation and "disappearance" of caffeine–citric-acid cocrystal polymorphs, Cryst. Growth Des. 20 (2020) 1119–1129.
[27] M. Mukaida, Y. Watanabe, K. Sugano, K. Terada, Identification and physicochemical characterization of caffeine–citric acid co-crystal polymorphs, Eur. J. Pharm. Sci. 79 (2015) 61–66.
[28] S. Aitipamula, P.S. Chow, R.B.H. Tan, Trimorphs of a pharmaceutical cocrystal involving two active pharmaceutical ingredients: potential relevance to combination drugs, CrystEngComm 11 (2009) 1823–1827.
[29] H. Kulla, C. Becker, A.A.L. Michalchuk, K. Linberg, B. Paulus, F. Emmerling, Tuning the apparent stability of polymorphic cocrystals through mechanochemistry, Cryst. Growth Des. 19 (2019) 7271–7279.
[30] K.K. Sarmah, T. Rajbongshi, A. Bhuyan, R. Thakuria, Effect of solvent polarity in mechanochemistry: preparation of a conglomerate vs. racemate, Chem. Commun. 55 (2019) 10900–10903.
[31] M.D. Eddleston, S. Sivachelvam, W. Jones, Screening for polymorphs of cocrystals: a case study, CrystEngComm 15 (2013) 175–181.
[32] A.A.L. Michalchuk, I.A. Tumanov, E.V. Boldyreva, Complexities of mechanochemistry: elucidation of processes occurring in mechanical activators via implementation of a simple organic system, CrystEngComm 15 (2013) 6403–6412.
[33] A.A.L. Michalchuk, I.A. Tumanov, V.A. Drebushchak, E.V. Boldyreva, Advances in elucidating mechanochemical complexities via implementation of a simple organic system, Faraday Discuss 170 (2014) 311–335.
[34] L.S. Germann, M. Arhangelskis, M. Etter, R.E. Dinnebier, T. Friščić, Challenging the Ostwald rule of stages in mechanochemical cocrystallisation, Chem. Sci. 11 (2020) 10092–10100.
[35] E. Ahmed, D.P. Karothu, M. Warren, P. Naumov, Shape-memory effects in molecular crystals, Nat. Commun. 10 (2019) 3723.
[36] A. Mondal, B. Bhattacharya, S. Das, S. Bhunia, R. Chowdhury, S. Dey, C.M. Reddy, Metal-like ductility in organic plastic crystals: role of molecular shape and dihydrogen bonding interactions in aminoboranes, Angew. Chem. 132 (2020) 11064–11073.
[37] S. Takamizawa, Y. Takasaki, Superelastic shape recovery of mechanically twinned 3,5-difluorobenzoic acid crystals, Angew. Chem. Int. Ed. 54 (2015) 4815–4817.
[38] S.H. Mir, Y. Takasaki, E.R. Engel, S. Takamizawa, Ferroelasticity in an organic crystal: a macroscopic and molecular level study, Angew. Chem. Int. Ed. 56 (2017) 15882–15885.
[39] S.H. Mir, Y. Takasaki, E.R. Engel, S. Takamizawa, Controllability of coercive stress in organoferroelasticity by the incorporation of a bulky flipping moiety in molecular crystals, CrystEngComm 20 (2018) 3807–3811.

[40] K.J. Ramos, D.F. Bahr, D.E. Hooks, Defect and surface asperity dependent yield during contact loading of an organic molecular single crystal, Philos. Mag. 91 (2011) 1276–1285.

[41] F. Liu, D.E. Hooks, N. Li, J.F. Rubinson, J.N. Wacker, J.A. Swift, Molecular crystal mechanical properties altered via dopant inclusion, Chem. Mater. 32 (2020) 3952–3959.

[42] C. Ouvrard, S.L. Price, Toward crystal structure prediction for conformationally flexible molecules: the headaches illustrated by aspirin, Cryst. Growth Des. 4 (2004) 1119–1127.

[43] T. Beyer, G.M. Day, S.L. Price, The prediction, morphology, and mechanical properties of the polymorphs of paracetamol, J. Am. Chem. Soc. 123 (2001) 5086–5094.

[44] K.S. Khomane, P.K. More, G. Raghavendra, A.K. Bansal, Molecular understanding of the compaction behavior of indomethacin polymorphs, Mol. Pharmaceutics 10 (2013) 631–639.

[45] A.B. Singaraju, K. Nguyen, A. Jain, R.V. Haware, L.L. Stevens, Aggregate elasticity, crystal structure, and tableting performance for p-aminobenzoic acid and a series of its benzoate esters, Mol. Pharmaceutics 13 (2016) 3794–3806.

[46] C.C. Sun, H. Hou, Improving mechanical properties of caffeine and methyl gallate crystals by cocrystallization, Cryst. Growth Des. 8 (2008) 1575–1579.

[47] J. Timmermans, Plastic crystals: a historical review, J. Phys. Chem. Solids 18 (1961) 1–8.

[48] J.N. Sherwood, The Plastically Crystalline State: Orientationally Disordered Crystals, Wiley, 1979.

[49] C.M. Reddy, M.T. Kirchner, R.C. Gundakaram, K.A. Padmanabhan, G.R. Desiraju, Isostructurality, polymorphism and mechanical properties of some hexahalogenated benzenes: the nature of Halogen···Halogen interactions, Chem. Eur. J. 12 (2006) 2222–2234.

[50] P.P. Bag, M. Chen, C.C. Sun, C.M. Reddy, Direct correlation among crystal structure, mechanical behaviour and tabletability in a trimorphic molecular compound, CrystEngComm 14 (2012) 3865–3867.

[51] G.R. Krishna, L. Shi, P.P. Bag, C.C. Sun, C.M. Reddy, Correlation among crystal structure, mechanical behavior, and tabletability in the co-crystals of vanillin isomers, Cryst. Growth Des. 15 (2015) 1827–1832.

[52] S. Ghosh, C. Malla Reddy, Co-crystals of caffeine with substituted nitroanilines and nitrobenzoic acids: structure–mechanical property and thermal studies, CrystEngComm 14 (2012) 2444–2453.

[53] S. Kakkar, B. Bhattacharya, C.M. Reddy, S. Ghosh, Tuning mechanical behaviour by controlling the structure of a series of theophylline co-crystals, CrystEngComm 20 (2018) 1101–1109.

[54] T. Sasaki, Y. Miyamoto, S. Takamizawa, Strictly regulated two-dimensional slippage in a lamellar single crystal of 5-fluorouracil, Cryst. Growth Des. 20 (2020) 4779–4782.

[55] O.V. Shishkin, V.V. Medvediev, R.I. Zubatyuk, Supramolecular architecture of molecular crystals possessing shearing mechanical properties: columns versus layers, CrystEngComm 15 (2013) 160–167.

[56] S. Ghosh, A. Mondal, M.S.R.N. Kiran, U. Ramamurty, C.M. Reddy, The role of weak interactions in the phase transition and distinct mechanical behavior of two structurally

similar caffeine co-crystal polymorphs studied by nanoindentation, Cryst. Growth Des. 13 (2013) 4435–4441.
[57] C.M. Reddy, R.C. Gundakaram, S. Basavoju, M.T. Kirchner, K.A. Padmanabhan, G.R. Desiraju, Structural basis for bending of organic crystals, Chem. Commun. (2005) 3945–3947.
[58] R. Devarapalli, S.B. Kadambi, C.-T. Chen, G.R. Krishna, B.R. Kammari, M.J. Buehler, U. Ramamurty, C.M. Reddy, Remarkably distinct mechanical flexibility in three structurally similar semiconducting organic crystals studied by nanoindentation and molecular dynamics, Chem. Mater. 31 (2019) 1391–1402.
[59] S. Saha, G.R. Desiraju, Crystal engineering of hand-twisted helical crystals, J. Am. Chem. Soc. 139 (2017) 1975–1983.
[60] S. Hu, M.K. Mishra, C.C. Sun, Twistable pharmaceutical crystal exhibiting exceptional plasticity and tabletability, Chem. Mater. 31 (2019) 3818–3822.
[61] C. Sun, D.J.W. Grant, Improved tableting properties of p-hydroxybenzoic acid by water of crystallization: a molecular insight, Pharm. Res. (N. Y.) 21 (2004) 382–386.
[62] F. Liu, D.E. Hooks, N. Li, N.A. Mara, J.A. Swift, Mechanical properties of anhydrous and hydrated uric acid crystals, Chem. Mater. 30 (2018) 3798–3805.
[63] M.K. Panda, K. Bhaskar Pal, G. Raj, R. Jana, T. Moriwaki, G.D. Mukherjee, B. Mukhopadhyay, P. Naumov, Flexibility in a molecular crystal accomplished by structural modulation of carbohydrate epimers, Cryst. Growth Des. 17 (2017) 1759–1765.
[64] S.-Y. Chang, C.C. Sun, Superior plasticity and tabletability of theophylline monohydrate, Mol. Pharmaceutics 14 (2017) 2047–2055.
[65] U.B.R. Khandavilli, M. Lusi, P.J. Frawley, Plasticity in zwitterionic drugs: the bending properties of Pregabalin and Gabapentin and their hydrates, IUCrJ 6 (2019) 630–634.
[66] S.A. Rather, B.K. Saha, Thermal expansion study as a tool to understand the bending mechanism in a crystal, Cryst. Growth Des. 18 (2018) 2712–2716.
[67] A.J. Brock, J.J. Whittaker, J.A. Powell, M.C. Pfrunder, A. Grosjean, S. Parsons, J.C. McMurtrie, J.K. Clegg, Elastically flexible crystals have disparate mechanisms of molecular movement induced by strain and heat, Angew. Chem. Int. Ed. 57 (2018) 11325–11328.
[68] S.P. Thomas, M.W. Shi, G.A. Koutsantonis, D. Jayatilaka, A.J. Edwards, M.A. Spackman, The elusive structural origin of plastic bending in dimethyl sulfone crystals with quasi-isotropic crystal packing, Angew. Chem. Int. Ed. 56 (2017) 8468–8472.
[69] G.R. Krishna, R. Devarapalli, G. Lal, C.M. Reddy, Mechanically flexible organic crystals achieved by introducing weak interactions in structure: supramolecular shape synthons, J. Am. Chem. Soc. 138 (2016) 13561–13567.
[70] S. Saha, G.R. Desiraju, A hand-twisted helical crystal based solely on hydrogen bonding, Chem. Commun. 53 (2017) 6371–6374.
[71] U.B. Rao Khandavilli, B.R. Bhogala, A.R. Maguire, S.E. Lawrence, Symmetry assisted tuning of bending and brittle multi-component forms of probenecid, Chem. Commun. 53 (2017) 3381–3384.
[72] N.K. Nath, M. Hazarika, P. Gupta, N.R. Ray, A.K. Paul, E. Nauha, Plastically bendable crystals of probenecid and its cocrystal with 4,4′-bipyridine, J. Mol. Struct. 1160 (2018) 20–25.
[73] T.V. Joshi, A.B. Singaraju, H.S. Shah, K.R. Morris, L.L. Stevens, R.V. Haware, Structure–mechanics and compressibility profile study of flufenamic acid:nicotinamide cocrystal, Cryst. Growth Des. 18 (2018) 5853–5865.

[74] J.-R. Wang, M. Li, Q. Yu, Z. Zhang, B. Zhu, W. Qin, X. Mei, Anisotropic elasticity and plasticity of an organic crystal, Chem. Commun. 55 (2019) 8532–8535.
[75] U.B.R. Khandavilli, M. Yousuf, B.E. Schaller, R.R.E. Steendam, L. Keshavarz, P. McArdle, P.J. Frawley, Plastically bendable pregabalin multi-component systems with improved tabletability and compressibility, CrystEngComm 22 (2020) 412–415.
[76] P. Gupta, S.A. Rather, B.K. Saha, T. Panda, D.P. Karothu, N.K. Nath, Mechanical flexibility of molecular crystals achieved by exchanging hydrogen bonding synthons, Cryst. Growth Des. 20 (2020) 2847–2852.
[77] N.K. Nath, P. Gupta, P.J. Hazarika, N. Deka, A. Mukherjee, G.K. Dutta, Plastically deformable and exfoliating molecular crystals of a 2-D coordination polymer and its ligand, Cryst. Growth Des. 19 (2019) 6033–6038.
[78] S. Ghosh, C.M. Reddy, Elastic and bendable caffeine cocrystals: implications for the design of flexible organic materials, Angew. Chem. Int. Ed. 124 (2012) 10465–10469.
[79] S. Dey, S. Das, S. Bhunia, R. Chowdhury, A. Mondal, B. Bhattacharya, R. Devarapalli, N. Yasuda, T. Moriwaki, K. Mandal, G.D. Mukherjee, C.M. Reddy, Mechanically interlocked architecture aids an ultra-stiff and ultra-hard elastically bendable cocrystal, Nat. Commun. 10 (2019) 3711.
[80] S. Ghosh, M.K. Mishra, S.B. Kadambi, U. Ramamurty, G.R. Desiraju, Designing elastic organic crystals: highly flexible polyhalogenated N-benzylideneanilines, Angew. Chem. Int. Ed. 127 (2015) 2712–2716.
[81] A. Mukherjee, G.R. Desiraju, Halogen bonds in some dihalogenated phenols: applications to crystal engineering, IUCrJ 1 (2014) 49–60.
[82] S. Ghosh, M.K. Mishra, S. Ganguly, G.R. Desiraju, Dual stress and thermally driven mechanical properties of the same organic crystal: 2,6-dichlorobenzylidene-4-fluoro-3-nitroaniline, J. Am. Chem. Soc. 137 (2015) 9912–9921.
[83] S. Saha, G.R. Desiraju, σ-hole and π-hole synthon mimicry in third-generation crystal engineering: design of elastic crystals, Chem. Eur. J. 23 (2017) 4936–4943.
[84] S. Saha, G.R. Desiraju, Using structural modularity in cocrystals to engineer properties: elasticity, Chem. Commun. 52 (2016) 7676–7679.
[85] S. Saha, G.R. Desiraju, Acid···Amide supramolecular synthon in cocrystals: from spectroscopic detection to property engineering, J. Am. Chem. Soc. 140 (2018) 6361–6373.
[86] S. Hayashi, T. Koizumi, Elastic organic crystals of a fluorescent π-conjugated molecule, Angew. Chem. Int. Ed. 128 (2016) 2751–2754.
[87] S. Hayashi, S.-y. Yamamoto, D. Takeuchi, Y. Ie, K. Takagi, Creating elastic organic crystals of π-conjugated molecules with bending mechanofluorochromism and flexible optical waveguide, Angew. Chem. Int. Ed. 57 (2018) 17002–17008.
[88] S. Hayashi, A. Asano, N. Kamiya, Y. Yokomori, T. Maeda, T. Koizumi, Fluorescent organic single crystals with elastic bending flexibility: 1,4-bis(thien-2-yl)-2,3,5,6-tetrafluorobenzene derivatives, Sci. Rep. 7 (2017) 9453.
[89] S. Hayashi, T. Koizumi, N. Kamiya, Elastic bending flexibility of a fluorescent organic single crystal: new aspects of the commonly used building block 4,7-Dibromo-2,1,3-benzothiadiazole, Cryst. Growth Des. 17 (2017) 6158–6162.
[90] S. Hayashi, T. Koizumi, Mechanically induced shaping of organic single crystals: facile fabrication of fluorescent and elastic crystal fibers, Chem. Eur. J. 24 (2018) 8507–8512.

[91] A. Worthy, A. Grosjean, M.C. Pfrunder, Y. Xu, C. Yan, G. Edwards, J.K. Clegg, J.C. McMurtrie, Atomic resolution of structural changes in elastic crystals of copper(II) acetylacetonate, Nat. Chem. 10 (2018) 65–69.
[92] A.K. Saini, K. Natarajan, S.M. Mobin, A new multitalented azine ligand: elastic bending, single-crystal-to-single-crystal transformation and a fluorescence turn-on Al(iii) sensor, Chem. Commun. 53 (2017) 9870–9873.
[93] H. Liu, Z. Bian, Q. Cheng, L. Lan, Y. Wang, H. Zhang, Controllably realizing elastic/plastic bending based on a room-temperature phosphorescent waveguiding organic crystal, Chem. Sci. 10 (2019) 227–232.
[94] M.K. Mishra, S.B. Kadambi, U. Ramamurty, S. Ghosh, Elastic flexibility tuning via interaction factor modulation in molecular crystals, Chem. Commun. 54 (2018) 9047–9050.
[95] L. Yuan, M. Xing, F. Pan, Polymorphs of 2,4,6-tris(4-pyridyl)-1,3,5-triazine and their mechanical properties, Acta Crystallogr., Sect. B 75 (2019) 987–993.
[96] H. Liu, Z. Lu, Z. Zhang, Y. Wang, H. Zhang, Highly elastic organic crystals for flexible optical waveguides, Angew. Chem. Int. Ed. 57 (2018) 8448–8452.
[97] B. Liu, Q. Di, W. Liu, C. Wang, Y. Wang, H. Zhang, Red-emissive organic crystals of a single-benzene molecule: elastically bendable and flexible optical waveguide, J. Phys. Chem. Lett. 10 (2019) 1437–1442.
[98] R. Huang, C. Wang, Y. Wang, H. Zhang, Elastic self-doping organic single crystals exhibiting flexible optical waveguide and amplified spontaneous emission, Adv. Mater. 30 (2018) 1800814.
[99] P. Gupta, D.P. Karothu, E. Ahmed, P. Naumov, N.K. Nath, Thermally twistable, photobendable, elastically deformable, and self-healable soft crystals, Angew. Chem. 130 (2018) 8634–8638.
[100] K. Wang, M.K. Mishra, C.C. Sun, Exceptionally elastic single-component pharmaceutical crystals, Chem. Mater. 31 (2019) 1794–1799.
[101] S. Saha, G.R. Desiraju, Trimorphs of 4-bromophenyl 4-bromobenzoate. Elastic, brittle, plastic, Chem. Commun. 54 (2018) 6348–6351.
[102] J.J. Whittaker, K.C. Jack, et al., Comment on "Trimorphs of 4-bromophenyl 4-bromobenzoate. Elastic, brittle, plastic" by S. Saha and G. R. Desiraju, Chem. Commun., 2018, 54, 6348, Chem. Commun. (2021), https://doi.org/10.1039/D0CC07668F. In press.
[103] S. Saha, G.R. Desiraju, Reply to the 'Comment on "Trimorphs of 4-bromophenyl 4-bromobenzoate. Elastic, brittle, plastic"' by J. J. Whittaker, A. J. Brock, A. Grosjean, M. C. Pfrunder, J. C. McMurtrie and J. K. Clegg, Chem. Commun., 2021, 57, DOI: 10.1039/D0CC07668F, Chem. Commun. (2021), https://doi.org/10.1039/D1CC00159K. In press.
[104] A.E. Masunov, M. Wiratmo, A.A. Dyakov, Y.V. Matveychuk, E.V. Bartashevich, Virtual tensile test for brittle, plastic, and elastic polymorphs of 4-bromophenyl 4-bromobenzoate, Cryst. Growth Des. 20 (2020) 6093–6100.
[105] P. Gupta, T. Panda, S. Allu, S. Borah, A. Baishya, A. Gunnam, A. Nangia, P. Naumov, N.K. Nath, Crystalline acylhydrazone photoswitches with multiple mechanical responses, Cryst. Growth Des. 19 (2019) 3039–3044.
[106] K. Yadava, G. Gallo, S. Bette, C.E. Mulijanto, D.P. Karothu, I.-H. Park, R. Medishetty, P. Naumov, R.E. Dinnebier, J.J. Vittal, Extraordinary anisotropic thermal expansion in photosalient crystals, IUCrJ 7 (2020) 83–89.
[107] Y. Nakagawa, M. Morimoto, N. Yasuda, K. Hyodo, S. Yokojima, S. Nakamura, K. Uchida, Photosalient effect of diarylethene crystals of thiazoyl and thienyl derivatives, Chem. Eur. J. 25 (2019) 7874–7880.

[108] T. Seki, K. Sakurada, M. Muromoto, H. Ito, Photoinduced single-crystal-to-single-crystal phase transition and photosalient effect of a gold(i) isocyanide complex with shortening of intermolecular aurophilic bonds, Chem. Sci. 6 (2015) 1491−1497.

[109] Y. Duan, S. Semin, P. Tinnemans, H. Cuppen, J. Xu, T. Rasing, Robust thermoelastic microactuator based on an organic molecular crystal, Nat. Commun. 10 (2019) 4573.

[110] T. Seki, T. Mashimo, H. Ito, Anisotropic strain release in a thermosalient crystal: correlation between the microscopic orientation of molecular rearrangements and the macroscopic mechanical motion, Chem. Sci. 10 (2019) 4185−4191.

[111] T. Seki, C. Feng, K. Kashiyama, S. Sakamoto, Y. Takasaki, T. Sasaki, S. Takamizawa, H. Ito, Photoluminescent ferroelastic molecular crystals, Angew. Chem. 132 (2020) 8924−8928.

[112] Ž. Skoko, S. Zamir, P. Naumov, J. Bernstein, The thermosalient phenomenon. "Jumping crystals" and crystal chemistry of the anticholinergic agent oxitropium bromide, J. Am. Chem. Soc. 132 (2010) 14191−14202.

[113] R. Medishetty, S.C. Sahoo, C.E. Mulijanto, P. Naumov, J.J. Vittal, Photosalient behavior of photoreactive crystals, Chem. Mater. 27 (2015) 1821−1829.

[114] M.I. Tamboli, D.P. Karothu, M.S. Shashidhar, R.G. Gonnade, P. Naumov, Effect of crystal packing on the thermosalient effect of the pincer-type diester naphthalene-2,3-diyl-bis(4-fluorobenzoate): A new class II thermosalient solid, Chem. Eur. J. 24 (2018) 4133−4139.

[115] S. Schramm, M.B. Al-Handawi, D.P. Karothu, A. Kurlevskaya, P. Commins, Y. Mitani, C. Wu, Y. Ohmiya, P. Naumov, Mechanically assisted bioluminescence with natural luciferase, Angew. Chem. 132 (2020) 16627−16631.

[116] P. Naumov, D.P. Karothu, E. Ahmed, L. Catalano, P. Commins, J. Mahmoud Halabi, M.B. Al-Handawi, L. Li, The rise of the dynamic crystals, J. Am. Chem. Soc. 142 (2020) 13256−13272.

[117] S. Takamizawa, Y. Miyamoto, Superelastic organic crystals, Angew. Chem. Int. Ed. 53 (2014) 6970−6973.

[118] S. Sakamoto, T. Sasaki, A. Sato-Tomita, S. Takamizawa, Shape rememorization of an organosuperelastic crystal through superelasticity−ferroelasticity interconversion, Angew. Chem. Int. Ed. 58 (2019) 13722−13726.

[119] M. Owczarek, K.A. Hujsak, D.P. Ferris, A. Prokofjevs, I. Majerz, P. Szklarz, H. Zhang, A.A. Sarjeant, C.L. Stern, R. Jakubas, S. Hong, V.P. Dravid, J.F. Stoddart, Flexible ferroelectric organic crystals, Nat. Commun. 7 (2016) 13108.

[120] T. Sasaki, S. Sakamoto, S. Takamizawa, Flash shape-memorization processing and inversion of a polar direction in a chiral organosuperelastic crystal of 1,3,5-tricyanobenzene, Cryst. Growth Des. 20 (2020) 4621−4626.

[121] S. Takamizawa, Y. Takasaki, Shape-memory effect in an organosuperelastic crystal, Chem. Sci. 7 (2016) 1527−1534.

[122] Y. Takasaki, T. Sasaki, S. Takamizawa, Temperature-diversified anisotropic superelasticity and ferroelasticity in a 3-methyl-4-nitrobenzoic acid crystal, Cryst. Growth Des. 20 (2020) 6211−6216.

[123] S.H. Mir, Y. Takasaki, E.R. Engel, S. Takamizawa, Enhancement of dissipated energy by large bending of an organic single crystal undergoing twinning deformation, RSC Adv. 8 (2018) 21933−21936.

[124] T. Mutai, T. Sasaki, S. Sakamoto, I. Yoshikawa, H. Houjou, S. Takamizawa, A superelastochromic crystal, Nat. Commun. 11 (2020) 1824.

[125] T. Sasaki, S. Sakamoto, Y. Takasaki, S. Takamizawa, A multidirectional superelastic organic crystal by versatile ferroelastical manipulation, Angew. Chem. Int. Ed. 59 (2020) 4340–4343.
[126] E.R. Engel, Y. Takasaki, S.H. Mir, S. Takamizawa, Twinning ferroelasticity facilitated by the partial flipping of phenyl rings in single crystals of 4,4′-dicarboxydiphenyl ether, R. Soc. Open Sci. 5 (2018) 171146.
[127] T. Sasaki, S. Sakamoto, S. Takamizawa, Twinning organosuperelasticity of a fluorinated cyclophane single crystal, Cryst. Growth Des. 19 (2019) 5491–5493.
[128] S. Takamizawa, Y. Takasaki, Versatile shape recoverability of odd-numbered saturated long-chain fatty acid crystals, Cryst. Growth Des. 19 (2019) 1912–1920.
[129] T. Sasaki, S. Sakamoto, S. Takamizawa, Organoferroelastic crystal prepared by supramolecular synthesis, *Cryst*. Growth Des. 20 (2020) 1935–1939.
[130] T. Sasaki, S. Sakamoto, K. Nishizawa, S. Takamizawa, Ferroelasticity with a biased hysteresis loop in a co-crystal of pimelic acid and 1,2-di(4-pyridyl)ethane, Cryst. Growth Des. 20 (2020) 3913–3917.
[131] E.R. Engel, S. Takamizawa, Versatile ferroelastic deformability in an organic single crystal by twinning about a molecular zone axis, Angew. Chem. Int. Ed. 57 (2018) 11888–11892.
[132] S.H. Mir, Y. Takasaki, S. Takamizawa, An organoferroelasticity driven by molecular conformational change, Phys. Chem. Chem. Phys. 20 (2018) 4631–4635.
[133] T. Sasaki, S. Takamizawa, Organoferroelasticity mediated by water of crystallization, Cryst. Growth Des. 20 (2020) 6990–6994.
[134] S. Takamizawa, Y. Takasaki, T. Sasaki, N. Ozaki, Superplasticity in an organic crystal, Nat. Commun. 9 (2018) 3984.
[135] S. Varughese, M.S.R.N. Kiran, U. Ramamurty, G.R. Desiraju, Nanoindentation in crystal engineering: quantifying mechanical properties of molecular crystals, Angew. Chem. Int. Ed. 52 (2013) 2701–2712.
[136] S. Guo, B.B. Akhremitchev, Investigation of mechanical properties of insulin crystals by atomic force microscopy, Langmuir 24 (2008) 880–887.
[137] O. Sahin, S. Magonov, C. Su, C.F. Quate, O. Solgaard, An atomic force microscope tip designed to measure time-varying nanomechanical forces, Nat. Nanotechnol. 2 (2007) 507–514.
[138] V.M. Masterson, X. Cao, Evaluating particle hardness of pharmaceutical solids using AFM nanoindentation, Int. J. Pharm. 362 (2008) 163–171.
[139] I. Azuri, E. Meirzadeh, D. Ehre, S.R. Cohen, A.M. Rappe, M. Lahav, I. Lubomirsky, L. Kronik, Unusually large young's moduli of amino acid molecular crystals, Angew. Chem. Int. Ed. 54 (2015) 13566–13570.
[140] A.B. Singaraju, D. Bahl, L.L. Stevens, Brillouin light scattering: development of a near century-old technique for characterizing the mechanical properties of materials, AAPS PharmSciTech 20 (2019) 109.
[141] C.-T. Chen, S. Ghosh, C. Malla Reddy, M.J. Buehler, Molecular mechanics of elastic and bendable caffeine co-crystals, Phys. Chem. Chem. Phys. 16 (2014) 13165–13171.
[142] M.J. Turner, S.P. Thomas, M.W. Shi, D. Jayatilaka, M.A. Spackman, Energy frameworks: insights into interaction anisotropy and the mechanical properties of molecular crystals, Chem. Commun. 51 (2015) 3735–3738.
[143] M.S.R.N. Kiran, S. Varughese, C.M. Reddy, U. Ramamurty, G.R. Desiraju, Mechanical anisotropy in crystalline saccharin: nanoindentation studies, Cryst. Growth Des. 10 (2010) 4650–4655.

[144] M. Butters, J. Ebbs, S.P. Green, J. MacRae, M.C. Morland, C.W. Murtiashaw, A.J. Pettman, Process development of Voriconazole: A novel broad-spectrum triazole antifungal agent, Org. Process Res. Dev. 5 (2001) 28–36.
[145] D.-K. Bučar, J.A. Elliott, M.D. Eddleston, J.K. Cockcroft, W. Jones, Sonocrystallization yields monoclinic paracetamol with significantly improved compaction behavior, Angew. Chem. Int. Ed. 54 (2015) 249–253.
[146] S. Karki, T. Friščić, L. Fábián, P.R. Laity, G.M. Day, W. Jones, Improving mechanical properties of crystalline solids by cocrystal formation: new compressible forms of paracetamol, Adv. Mater. 21 (2009) 3905–3909.
[147] Y. Hu, A. Erxleben, B.K. Hodnett, B. Li, P. McArdle, Å.C. Rasmuson, A.G. Ryder, Solid-state transformations of sulfathiazole polymorphs: the effects of milling and humidity, Cryst. Growth Des. 13 (2013) 3404–3413.
[148] U. Ramamurty, J.-i. Jang, Nanoindentation for probing the mechanical behavior of molecular crystals—a review of the technique and how to use it, CrystEngComm 16 (2014) 12–23.
[149] B.P.A. Gabriele, C.J. Williams, M.E. Lauer, B. Derby, A.J. Cruz-Cabeza, Nanoindentation of molecular crystals: lessons learned from aspirin, Cryst. Growth Des. 20 (2020) 5956–5966.
[150] P. Manimunda, S.A. Syed Asif, M.K. Mishra, Probing stress induced phase transformation in aspirin polymorphs using Raman spectroscopy enabled nanoindentation, Chem. Commun. 55 (2019) 9200–9203.
[151] M.K. Mishra, G.R. Desiraju, U. Ramamurty, A.D. Bond, Studying microstructure in molecular crystals with nanoindentation: intergrowth polymorphism in Felodipine, Angew. Chem. *126* (2014) 13318–13321.
[152] M.K. Mishra, U. Ramamurty, G.R. Desiraju, Solid solution hardening of molecular crystals: tautomeric polymorphs of omeprazole, J. Am. Chem. Soc. 137 (2015) 1794–1797.
[153] M.K. Mishra, C.C. Sun, Conformation directed interaction anisotropy leading to distinct bending behaviors of two ROY polymorphs, Cryst. Growth Des. 20 (2020) 4764–4769.
[154] K.B. Raju, S. Ranjan, V.S. Vishnu, M. Bhattacharya, B. Bhattacharya, A.K. Mukhopadhyay, C.M. Reddy, Rationalizing distinct mechanical properties of three polymorphs of a drug adduct by nanoindentation and energy frameworks analysis: role of slip layer topology and weak interactions, Cryst. Growth Des. 18 (2018) 3927–3937.
[155] R. Chinnasamy, A. Arul, A. AlMousa, M.S.R.N. Kiran, P. Das, A.S. Jalilov, A.M.P. Peedikakkal, S. Ghosh, Structure property correlation of a series of halogenated Schiff base crystals and understanding of the molecular basis through nanoindentation, Cryst. Growth Des. 19 (2019) 6698–6707.
[156] M.K. Panda, S. Ghosh, N. Yasuda, T. Moriwaki, G.D. Mukherjee, C.M. Reddy, P. Naumov, Spatially resolved analysis of short-range structure perturbations in a plastically bent molecular crystal, Nat. Chem. 7 (2015) 65–72.
[157] E. Ahmed, D.P. Karothu, L. Pejov, P. Commins, Q. Hu, P. Naumov, From mechanical effects to mechanochemistry: softening and depression of the melting point of deformed plastic crystals, J. Am. Chem. Soc. 142 (2020) 11219–11231.
[158] R. Thakuria, M.D. Eddleston, E.H.H. Chow, L.J. Taylor, B.J. Aldous, J.F. Krzyzaniak, W. Jones, Comparison of surface techniques for the discrimination of polymorphs, CrystEngComm 18 (2016) 5296–5301.

[159] M.K. Mishra, K. Mishra, A. Narayan, C.M. Reddy, V.R. Vangala, Structural basis for mechanical anisotropy in polymorphs of a caffeine–glutaric acid cocrystal, Cryst. Growth Des. 20 (2020) 6306–6315.

[160] P. Sanphui, M.K. Mishra, U. Ramamurty, G.R. Desiraju, Tuning mechanical properties of pharmaceutical crystals with multicomponent crystals: voriconazole as a case study, Mol. Pharmaceutics 12 (2015) 889–897.

[161] J.P. Yadav, R.N. Yadav, P. Sihota, H. Chen, C. Wang, C.C. Sun, N. Kumar, A. Bansal, S. Jain, Single-crystal plasticity defies bulk-phase mechanics in isoniazid cocrystals with analogous coformers, Cryst. Growth Des. 19 (2019) 4465–4475.

[162] J.P. Yadav, R.N. Yadav, P. Uniyal, H. Chen, C. Wang, C.C. Sun, N. Kumar, A.K. Bansal, S. Jain, Molecular interpretation of mechanical behavior in four basic crystal packing of isoniazid with homologous cocrystal formers, Cryst. Growth Des. 20 (2020) 832–844.

[163] P.K. Mondal, S. Bhandary, M.G. Javoor, A. Cleetus, S.R.N.K. Mangalampalli, U. Ramamurty, D. Chopra, Probing the distinct nanomechanical behaviour of a new co-crystal and a known solvate of 5-fluoroisatin and identification of a new polymorph, CrystEngComm 22 (2020) 2566–2572.

[164] S. Chattoraj, L. Shi, M. Chen, A. Alhalaweh, S. Velaga, C.C. Sun, Origin of deteriorated crystal plasticity and compaction properties of a 1:1 cocrystal between piroxicam and saccharin, Cryst. Growth Des. 14 (2014) 3864–3874.

[165] E.H.H. Chow, D.-K. Bučar, W. Jones, New opportunities in crystal engineering – the role of atomic force microscopy in studies of molecular crystals, Chem. Commun. 48 (2012) 9210–9226.

[166] M.D. Ward, Bulk crystals to surfaces: combining X-ray diffraction and atomic force microscopy to probe the structure and formation of crystal interfaces, Chem. Rev. 101 (2001) 1697–1726.

[167] D. Tranchida, Z. Kiflie, S. Acierno, S. Piccarolo, Nanoscale mechanical characterization of polymers by atomic force microscopy (AFM) nanoindentations: viscoelastic characterization of a model material, Meas. Sci. Technol. 20 (2009) 095702.

[168] C. Karunatilaka, D.K. Bučar, L.R. Ditzler, T. Friščić, D.C. Swenson, L.R. MacGillivray, A.V. Tivanski, Softening and hardening of macro-and nano-sized organic cocrystals in a single-crystal transformation, Angew. Chem. 123 (2011) 8801–8805.

[169] T.P. Rupasinghe, K.M. Hutchins, B.S. Bandaranayake, S. Ghorai, C. Karunatilake, D.-K. Bučar, D.C. Swenson, M.A. Arnold, L.R. MacGillivray, A.V. Tivanski, Mechanical properties of a series of macro- and nanodimensional organic cocrystals correlate with atomic polarizability, J. Am. Chem. Soc. 137 (2015) 12768–12771.

[170] Y. Jing, Y. Zhang, J. Blendell, M. Koslowski, M.T. Carvajal, Nanoindentation method to study slip planes in molecular crystals in a systematic manner, Cryst. Growth Des. 11 (2011) 5260–5267.

[171] R. Thakuria, M.D. Eddleston, E.H.H. Chow, G.O. Lloyd, B.J. Aldous, J.F. Krzyzaniak, A.D. Bond, W. Jones, Use of in-situ atomic force microscopy to follow phase changes at crystal surfaces in real time, Angew. Chem. Int. Ed. 52 (2013) 10541–10544.

[172] W. Jones, M.D. Eddleston, Introductory lecture: mechanochemistry, a versatile synthesis strategy for new materials, Faraday Discuss 170 (2014) 9–34.

[173] J.-L. Do, T. Friščić, Mechanochemistry: a force of synthesis, ACS Cent. Sci. 3 (2017) 13–19.

[174] S.L. James, C.J. Adams, C. Bolm, D. Braga, P. Collier, T. Friščić, F. Grepioni, K.D.M. Harris, G. Hyett, W. Jones, A. Krebs, J. Mack, L. Maini, A.G. Orpen, I.P. Parkin, W.C. Shearouse, J.W. Steed, D.C. Waddell, Mechanochemistry: opportunities for new and cleaner synthesis, Chem. Soc. Rev. 41 (2012) 413–447.
[175] A. Delori, T. Friščić, W. Jones, The role of mechanochemistry and supramolecular design in the development of pharmaceutical materials, CrystEngComm 14 (2012) 2350–2362.
[176] T. Friščić, W. Jones, Recent advances in understanding the mechanism of cocrystal formation via grinding, Cryst. Growth Des. 9 (2009) 1621–1637.

CHAPTER

4 A combined theoretical and CSD perspective on σ-hole interactions with tetrels, pnictogens, chalcogens, halogens, and noble gases

Rosa M. Gomila[1], Tiddo J. Mooibroek[2], Antonio Frontera[1]

[1]*Universitat de les Illes Balears, Palma de Mallorca (Baleares), Spain;* [2]*van 't Hoff Institute for Molecular Sciences, Amsterdam, the Netherlands*

1. Introduction

Covalently bonded hydrogen (X—H) is well known to act as an electron-deficient partner to engage in hydrogen bonding (HB) interactions with some electron donor, especially if the X—H bond is strongly polarized. The term "hydrogen bond" is derived from the element that bears the positive electrostatic potential. It has become clear in recent years that covalently bonded elements of groups 13—18 also have a strong tendency to interact with electron-rich atoms or π-systems [1—13]. Inspired by the HB terminology, these interactions can be named according to the element bearing the electropositive potential. Alternatively, it has been proposed to designate the noncovalent interactions between electrophilic and electron-donating sites by using the name of the group of the Periodic Table the electrophilic atom belongs to [14,15]. In fact, the terms halogen bond (HaB, group 17) [16] and chalcogen bond (ChB, group 16) [17] are now recommended by the IUPAC, and these interactions are the most investigated of the p-block family. Moreover, the terms noble gas bonding (NgB, group 18) [12], pnictogen bonding (PnB, group 15) [18,19], tetrel bonding (TrB, group 14) [20], and triel bonding (TrlB, group 13) [7] are commonly used. Recent investigations have demonstrated that these interactions can be exploited in supramolecular chemistry, catalysis, drug design, membrane transport, and so on. Several review articles, accounts, and feature articles are available in the literature describing a variety of applications and fundamental properties [21—41], [42—62]. Moreover, several relevant manuscripts have compared these interactions to the ubiquitous HB [63—68].

The net attractive interaction between the Lewis acid p-block element Y covalently bound to some electron-withdrawing group (EWG) and the electron-donating site (:X) can be written as EWG—Y···:X. The physical nature of these

Hot Topics in Crystal Engineering. https://doi.org/10.1016/B978-0-12-818192-8.00001-9

interactions has been rationalized as the sum of electrostatics, charge transfer, orbital mixing, polarization, and dispersion forces [69,70]. Typically, the electrostatic term is described either by a σ-hole (a region of positive electrostatic potential located on the Y atom at the elongation of the EWG—Y bond) or a π-hole (a region of positive electrostatic potential located above and below the EWG-Y bond). Dispersion and polarization terms can be very important in certain cases, especially in heavier elements [37,71–75]. The orbital mixing term in σ-hole interactions is commonly designated as an $n \rightarrow \sigma^*$ or $\pi \rightarrow \sigma^*$ donation, indicating whether lone pair (n) or π-type orbitals interact with the antibonding σ* orbital of the EWG—Y bond. For the π-hole interaction, the π*(EWG—Y) orbital instead of σ*(EWG—Y) participates in the donor—acceptor interaction. The positive potential on an element is proportional to the strength of an interaction, and this property can be enhanced by increasing the electron-withdrawing ability of EGW. Moreover, the utilization of either heavier elements or strong EWGs decreases the energy level of the σ* or π* orbitals, thus making them better electron acceptors [70]. The utilization of more polarizable heavier atoms lowers the energy of the antibonding orbitals and also increases the contribution of the favorable dispersion term [76,77]. The triels present a special case because $TrlF_3$ structures have a formally empty p-orbital, making them classical Lewis acids (strong $n \rightarrow p$ donation). This vacant p-orbital forces $TrlR_3$ structures into a trigonal planar geometry, meaning that the electropositive sites are located "above"/"below" the molecule. As a consequence, interactions with saturated triels (e.g., $TrlF_3$) can be seen as π-hole interactions according to the given definition and will not be further discussed in this chapter.

The interest in π-hole [78] and σ-hole [79] interactions is growing very fast among the scientific community, thus allowing the understanding, developing, and exploitation of new lines of research [14,15,80]. The distribution of electron density is anisotropic in the elements of groups 15—17 due to the coexistence of lone pairs and σ-holes on the same element. The number of lone pairs varies from 0 to 3 and the number of σ-holes from 4 to 1 on going from group 14 to group 17 for saturated elements. The number and location of positive (σ-holes) and negative electrostatic potential regions at the surface of atoms is related to the number and position of the covalent bonds the atoms are engaged in. That is, the elements of groups 14, 15, 16, and 17 most frequently form four, three, two, and one covalent bonds that coincide with the number of σ-holes opposite to these bonds. Molecular entities capable of engaging in π-hole interactions always have a maximum of two π-holes due to their flat geometry (i.e., the colloquial "top" and "bottom").

In this chapter, σ-hole interactions involving elements of groups 14 to 18 are first analyzed and compared using a theoretical perspective from the novel vantage point: comparing properties of σ-hole donating atoms across the groups and rows of the Periodic Table. Secondly, illustrative examples retrieved from the Cambridge Structural Database (CSD) for the 20 elements of groups 14—18 and periods 2—5 are highlighted. A comparative statistical CSD survey of σ-hole interactions is presented in Sections 3, followed by a summary and some concluding remarks in Section 4.

2. Results and discussion

2.1 *Ab initio* study

In this section, some theoretical features of the p-block elements in groups 14–18 are analyzed. These include the elemental polarizabilities, the electropositivity at the σ-hole of F-saturated elements (e.g., CF_4), and a study of the orbital interactions of acetonitrile adducts (e.g., $CH_3CN\cdots CF_4$). For details on the computational methods and appropriate references, see Section 4. Table 4.1 summarizes the polarizabilities (and van der Waals radii) [81] of the 20 elements belonging to groups 14–18 and periods 2–5. The polarizability increases on going from periods 2–5 and also from groups 18–14. Therefore, the most polarizable element is Sn (37.3 a.u.), and the least one is Ne (1.2 a.u.). It is also worthy to comment that the difference in the atomic polarizabilities between elements of periods 2 and 3 is quite large (by a factor of approximately 3).

The molecular electrostatic potential (MEP) values at the σ-holes of the saturated fluoride derivatives of the 16 elements (from groups 14 to 17) are given in Table 4.2, and some of them are represented in Fig. 4.1. We have not used elements of group 18 because their fluoride derivatives (NgF_n, $n = 2, 4, 6$) do not present σ-holes on the Ng atoms (only π-hole), and consequently the values are not comparable. The MEP values of series of noble gas compounds are available in the

Table 4.1 Atomic polarizabilities (α, a.u.) of elements from groups 14–18 at the MP2/def2-TZVP level of theory and their van der Waals radii [81] in parenthesis.

	Group 14	Group 15	Group 16	Group 17	Group 18
Period 2	9.0 (1.70)	5.3 (1.55)	3.0 (1.52)	1.8 (1.47)	1.2 (1.54)
Period 3	26.1 (2.10)	16.9 (1.80)	11.8 (1.80)	8.4 (1.75)	6.3 (1.88)
Period 4	28.4 (2.11)	21.6 (1.85)	17.5 (1.90)	14.0 (1.89)	11.5 (2.02)
Period 5	37.3 (2.17)	30.8 (2.06)	25.9 (2.06)	22.6 (1.98)	19.9 (2.16)

Table 4.2 Molecular electrostatic potential energy (kcal/mol) at the σ-holes for saturated fluoride derivatives of indicated elements at the MP2/def2-TZVP level of theory.

	TrF_4	PnF_3	ChF_2	HaF
Period 2	18.6	15.9	16.8	15.9
Period 3	39.0	27.4	35.6	42.3
Period 4	50.2	38.5	44.9	49.7
Period 5	66.5	46.7	52.6	55.8

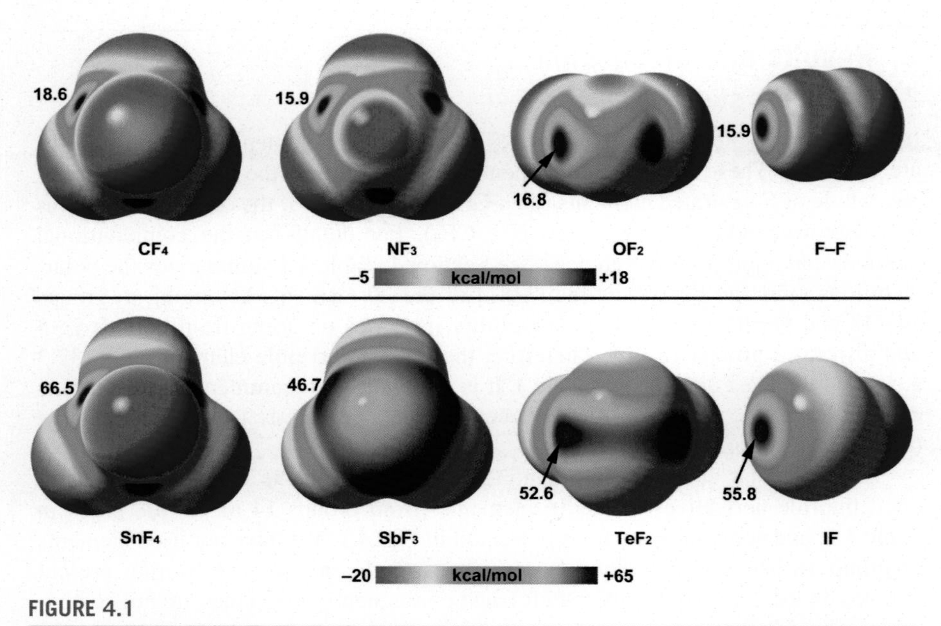

FIGURE 4.1

MEP surfaces of several fluorides of elements of groups 14–17 using the 0.001 a.u. isosurface and the MP2/def2-TZVP level of theory (energies in kcal/mol). *MEP*, molecular electrostatic potential.

literature [12]. The results gathered in Table 4.2 show that the MEP values increase when descending from period 2 to period 5, similarly to the behavior observed for the polarizability.

All MEP values in period 2 are roughly identical. The MEP values in groups 15–18 show a clearly increasing trend (15 → 18). That these values for elements in group 14 are larger than those of group 15 is likely due to the absence of free lone pairs on the tetrels (which could contribute some electron density to dampen the σ-holes).

As illustrative examples, the MEP surfaces of the second and fifth periods are shown in Fig. 4.1. It is interesting to comment that the main difference (apart from the MEP values) is found in the pnictogen and chalcogen derivatives. That is, while the location, size, and shape of the σ-hole are similar for the halogen (F–F and I–F) and tetrel (CF_4 and SnF_4) molecules, the size and shape significantly change in the chalcogen and pnictogen fluorides. The σ-holes are less defined, thus anticipating lesser directionality of ChBs and PnBs compared with TrBs and HaBs. Moreover, another significant difference is the fact that in TrF_4 molecules (also in the NF_3), the electrostatic attraction between the electron-rich site and the σ-hole can be compensated by the repulsion with the negative F-atoms (or the LP in NF_3).

The interaction energies of the fluorides of elements of groups 14–18 with acetonitrile as electron donor are summarized in Table 4.3 along with the van der

Table 4.3 Interaction energies (in kcal/mol) and van der Waals–corrected equilibrium distances (in Å) for the σ-hole complexes of CH_3CN with fluoride derivatives of elements of the p-block at the MP2/def2-TZVP level of theory. Both the energies and distances are basis set superposition error (BSSE) corrected.

Period	$CH_3CN\cdots TrF_4$	$CH_3CN\cdots PnF_3$	$CH_3CN\cdots ChF_2$	$CH_3CN\cdots HaF$	$CH_3CN\cdots NgF_2$
2	−0.9 (0.285)	−0.9 (0.199)	−1.1 (−0.040)	−1.1 (−0.193)	Not found
3	−2.8 (−0.545)	−2.5 (−0.173)	−3.9 (−0.493)	−5.5 (−0.731)	−2.3 (−0.244)
4	−5.5 (−1.242)	−4.6 (−0.406)	−6.3 (−0.718)	−8.1 (−0.922)	−2.6 (−0.247)
5	−15.5 (−1.391)	−7.6 (−0.636)	−9.3 (−0.918)	−10.8 (−0.962)	−2.7 (−0.191)

Waals–corrected equilibrium distances (i.e., the calculated distance minus the van der Waals radii of N and the electrophilic element, see also Table 4.1). Since acetonitrile is a weak lone pair donor (low basicity of the sp-hybridized lone pair), the interaction energies span from modest to moderately strong (−0.9 to −15.5 kcal/mol). Interestingly, the interaction energies are similar for the complexes of elements belonging to period 2; however, the amount of van der Waals overlap increases from none at all in $CH_3CN\cdots CF_4$ (0.285 Å) and $CH_3CN\cdots NF_3$ (0.199 Å) complexes to −0.040 Å in $CH_3CN\cdots OF_2$ and even −0.193 Å in $CH_3CN\cdots F_2$. These differences are likely due to the repulsion with the F-atoms/lone pair in former complexes. The HaB and ChB complexes present stronger energies and more van der Waals overlap, in line with the large σ-hole MEP values and absence of steric/repulsion effects with the substituents. The strongest complex corresponds to the $CH_3CN\cdots SnF_4$ due to the partial covalent character (−15.5 kcal/mol and van der Waals–corrected distance is −1.391 Å). The overall pattern of interaction energies as well as the trends in van der Waals overlap follows the pattern seen in MEP values (Table 4.2).

Fig. 4.2 shows the geometries of the complexes of the second and fifth periods, apart from the Ng series where the $F_2Ar\cdots NCCH_3$ complex (third period) is represented instead of the $F_2Ne\cdots NCCH_3$ that is not stable. The most important feature is that the high directionality observed in the σ-hole complexes of the second period is significantly lost in the ChB and PnB complexes of the fifth period, and conserved in the HaB and TrB complexes, in line with the MEP surface analysis commented on before. It is also worth noting that in case of SnF_4, the tetrahedral geometry around the Sn atom changes to a pseudo-trigonal pyramid, even in the presence of a very weak Lewis base, thus evidencing the strong tendency of heavier tetrels to expand their valence. Finally, in the SbF_3 complex, the orientation of acetonitrile suggests that the electron donor is the π-system of the CN triple bond instead of the lone pair.

Table 4.4 summarizes the orbital donor–acceptor interactions computed for the acetonitrile complexes to analyze the variation of the LP → σ* orbital contribution depending on the group and period of the periodic table. Apart from tetrels, the contribution increases on going from groups 15 to 17 and descending in the group. For Sn, the bond is considered as covalent in the NBO treatment, and for Ge, the contribution is larger than the pnictogen and chalcogen elements of the same period.

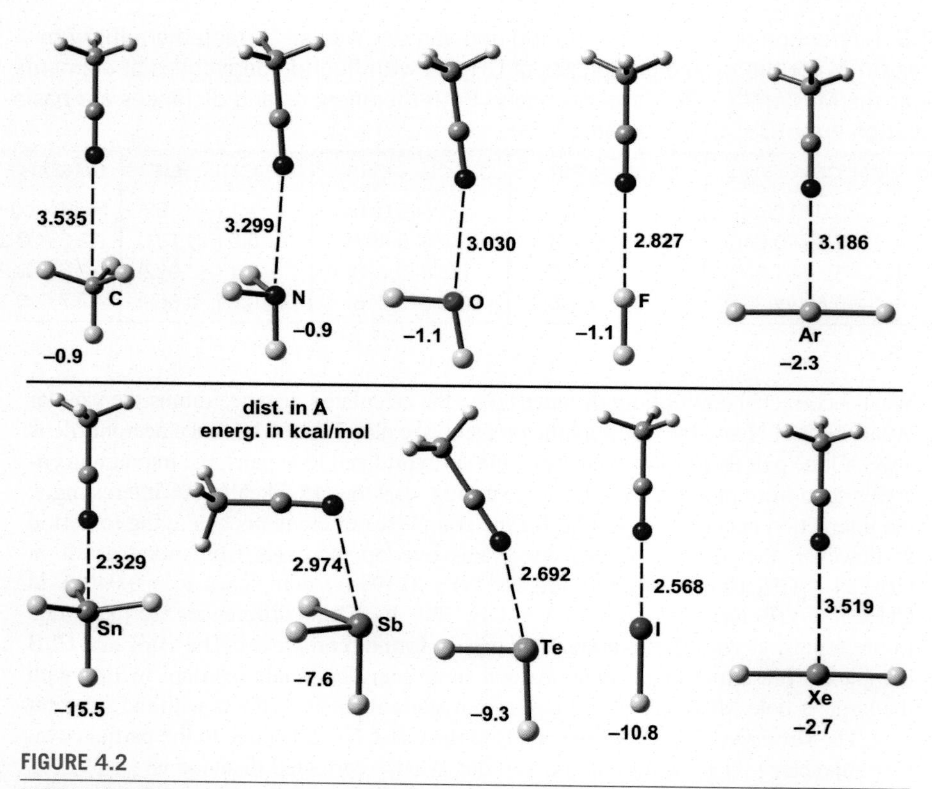

FIGURE 4.2

MP2/def2-TZVP complexes with indication of the BSSE-corrected energies and geometries.

Table 4.4 Orbital donor—acceptor [LP(N) → σ*(Y—F), Y = p-block element)] $E^{(2)}$ energies (kcal/mol) obtained from the second-order perturbation analysis for the σ-hole complexes of CH_3CN with fluoride derivatives of elements of the p-block at the MP2/def2-TZVP level of theory.

Period	HCN··· TrF_4	HCN···PnF_3	HCN···ChF_2	HCN···HaF
2	0.02	0.09	0.32	0.63
3	0.23	0.19	2.43	9.17
4	11.31[a]	1.73	6.44	18.21
5	Covalent	2.47[a]	12.26	25.16

[a] *Also includes a contribution of π(CN) → σ*(Tr/Pn—F).*

2.2 Cambridge Structural Database examples and analyses

2.2.1 Tetrel bonding

Tetrel bonds are rather different compared with other σ-hole interactions. That is, taking the HaB as an example, the approaching Lewis Base is separated by 180 degrees from the EWG that is covalently attached to the halogen atom. As a result, there is no steric repulsion at all. The situation is similar for ChBs and, in the case of PnBs (the R_3Pn molecule is trigonal pyramidal in shape), the electron-rich atom is still able to approach the Pn atom without substantial steric repulsion. However, the situation is clearly different in tetrel atoms since the tetrahedral spacing of its four substituents increases the steric crowding and increases the steric repulsions with the electron-rich entity. This issue is very important in TrBs and has been studied in detail by Scheiner [82] focusing on the size of the tetrel atom (from Si to Pb), the size of R in the R_3TrF σ-hole donor, the type of electron donor (neutral or anionic), and the electron-withdrawing ability of R.

2.2.1.1 Carbon

The ability of heavier tetrel atoms to act as Lewis acids suitable to accommodate electron-rich guests [20,28,83–86] has been recognized for decades. Surprisingly the investigations dedicated to studying the most abundant of Tr atoms, namely, the carbon atom [87] are much more recent. Some of these investigations have demonstrated that 1,1,2,2-tetracyanocyclopropane [88–92] is a convenient building block to form strong complexes with electron-rich entities by using the strong σ-hole located in the middle of the $(CN)_2C{-}C(CN)_2$ bond. Moreover, Mani and Arunan [93–96] have demonstrated the ability of the carbon atom in a methyl group or in an aliphatic chain (sp^3 hybridized) to participate in σ-hole interactions, leading to noncovalent "carbon bonding." This term should be avoided and replaced by tetrel bonding following the nomenclature proposed by the IUPAC.

Tetrel bonds (TrBs) formed in crystalline solids embracing sp^3-hybridized carbon atoms have been recently reviewed [2], where some functional groups particularly tailored to form TrBs were identified and described, including methyl and methylene groups bound to ammonium, pyridinium, and sulfonium. These cationic EWGs promote particularly short and directional TrBs. Neutral groups such as fluorine, nitro, and cyano substituents are also able to promote the formation of TrBs in solid-state X-ray structures. TrBs involving sp^3-hybridized carbon atom are also important in biological systems, in particular the inhibition of serine proteases and antagonists of the muscarinic acetylcholine receptor [97].

Fig. 4.3 shows the X-ray structures of selected examples where the TrBs involving carbon are relevant and very directional due to the presence ammonium cations, thus promoting charge-assisted TrBs. For instance, 9-cyano-10-methylacridinium [98] forms infinite 1D supramolecular assemblies in the solid state where the methyl group establishes a short contact with the N-atom of the cyano group of the adjacent molecule (3.189 Å, see Fig. 4.3a) that is located exactly at the elongation of the $^+N{-}CH_3$ bond (180 degrees). Fig. 4.3b shows that the

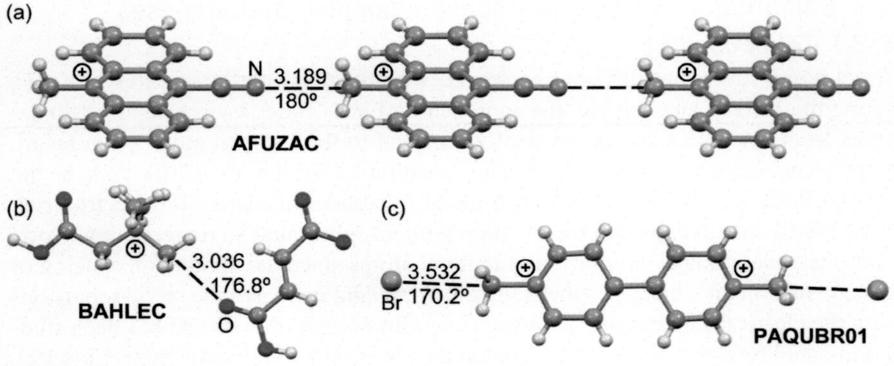

FIGURE 4.3

TrBs in the solid state of CSD refcodes AFUZAC (a), BAHLEC (b), and PAQUBR01 (c) X-ray structures (represented in ball and stick). The TrBs are represented using dashed lines. The values of distances (Å) and angles of interactions are indicated close to the contacts. *CSD*, Cambridge Structural Database; *TrBs*, tetrel bonds.

methyl group of *N,N,N*-trimethyl glycine fumarate [99] establishes a quite short and directional N—C···O TrB (Fig. 4.3b) with a C···O separation of 3.036 Å and N—C···O angle of 176.8 degrees. Finally, the carbon atom in *N*-methyl pyridinium derivatives is very well suited for establishing TrBs, as exemplified by the *N,N′*-dimethyl-4,4′-dipyridinuim bromide [100] represented in Fig. 4.3c. The bromide counteranions are located opposite to the $^{+}N-CH_3$ bonds, thus establishing two symmetrically equivalent TrBs with C···Br separation of 3.532 Å.

2.2.1.2 Silicon

Silicon is best prepared for establishing noncovalent TrBs than the rest of heavier tetrels because Ge and Sn can easily expand their valence, thus engaging in more traditional chemistry. In case of Pb, tetrahedral Pb(IV) compounds are quite rare, and both Pb(II) and Sn have a rich coordination chemistry [101—104] and are usually treated as metals. Nevertheless, it should be mentioned that hypervalent species of Si are rarer but also known [105—115]. TrB interactions involving Si were initially explored by Alkorta et al. [116] both theoretically in tetrahalosilanes and also experimentally in aminopropylsilanes [117]. Moreover, Scheiner et al. have compared TrBs in TrF_3 adducts (Tr = C, Si, and Ge) with charge neutral and protonated pyridine, furan, and NH_3 [118].

Fig. 4.4 shows two illustrative examples of relevant TrB interactions in the solid state between fluoride and chloride anions and Si where the typical nucleophilic attack (valence expansion) did not happen. It has been shown that the utilization of fluoride salts improves the synthesis of spherosilicates like the one shown in Fig. 4.4a (WAVYAV) [119]. This class of cage compounds incorporates the fluoride ion perfectly centered within the octasiloxane cage, establishing eight

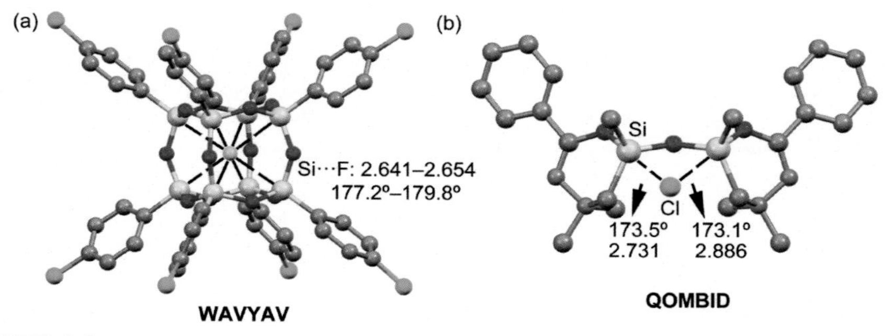

FIGURE 4.4

TrBs in the solid state of CSD refcodes WAVYAV (a) and QOMBID (b) X-ray structures (represented in ball and stick). The TrBs are represented using dashed lines. The values of distances or range of distances (Å) and angles or range of angles of interactions are indicated close to the contacts. *CSD*, Cambridge Structural Database; *TrBs*, tetrel bonds.

pseudosymmetric Si···F TrBs [120]. Directional TrBs have been also described in the disilolane derivatives like the one highlighted in Fig. 4.4b (QOMBID) [121] that is able to recognize chloride anions by establishing two short Si···Cl noncovalent contacts.

Fig. 4.5a shows the X-ray solid-state structure of Si_6Cl_{12} that exhibits a chair-like conformation (refcode AZAZUX) [122]. Quite remarkably, the addition of chloride anion induces planarity in the dodecachlorohexasilinane ring (Fig. 4.5b, refcode LECXIC) [123]. The unexpected formation of this “inverse sandwich” complex [123,124], where two anions are located above and below the hexasilinane ring, is due to the formation of six concurrent tetrel bonds. These TrBs are able to compensate the energy difference between the chain and planar conformation of the six-membered ring and also overcome the electronic repulsion of the two anions. The MEP surfaces of dodecachlorohexasilinane in the chair-like and planar conformations (Fig. 4.5c and d, respectively) evidence that the σ-holes are not accessible in the chair conformation. However, in the D_{6h} conformation, two strong and accessible positive regions (+85 kcal/mol) appear as a consequence of the superposition of six σ-holes above and below the Si_6 ring.

2.2.1.3 Germanium and tin

While most of the experimental and theoretical work on TrB interactions has been directed toward carbon and silicon [2,20,116,117], investigations of the heavier elements of group 14 have been reported [125]. Two excellent reviews are available in the literature where a large number of crystal structures are provided to demonstrate the ability of Ge and Sn to establish highly directional interactions [126,127]. These TrB interactions are very important governing their crystal packing. These reviews mainly focus on organic derivatives of Ge and Sn and demonstrate that TrBs are more directional than PnBs and ChBs and similar to HaBs, in good agreement with the aforementioned MEP surfaces (see Fig. 4.1).

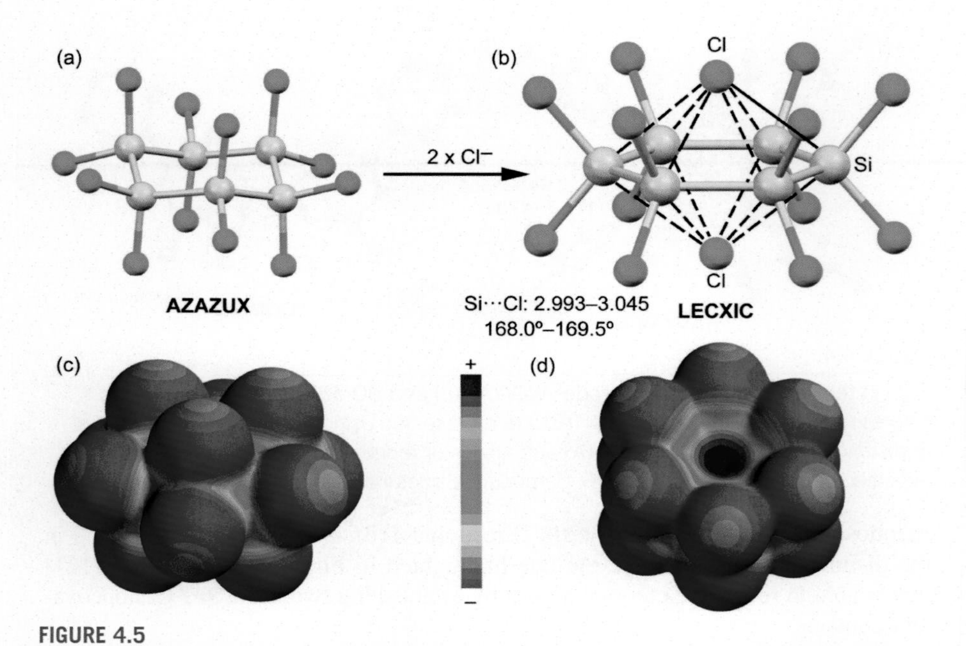

FIGURE 4.5

(a) X-ray structure of dodecachlorohexasilinane (refcode AZAZUX). (b) TrBs in the solid state of CSD refcode LECXIC X-ray structure (represented in ball and stick). The TrBs are represented using dashed lines. The values of range of distances (Å) and range of angles of TrB interactions are indicated. (c) MEP surface (0.001 a.u.) of dodecachlorohexasilinane in a chair conformation. (d) MEP surface (0.001 a.u.) of dodecachlorohexasilinane in a planar conformation (D_{6h}). The MEP value at the σ-hole is +85 kcal/mol. Color code for MEP surfaces: red negative and blue positive. *CSD*, Cambridge Structural Database; *MEP*, molecular electrostatic potential; *TrBs*, tetrel bonds.

Two relevant examples of Ge and Sn are highlighted herein to illustrate their ability to establish highly directional TrBs. Fig. 4.6a shows the X-ray structure of $[Ge_8O_{12}(OH)_8F]^-$ anion (refcode VUZVUH) [128] where the $Ge_8O_{12}(OH)_8$ moiety presents a cuboidal structure with the fluoride anion encapsulated in the center, in a situation similar to the silicon derivative commented on before (refcode WAVYAV). The other selected structure of germanium (Fig. 4.6b) presents tetrel-bonded infinite chains in its crystal packing (refcode QAHXIG) [129] where the most positive σ-hole on germanium (opposite to the O_2SC–Ge covalent bond) interacts with the sulfonyl oxygen atom. This interaction is particularly directional with a Ge···O separation of 3.520 Å and C–Ge···O angle of 179.8 degrees.

For tin, two X-ray structures characterized by the formation of infinite 1D supramolecular chains have been selected. In both cases, the electron-rich atom approaches the Sn atom opposite to the most polarized bond (Sn–X, X = Cl for

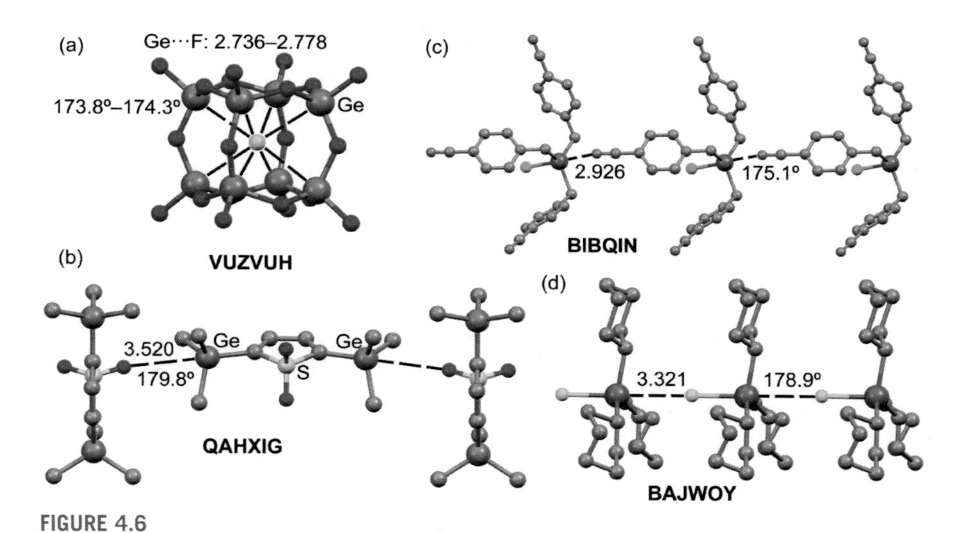

FIGURE 4.6

TrBs in the solid state of CSD refcodes VUZVUH (a), QAHXIG (b), BIBQIN (c), and BAJWOY (d) X-ray structures (represented in ball and stick). The TrBs are represented using dashed lines. The values of distances (Å) and angles of interactions are indicated close to the contacts. *CSD*, Cambridge Structural Database; *TrBs*, tetrel bonds.

BIBQN [130] and F for BAJWOY [131]). The Sn···N distance in the chlorotris(4-cyanobenzyl)tin structure is very short, especially taking into consideration that the electron donor is poor Lewis base (sp-hybridized N-atom). This is likely due to the large polarizability of the Sn atom, and it is in good agreement with the $F_4Sn\cdots NCCH_3$ complex described earlier in Section 2.1 (see Fig. 4.2). A similar behavior is also observed in structure BAJWOY, tricyclohexyltin fluoride [131], where the fluorine atom of one molecule establishes a short and remarkably linear TrB opposite to the covalent Sn—F bond of an adjacent molecule (thus generating the infinite chains).

2.2.2 Pnictogen bonding

While theoretical manuscripts dealing with PnBs are numerous [35,63,132–140], those presenting experimental evidences are much less common [141–143]. Several reviews have evidenced that PnBs are relevant in many single-crystal X-ray structures and can be used as a reliable tool in crystal engineering. Some examples are highlighted in this section [35,36,144].

2.2.2.1 Nitrogen

To our knowledge, there is no experimental evidence of σ-hole interactions involving nitrogen acting as Lewis acid (σ-hole donor). In stark contrast, there are numerous examples where the N atoms participate in π-hole interactions especially in nitro derivatives ($R—NO_2$) [145–149]. Nitrate anions can even act as π-hole donors in crystal structures when the negative charge is partially transferred to adjacent molecules by means of noncovalent interactions such as hydrogen bonding [150] or coordination to a metal ion [151]. Moreover, rotational spectroscopy experiments revealed that π-hole bonding geometry prevails in the gas phase for the complex between Me_3N and $EtNO_2$ [152]. Although it is not the main topic of this chapter, due to its relevance, several examples of X-ray structures to illustrate the ability of N atom to establish π-hole interactions are given in Fig. 4.7. It shows four examples, where in one of them (FBATNB [153]) in addition to the π-hole interaction established between the F atom and the nitro's N atom, an additional N···O contact is established, approximately opposite to the polarized F—N bond. It is difficult to assert whether this contact can be classified as an attractive σ-hole interaction or it is simply originated as a consequence of the minimization of the steric repulsion between the substituents of the aromatic ring. To this respect, Resnati et al. [144] analyzed the CSD and found four structures where a C—NF_2 moiety is present and in short contact with electron-rich atoms approximately on the extension of the N—F covalent bond. Although this data set is too small to draw any conclusion, it seems that the N atom is able to establish σ-hole interactions when it is bonded to the electron-withdrawing fluorine atom, also in agreement with the theoretical results (see Table 4.3).

Regarding the other three structures of Fig. 4.7, one corresponds to a nitro derivative (refcode NAPJAS [147]) and the other two to nitro esters (refcodes NAYHAX [154] and BEDSUA [155]). It is interesting to highlight that in the cocrystal dinitrobutadiene with dioxane (see Fig. 4.7b), the central dinitrobutadiene forms up to four O···N π-hole interactions in the solid state illustrating the ability of the nitro group

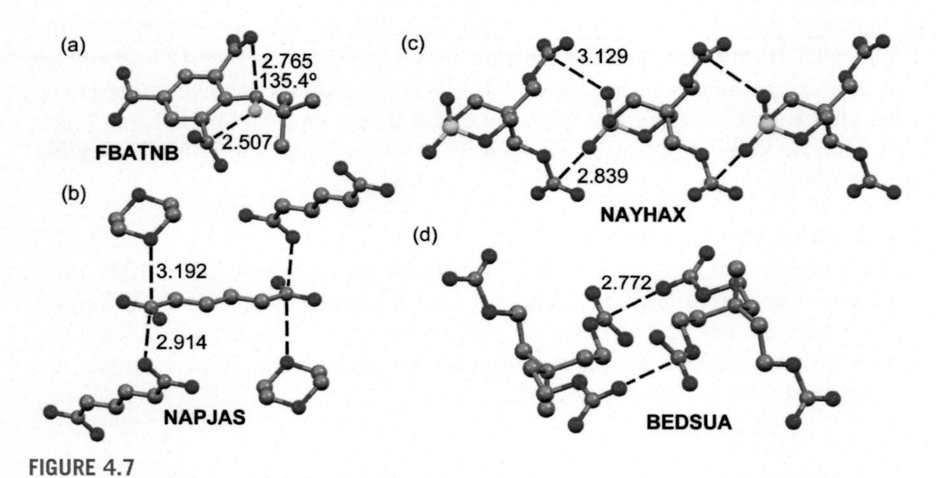

FIGURE 4.7

PnBs in the solid state of CSD refcodes FBATNB (a), NAPJAS (b), NAYHAX (c), and BEDSUA (d) X-ray structures (represented in ball and stick). The PnBs are represented using dashed lines. The values of distances (Å) are indicated close to the contacts. *CSD*, Cambridge Structural Database; *PnBs*, pnictogen bonds.

to act as a Lewis acid. The formation of infinite 1D supramolecular chains in NAYHAX structure is also remarkable, where two N···O PnBs are established between the O atoms of the sulfate ester group with the N atoms of the nitrate ester groups (see Fig. 4.7c). Finally, the BEDSUA structure forms self-assembled dimers in the solid state where the nitrate ester groups act as both donor and acceptor, forming two short N···O contacts responsible for the generation of the dimers.

2.2.2.2 Phosphorus

Some theoretical aspects of PnBs involving phosphorus were revised by Scheiner in 2013 [35]. The physical nature, effects of the substituents, and comparison with other interactions were described and discussed in detail [35]. Remarkably, it was demonstrated that PnBs can be more attractive than H bonds. Experimental evidences for intermolecular pnictogen bonding involving phosphorus are scarcely found in the literature. It is well known, however, that M-PR_3 bonds in complexes of late transition metals (M) and phosphane ligands (PR_3) can be stabilized by so-called "back-bonding" interactions, where filled metal d-orbitals donate electron density into the σ^* orbitals of a P—R bond [156]. This back-bonding can thus be seen as a special case of a σ-hole interaction.

It has been recently demonstrated using matrix isolation infrared spectroscopy that P···O interactions occur in PCl_3—CH_3OH adducts [142]. Similarly, P···O and P···π interactions have been described in PCl_3—H_2O and PCl_3—C_6H_6 systems where PnBs dominate over conventional HBs [157,158].

While there are many examples in the literature where PnBs are recognized as important structure directing interactions in the solid state of arsenic [159—161], antimony [162—166], or bismuth [167—175], those involving phosphorus are less common [176]. Fig. 4.8 shows three X-ray structures where directional PnBs are formed. Those shown in Fig. 4.8a and b are tetraphosphines of formula *p*-$C_6H_4[N(PX_2)_2]_2$, where X = F for LOGSIK and X = Cl for LOGSEG [177]. It is interesting to highlight that their X-ray solid-state architectures are totally different, thus revealing that the substitution of F by Cl has a great impact on the crystal packing. In fact, the LOGSEG structure forms 2D sheets in the solid state [177] governed by P···P PnBs where the lone pair of one P-atom is located opposite to the N—P bond. In sharp contrast, the LOGSIK structure forms P···F contacts where the electron-rich F atom is located opposite to the most polarized P—F bond. In both cases, the directionality of the interaction is similar (around 170 degrees). Finally, Fig. 4.8c shows the X-ray cocrystal of 1,4-dimethylpiperazine and PBr_3 (MEWYES) where two highly directional P···N bonds are formed with the axial lone pairs pointing exactly to the elongation of the Br—P bonds [178].

2.2.2.3 Arsenic and antimony

As commented earlier, As and Sb are more prone to establish PnBs than the lighter N and P. In fact, they have recently been used in catalysis [179] and anion recognition [180] and transmembrane anion transport [59]. Among the heavier pnictogens, Sb(III) is the most promising to be used in supramolecular chemistry, particularly

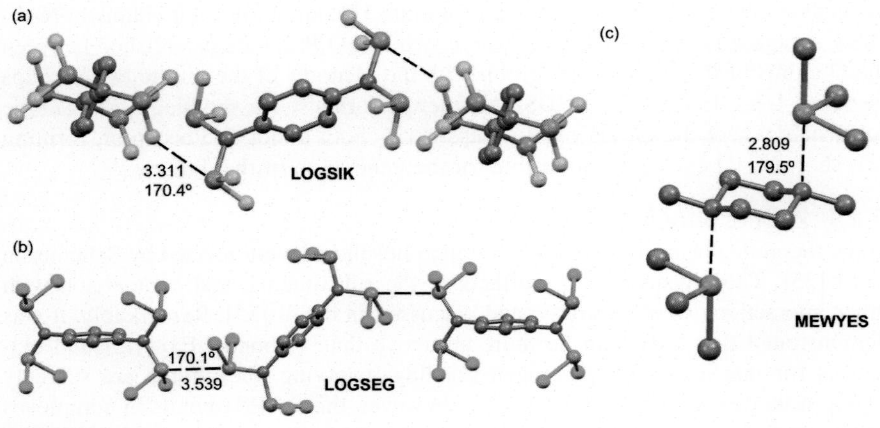

FIGURE 4.8

PnBs in the solid state of CSD refcodes LOGSIK (a), LOGSEG (b), and MEWYES (c) X-ray structures (represented in ball and stick). The PnBs are represented using dashed lines. The values of distances (Å) and angles are indicated close to the contacts. *CSD*, Cambridge Structural Database; *PnBs*, pnictogen bonds.

in the recognition of anions. For instance, Gabbaï's group has used bis-stilbonium to build a host [181] that includes two cationic Sb(III) centers that establish several PnB interactions with the oxygen atoms of the triflate counteranion. Cozzolino and coworkers have also described bidentate PnB anion-binding motifs containing two Sb(III) centers bridged by either oxygen or sulfur atoms [182].

Fig. 4.9 gathers several examples of As and Sb structures to illustrate the ability of As(III) and Sb(III) derivatives to form PnBs. For instance, in the inorganic compounds, triazide arsenic and triazide antimony (ICSD413360 and ICSD413359 [183], respectively), the three σ-holes located at the extension of the N—Pn bonds interact with the electron-rich azide group of three adjacent molecules to establish three concurrent PnBs (identical in the $Sb(N_3)_3$ compound). Similarly, tricyano-arsenic(III) also forms three PnBs with the adjacent molecules in the crystal structure (refcode USEPUF [184]). In case of the tricyano-antimony(III) cocrystallized with 2,2′-bipyridine (refcode USEQAM [184], Fig. 4.9e), the N(bpy)—Sb PnBs are very short (2.724 and 2.560 Å), and the remaining σ-hole establishes a longer contact with the N atom of the cyano group of the adjacent molecule. This leads to the formation of infinite 1D chains in the solid state (not shown in Fig. 4.9). Two additional structures are included in Fig. 4.9 as examples of Pn···O contacts. In UROZIL [185], the As atom establishes three concurrent PnBs with the neighboring phosphine oxide groups. Interestingly, the UROYUW [185] solid-state structure reveals that two pyramidal SbF_3 units interact with two bridging diphosphine dioxides to generate a 12-membered supramolecular ring. In general, the angles observed for the pnictogen bonding interactions are less directional than those commented above for the tetrel bonding interactions. This agrees well with the MEP

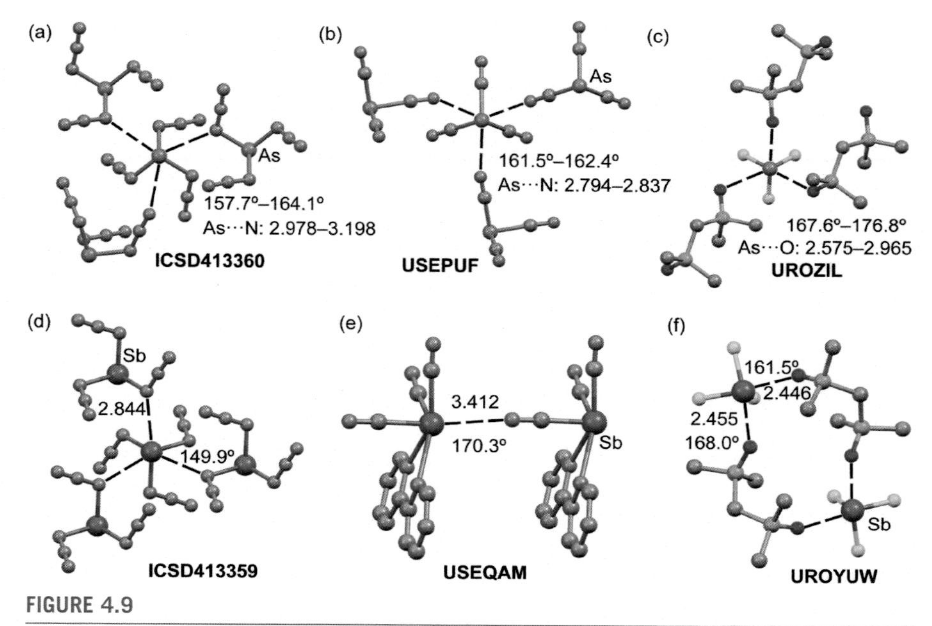

FIGURE 4.9

PnBs in the solid state of ICSD and CSD refcodes 413360 (a), USEPUF (b), UROZIL (c), 413359 (d), USEQAM (e), and UROYUW (f) X-ray structures (represented in ball and stick). The PnBs are represented using dashed lines. The values of distances (Å) and angles are indicated close to the contacts. *CSD*, Cambridge Structural Database; *ICSD*, Inorganic Crystal Structural Database; *PnBs*, pnictogen bonds.

surface analysis described in Section 2.1 (see Fig. 4.1 and Table 4.2) where the size and shape of the σ-hole anticipated that the most directional interactions are the tetrel and halogen bonds.

2.2.3 Chalcogen bonding

In recent years, a significant progress has been accomplished in the field of chalcogen bonding recognition, particularly in the control and tuning of new chemical systems for applications in crystal engineering, supramolecular chemistry, catalysis, transport of anions, biosensors, and functional materials [3,9,10,34,57–62,178]. Some intriguing examples of ChBs in the solid state are commented on in this section.

2.2.3.1 Oxygen

The oxygen atom is a very poor Ch-bond donor due to its low polarizability and large electronegativity. Several works [186,187] have investigated and evidenced attractive chalcogen bonds in $O(CN)_2$ and OF_2 compounds. Moreover, a recent comprehensive theoretical study reported on chalcogen bonds involving O-atom [188] using several σ-hole donor and acceptor molecules.

To our knowledge, there is no strong experimental evidence for the existence of σ-hole ChBs involving O-atom as Lewis acid. Scilabra et al. [189] have proposed their participation in the crystal packing of 2-methylsulfonyl-3-(2-chloro-5-nitrophenyl)oxaziridine and several oxynitrobenzodifuroxan derivatives (one represented in Fig. 4.10). In this structure (XERPUG [190]), oxynitrobenzodifuroxan forms infinite 1D supramolecular polymers in the solid state where short and directional O···O interactions are present. In the case of FIVJEZ [191] (Fig. 4.10), the electron donor corresponds to the negative belt at the Cl atom that is located opposite to the N—O bond of the oxaziridine ring.

2.2.3.2 Sulfur

ChBs involving sulfur are not very competitive in solution, especially compared with the heavier chalcogens [192—195]. However, ChBs involving sulfur in the solid state are commonly described, and the typical σ-hole donor compounds encompass sulfides, disulfides, trisulfides, and sulfur-containing heteroaromatics [189]. Fig. 4.11 shows one example for each category, to illustrate the importance of ChBs involving S in crystal structures. As example for sulfide, we have selected 4-nitrobenzylthiocyanate that forms chains in the solid state where the electron-rich O-atom is located opposite to the most polarized S—C bond (refcode CIGGII [196], see Fig. 4.11). Fig. 4.11 shows the structure of the 4-(4-nitrophenyl)-1,2,3,5-dithiadiazolyl radical where the O atoms of the nitro group are located opposite to the S—N bonds, thus promoting the formation of 1D supramolecular polymers by means of two concurrent S···O ChBs [197]. 6-Trifluoromethyl-4-nitrobenzotrithiole has been selected to illustrate ChBs in trisulfides. The O atom of the nitro group is located opposite to the S—S bond of the adjacent molecule (refcode DAHDOF [198], Fig. 4.11). Finally, the EFUNEA [199] structure exemplifies ChBs in heteroaromatic rings. This molecule forms chains in the solid state where the sp^2-hybridized lone pair is located opposite to the C—S bond, thus establishing a quite directional ChB (174 degrees).

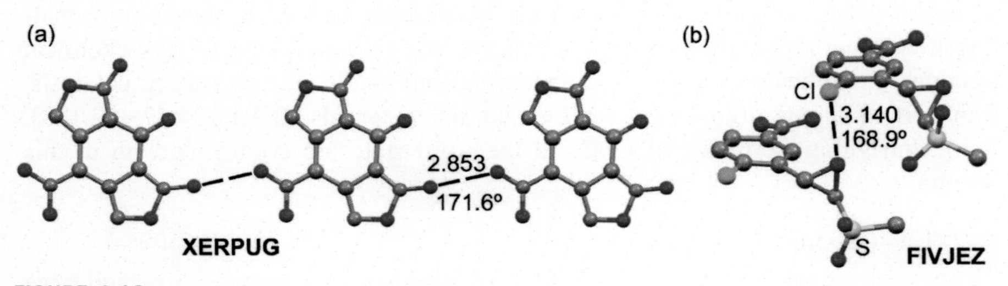

FIGURE 4.10

ChBs in the solid state of CSD refcodes XERPUG (a) and FIVJEZ (b). X-ray structures (represented in ball and stick). The PnBs are represented using dashed lines. The values of distances (Å) and angles are indicated close to the contacts. *ChBs*, chalcogen bonds; *CSD*, Cambridge Structural Database; *PnBs*, pnictogen bonds.

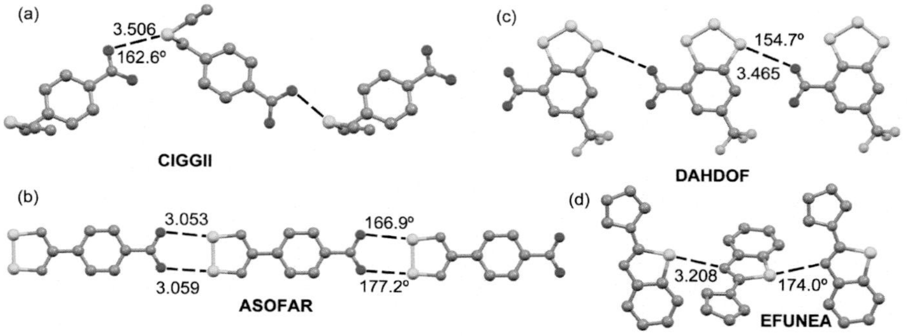

FIGURE 4.11

ChBs in the solid state of CSD refcodes CIGGII (a), ASOFAR (b), DAHDOF (c), and EFUNEA (d) X-ray structures (represented in ball and stick). The ChBs are represented using dashed lines. The values of distances (Å) and angles are indicated close to the contacts. *ChBs*, chalcogen bonds; *CSD*, Cambridge Structural Database.

2.2.3.3 Selenium and tellurium

Since selenium is an essential trace element [200,201], it is continuously under investigation due to the importance of its derivatives in the pharmaceutical industry. In fact, there are several reviews [9,32] devoted to the study of organoselenium compounds and their interactions with electron-rich species. In addition, there are plenty of examples in the literature where organic diselenides establish both inter- and intramolecular ChB with Lewis bases, which are stronger in the prolongation of the Se—Se bond [9]. In the case of organotellurium compounds like tellurophene and bis(tellurophene) derivatives, they are gaining interest to study in solution as both catalysts and anion receptors [11,202,203]. Moreover, an excellent review providing a survey of tellurium-centered secondary-bonding supramolecular synthons is available in the literature [204]. Several theoretical works have demonstrated that Se and Te are able to form very strong ChBs [205,206].

Fig. 4.12 shows four selected examples of ChBs involving Se and Te. The 2,2′-(benzo[1,2-c:4,5-c′]bis([1,2,5]selenadiazole)-4,8-diylidene)dimalononitrile molecule (refcode GEFVOC10 [207]) is planar, and the selenium atoms act as bifurcated ChB donors toward the N atoms of the adjacent cyano groups with moderate directionality (169.5 degrees). These Se···N contacts connect the molecule with four neighbors (Fig. 4.12), leading to the formation of infinite 2D layers. Fig. 4.12 shows the cocrystal of selenium dicyanide (refcode QUHYAV [208]) and 18-crown-6 where two short and moderately directional ChBs are established opposite to both electron-withdrawing cyano groups. An interesting head-to-tail arrangement is observed in the solid-state structure of bis(thiobenzoato-*S*)-tellurium (refcode CISPUP [209]), where each Te atom establishes two ChBs with the S atoms of the adjacent molecule, generating infinite 1D supramolecular chains. Finally, phenanthro(9,10-c)-1,2,5-telluradiazole (refcode FELPUH [210], Fig. 4.12) forms

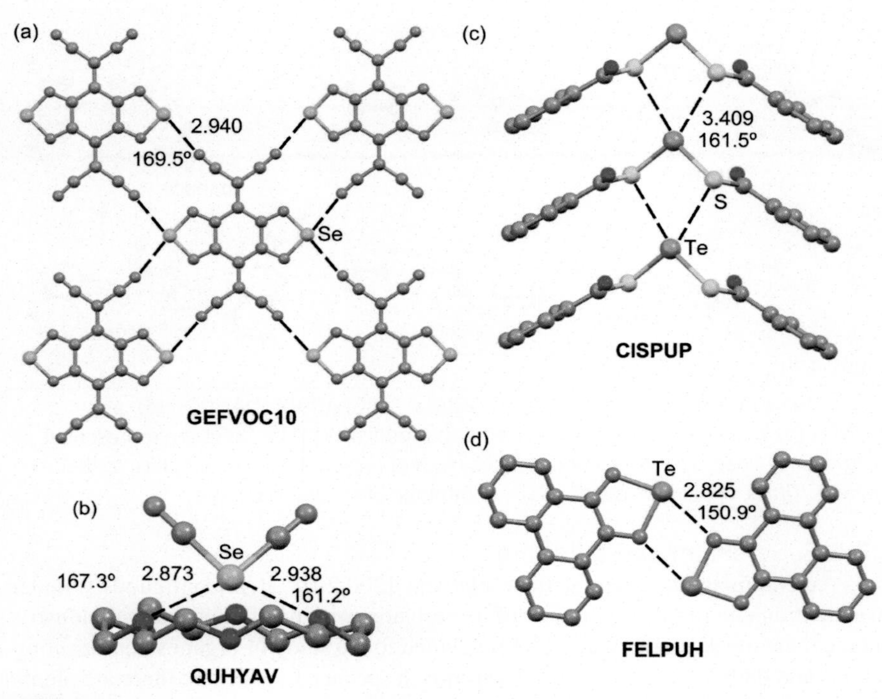

FIGURE 4.12

ChBs in the solid state of CSD refcodes GEFVOC10 (a), QUHYAV (b), CISPUP (c), and FELPUH (d) X-ray structures (represented in ball and stick). The ChBs are represented using dashed lines. The values of distances (Å) and angles are indicated close to the contacts. *ChBs*, chalcogen bonds; *CSD*, Cambridge Structural Database.

self-assembled dimers in its crystal packing where two short and symmetrically equivalent Te···N ChBs are established. The directionality of ChBs is generally moderate in the structures represented in Fig. 4.12. This fact agrees with the MEP surfaces commented earlier, where the size and intensity of the σ-hole is very large in the heavier chalcogen and pnictogen elements, thus allowing larger readjustments of the geometries in these molecules due to packing effects without weakening the ChB to a large extent.

2.2.4 Halogen bonding

Halogen bonding interactions have been extensively studied and reviewed [29–31]. Remarkably, HaB has had a seminal and inspirational role in rationalizing the ability of the other elements of the p-block of the periodic table to form the analogous ChBs, PnBs, and TrBs. Moreover, it has also promoted the generalization of the σ-hole concept. Since there is a great deal of information in the form of reviews and book chapter devoted to HaB [29–31], this interaction is briefly described in this section.

2.2.4.1 Fluorine

Although it is rather uncommon and occasionally named fluorine bond [211–213], in some circumstances, fluorine can act as HaB donor [214]. Several theoretical studies have demonstrated that fluorine has the capability of forming HaBs if the EWG bonded to F is sufficiently electron withdrawing [213,215–218]. Moreover, Hardegger et al. in 2011 [219] analyzed both the CSD and PDB databases and suggested that it is possible to form an HaB between organofluorine compounds and electronegative atoms.

Recently, experimental evidence that F atom can serve as a σ-hole donor to the O atom has been reported [220]. As shown in Fig. 4.13, one F atom of the aliphatic chain forms short C–F···O HaBs in both TOVNOK and TOVNUQ structures [220] (significantly shorter than the sum of their van der Waals radii, i.e., 3.020 Å), although its directionality is quite poor compared with the typical HaBs [29–31]. In the solid state, these molecules form herringbone-type supramolecular polymers promoted by F···O HaBs as highlighted in Fig. 4.13.

2.2.4.2 Chlorine

Compared with the heavier halogens, chlorine is much less used in molecular recognition, crystal engineering, and catalysis because its HaB interactions are significantly weaker and less competitive in polar media. However, HaBs involving chlorine are abundant in crystal structures. The cocrystallization of halogenated solvents is favored whenever lone pair donors (HaB acceptors) are present in the crystal components. Solvates of $CHCl_3$ and CH_2Cl_2 [221–240] are frequently obtained while those of CCl_4 and CCl_2CCl_2 [241,242] are less frequent.

As examples of solvates, Fig. 4.14 represents the X-ray structures of refcodes FIYLEG [243], CABZOU11 [237], and NAXYIW [241] corresponding to CH_2Cl_2, $CHCl_3$ and CCl_4 cocrystals, respectively. In all cases, the interaction is very directional as common in HaBs. It is interesting to highlight the structure of

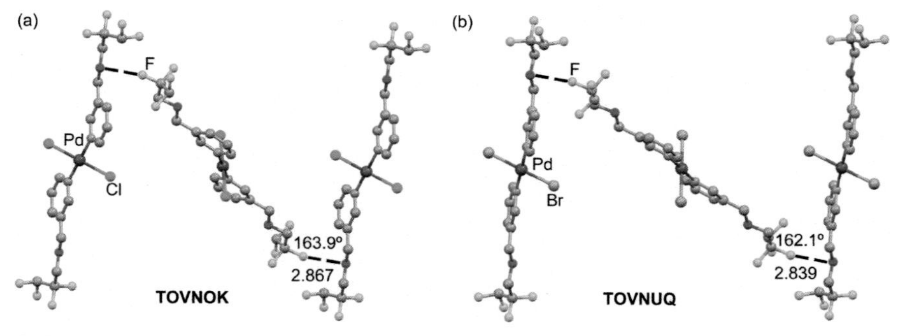

FIGURE 4.13

HaBs in the solid state of CSD refcodes TOVNOK (a) and TOVNUQ (b) X-ray structures (represented in ball and stick). The HaBs are represented using dashed lines. The values of distances (Å) and angles are indicated close to the contacts. *CSD*, Cambridge Structural Database; *HaBs*, halogen bonds.

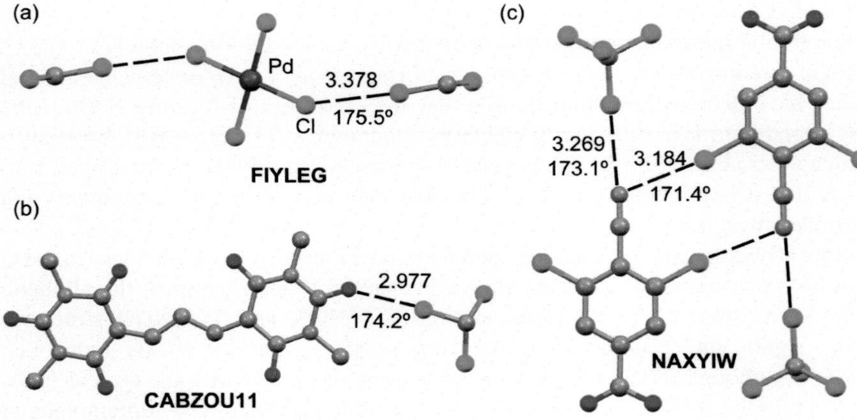

FIGURE 4.14

HaBs in the solid state of three solvates, CSD refcodes FIYLEG (a), CABZOU11 (b), and NAXYIW (c) X-ray structures (represented in ball and stick). The HaBs are represented using dashed lines. The values of distances (Å) and angles are indicated close to the contacts. *CSD*, Cambridge Structural Database; *HaBs*, halogen bonds.

3,5-dichloro-4-cyanobenzoic acid tetrachloromethane solvate (Fig. 4.14c) where the 3,5-dichloro-4-cyanobenzoic acid forms self-assembled dimers in the solid state governed by two symmetrically equivalent Cl···N HaBs, and moreover, each dimer interacts with two adjacent CCl_4 molecules.

2.2.4.3 Bromine and iodine

As previously stated, HaBs involving bromine and iodine are commonly used in crystal engineering, catalysis, supramolecular chemistry, molecular recognition of anions, and membrane transport. Therefore, in this section, less studied multivalent halogen bonds are discussed.

Elements from groups 14 to 16 form more than one covalent bond and consequently have more than one σ-hole. Therefore, they can form more than one-directional interactions with electron-rich sites. Similarly, hypervalent halogens, which also form more than one covalent bond, may also have more than one σ-hole and form more than one HaB [244]. Four interesting examples with multiple HaBs are given in Fig. 4.15. It is quite interesting that in the YATMEO [244] structure where a pentameric assembly is formed, six symmetrically equivalent Br···Cl HaBs are present, connecting two chloride anions to the bis(4-fluorophenyl)bromonium cations. The crystal lattice of phenyl-2-methoxycarbonylphenylbromonium tetrafluoroborate (YATMIS) [244] shows an intramolecular Br···O contact between the carbonyl oxygen and one of the bromine σ-holes (Fig. 4.15b). The other σ-hole on the bromine interacts with one fluorine atom of the counteranion. The HaB interactions in both YATMEO and YATMIS structures are quite directional, in agreement with the existence of two σ-holes on the Br atom.

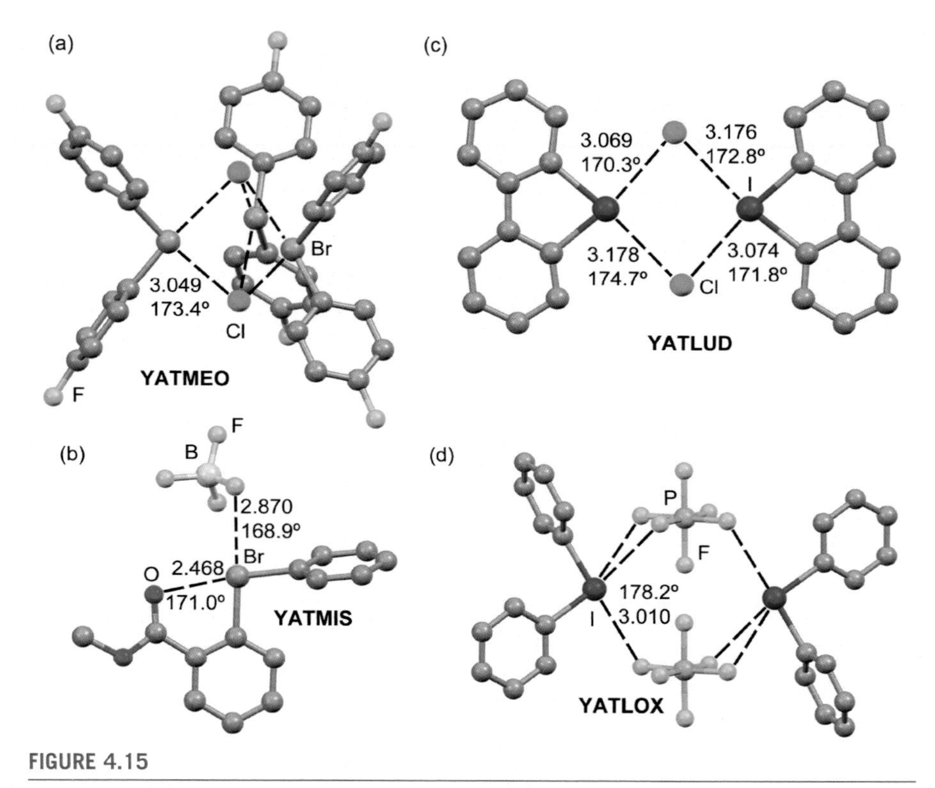

FIGURE 4.15

Multivalent HaBs in the solid state of CSD refcodes YATMEO (a), YATMIS (b), YATLUD (c), and YATLOX (d) X-ray structures (represented in ball and stick). The HaBs are represented using dashed lines. The values of distances (Å) and angles are indicated close to the contacts. *CSD*, Cambridge Structural Database; *HaBs*, halogen bonds.

Fig. 4.15c shows the tetrameric structure of dibenzo[b,d]iodol-5-ium chloride, where the anions are located opposite to the C–I bonds thus interacting with both σ-holes. In the structure of diphenyliodonium hexafluorophosphate (YATLOX) [244], a similar tetrameric adduct is formed where each diphenyliodonium establishes one conventional HaB with one F atom of the anion using one σ-hole and one bifurcated HaB involving two F atoms of the anion by using the other σ-hole (Fig. 4.15d).

2.2.5 Noble gas bonding

Noble gas bonding, usually termed aerogen bonding in the literature [245], has been analyzed both experimental [245–248] and theoretically [249–251]. Although the theoretical studies comprise Ar and Kr, X-ray structures are only available for Xe, which are discussed in the following and subdivided depending on the oxidation state of Xe and limited to Xenon fluorides, which have been scarcely discussed in the literature in terms of NgB donors. A recent review analyzes NgBs in organoxenon and xenon trioxide cocrystals [12].

2.2.5.1 XeF_2

There are several X-ray structures in the Inorganic Crystal Structural Database (ICSD) containing XeF_2, which are represented in Fig. 4.16. It can be observed that all X-ray structures exhibit short Xe···F contacts with F–Xe···F angles smaller than 90 degrees likely due to the presence of three lone pairs at Xe located perpendicular to the F–Xe–F axis. The XeF_2 crystal structure (ICSD26626 [252]) forms 1D supramolecular polymers where two symmetrically equivalent Xe···F contacts are established (see Fig. 4.16a). In the case of ICSD18128 [253] (a cocrystal of XeF_2 and XeF_4), the Xe establishes four NgBs, with the adjacent XeF_4 and XeF_2 molecules (see Fig. 4.16b). The ICSD42962 cocrystal [254] forms trimers in the solid state where the central XeF_2 molecule establishes two symmetric NgBs with the XeF_4O molecules, which present square pyramid geometry (see Fig. 4.16c).

2.2.5.2 XeF_4

The XeF_4 was the first fluoride of xenon discovered; however, it is the most difficult to synthesize [255]. Therefore, there are few examples of X-ray structures including the XeF_4 moiety in the literature, which are represented in Fig. 4.17. One of those is the $XeF_2 \cdot XeF_4$ adduct already commented before [253]. The XeF_4 participates in two short and symmetrically equivalent Xe···F NgB interactions with the adjacent XeF_2 molecules (ICSD18128, see Fig. 4.17a). A similar arrangement is observed in the X-ray structure of XeF_4 (ICSD27467 [256]), where two symmetric NgBs above and below the molecular plane (see Fig. 4.17b) are formed. This type of arrangement is also observed in the ICSD71592 structure [257], where the distorted CrF_6-

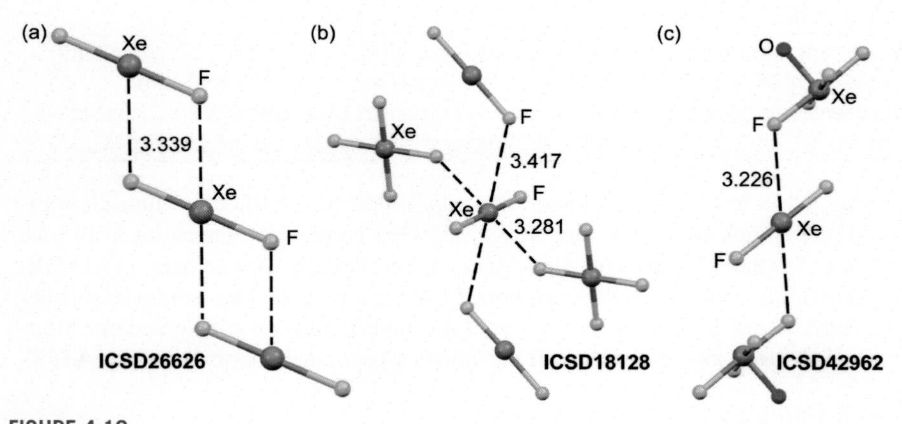

FIGURE 4.16

NgBs in the solid state of ICSD refcodes ICSD26626 (a), ICSD18128 (b), and ICSD42962 (c) X-ray structures (represented in ball and stick). The NgBs are represented using dashed lines. The values of distances (Å) are indicated close to the contacts. *ICSD*, Inorganic Crystal Structural Database; *NgBs*, noble gas bonds.

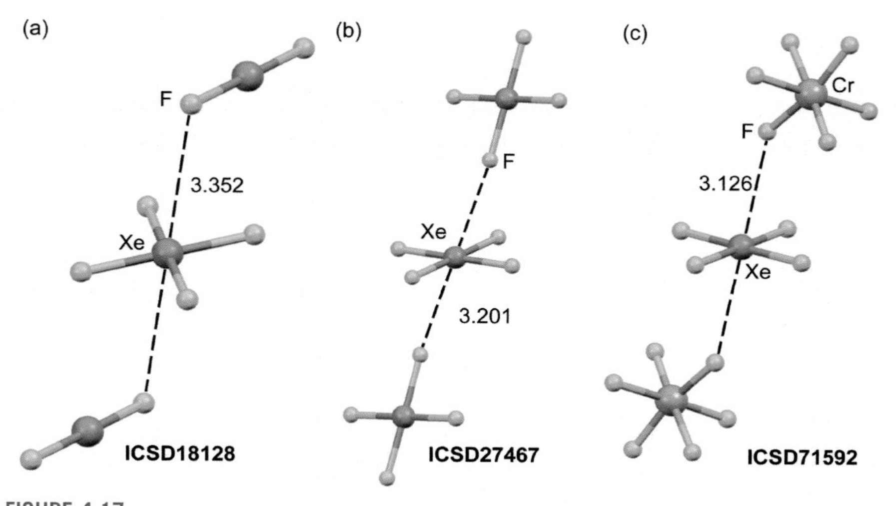

FIGURE 4.17

NgBs in the solid state of ICSD refcodes ICSD18128 (a), ICSD27467 (b) and ICSD71592 (c) X-ray structures (represented in ball and stick). The NgBs are represented using dashed lines. The values of distances (Å) are indicated close to the contacts. *ICSD*, Inorganic Crystal Structural Database; *NgBs*, noble gas bonds.

octahedra are connected to the XeF_4 via two equivalent Xe···F NgBs. It is noteworthy that the stereoactive electron lone pairs are located above and below the XeF_4 plane; therefore the approximation of the electron-rich F-atom deviates from the C_4 axis.

2.2.5.3 XeF_6

The synthesis and X-ray characterization of two adducts of XeF_6 with acetonitrile of composition $F_6Xe(NCCH_3)$ and $F_6Xe(NCCH_3)_2 \cdot CH_3CN$ were reported in 2015 [258], and they are the only available structures where the NgB interaction involves nitrogen as electron donor (see Fig. 4.18) and the oxidation state of xenon is +6 (XeF_6). However, this type of Xe···N NgB interaction has been also reported for several xenon oxides like $F_2Xe = O$ [259] and XeO_3 [246]. Fig. 4.18a and b clearly shows that in the XeF_6·acetonitrile adducts, the XeF_6 units are not octahedral, thus facilitating the approximation of the lone pair of ACN to the Xe atom. The Xe···distances are similar in both adducts.

According to several experimental evidences, XeF_6 exists in at least six different polymorphs, depending on the temperature [260]. At high temperature, the polymorph corresponds to a tetrameric $(XeF_5^+F^-)_3 \cdot XeF_6$ assembly. A partial view of this tetramer is represented in Fig. 4.18c, and it can be observed that two fluoride anions bridge the $[XeF_5]^+$ cation and the XeF_6 units by means of four Xe···F NgBs. As expected, those involving the $[XeF_5]^+$ cation are shorter than those involving the neutral XeF_6 that maintains a pseudo-octahedral geometry.

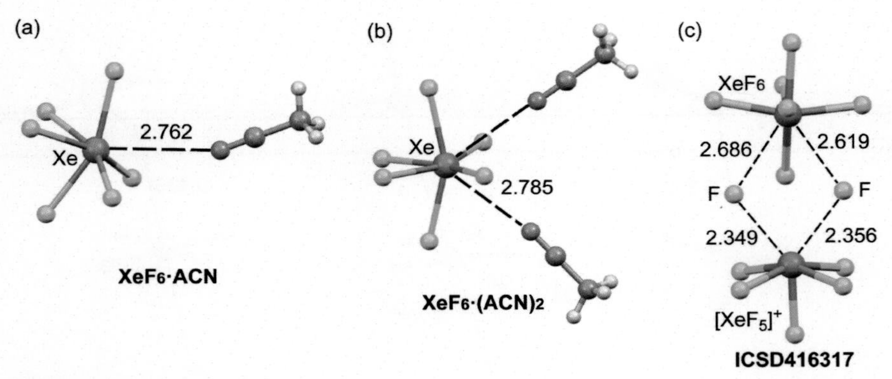

FIGURE 4.18

NgBs in the solid state of $XeF_6 \cdot NCCH_3$ (a), $XeF_6 \cdot (NCCH_3)_2$ (b) and ICSD416317 (c) X-ray structures (represented in ball and stick). The NgBs are represented using dashed lines. The values of distances (Å) are indicated close to the contacts. *ICSD*, Inorganic Crystal Structural Database; *NgBs*, noble gas bonds.

2.3 Comparative Crystal Structural Database analysis

2.3.1 General comments

A comprehensive overview of statistical surveys involving specific σ-hole interactions has been provided in 2015 (see Section 4 in Ref. [261]). Nearly all studies of σ-hole interactions after that year involve a detailed computational investigation together with anecdotal structural evidence extracted from the CSD. Noticeable exceptions involve the CSD analysis of σ-hole interactions with ethylene derivatives [262] and the CSD and the PDB analyses of σ-hole interactions with a methyl group [263,264]. To complement the existing literature and provide some comparative data, we here present a simplified CSD analysis of all the σ-hole interactions highlighted in previous sections.

2.3.2 Methodology

The CSD version 5.41 was inspected using ConQuest version 2.0.4 on April 1, 2020. All queries were restricted to single-crystal structures containing 3D coordinates with an R value of 0.1 or less. The queries used are shown in Fig. 4.19.

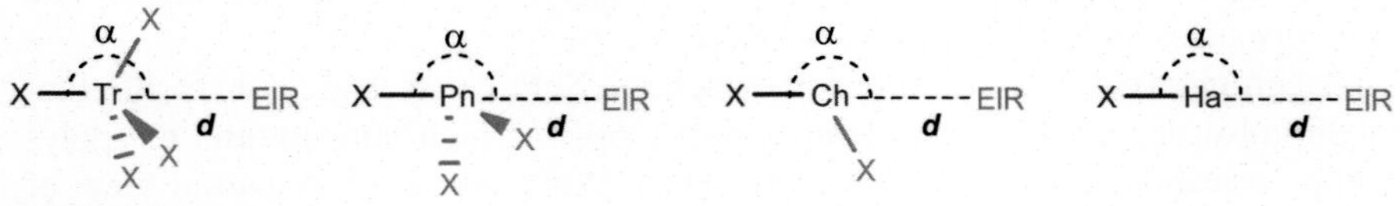

FIGURE 4.19

Overview of the four queries used to evaluate the CSD. $\alpha = 150–180$ degrees and $d \leq$ van der radius Tr/Pn/Ch/Ha + EIR + 0.5 Å. See Table 4.5 and Figs. 4.20 and 4.21 for results.

The bonds with Tr/Pn/Ch/Ha in each query were set as single bonds, and the number of bonded atoms was restricted to the number of X-substituents drawn to exclude hypervalent, coordinated, and protonated species (X = any atom). While the Tr/Pn/Ch/Ha atoms were considered as σ-hole sites, an electron-rich atom was specified as potential electron donor (ElR = N, O, F, P, S, Cl, As, Se, Br, Te, I, At). The X–Tr/Pn/Ch/Ha–ElR angle (α) was always restricted to 150–180 degrees to exclude H-bonding geometries with one of the X-substituents (i.e., when X = H). The intermolecular distance ***d*** was set to the sum of the van der Waals radius of ElR + Tr/Pn/Ch/Ha + 0.5 Å. This resulted in the data sets as specified in Table 4.5. The van der Waals–corrected distance ***d'*** has been used to generate the plots shown in Fig. 4.20. This analysis does not take into account the more detailed chemical context of the elements involved, does not reveal angular dependencies, and is not a measure of directionality. Nevertheless, for current purposes, the relative amounts of van der Waals overlap within the tightly confined area (α = 150–180 degrees and ***d'*** ≤ 0.5 Å) can be taken as a reflection of the relative strength of a certain σ-hole interaction (see also Section 2).

Table 4.5 Numerical overview of the comparative CSD evaluation using the queries shown in Fig. 4.19. See Figs. 4.20 and Fig. 4.21 for further details.

Entry	σ-hole		A.P.[a]	CIFs[b]	Hits[c]	vdW[d]	*d'*(m)[e]
1	Tr	C	11.5	461,501	1,521,142	9.3	0.265
2		Si	38.1	3390	5472	13.1	0.306
3		Ge	40.3	383	730	12.2	0.297
4		Sn	55.6	992	1725	44.5	0.068
5	Pn	N	7.1	29,913	51,418	19.3	0.251
6		P	25.0	1183	2573	24.8	0.188
7		As	29.7	541	1656	40.8	0.084
8		Sb	43.3	474	1303	71.8	−0.261
9	Ch	O	4.9	78,504	134,622	22.5	0.225
10		S	19.3	30,973	123,529	28.1	0.157
11		Se	25.4	2605	6038	53.9	−0.028
12		Te	38.3	523	1217	65.7	−0.176
13	Ha	F	3.4	51,895	148,776	25.6	0.163
14		Cl	14.3	49,295	89,453	32.8	0.107
15		Br	20.5	17,080	27,711	49.5	0.003
16		I	32.3	9280	17,370	61.3	−0.116
17	Ng	All	–	10	–	–	–

CSD, Crystal Structural Database
[a] *Atomic polarizability in a.u., taken from Table 4.1.*
[b] *Crystallographic Information Files (CIFs) found to contain the query.*
[c] *Number of individual instances that comply to the geometric criteria of the query. Note that a single CIF file can contain multiple hits.*
[d] *Percentage of hits that display van der Waals overlap.*
[e] *Median* ***d'*** *distance in Å.*

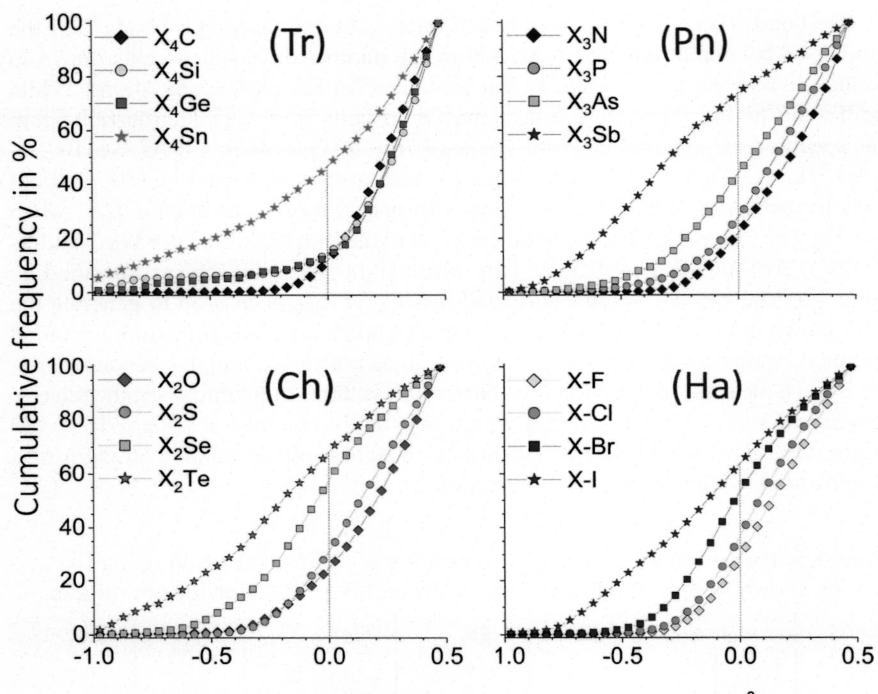

FIGURE 4.20

Plots of the cumulative frequencies as a function of the van der Waals–corrected Tr/Pn/Ch/Ha–EIR distance (*d'*) resulting for the queries shown in Fig. 4.19. See Table 4.5 for a numerical overview, including the total amount (in percentages) of van der Waals overlap for each search. The vertical dashed line is a guide to the eye for the van der Waals benchmark.

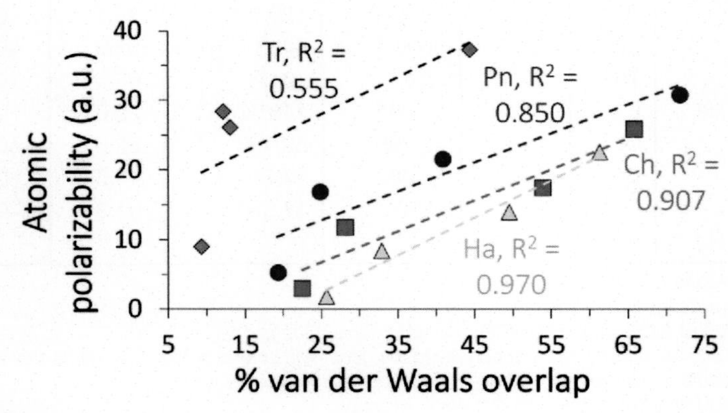

FIGURE 4.21

Comparative plots of the atomic polarizability (in a.u., see Table 4.1) as a function of the relative amounts (in %) of van der Waals overlap observed for each data set (see also Table 4.5).

2.3.3 Results and discussion

Shown in Fig. 4.20 are plots of the cumulative frequencies (in %) as a function of the van der Waals–corrected Tr/Pn/Ch/Ha···ElR distances (***d'***) grouped by element family.

It is clear from Fig. 4.20 (Tr) that the heavier and more polarizable Sn (star symbol) displays the shortest van der Waals–corrected distances (***d'***), as can also be quantified by the median ***d'*** of 0.07 Å versus ~0.3 Å for Ge, Si, and C. Sn is also involved in the largest amount of van der Waals overlap and the trend is Sn (44.5%) >> Ge (12.2%) ≈ Si (13.1%) > C (9.3%) (see also entries 1–4 in Table 4.5). The heavier elements (star symbol) of the pnictogen, chalcogen, and halogen families are clearly also involved in shorter ***d'*** distances and more van der Waals overlap. For all these three families, there is a clear trend that lighter elements have longer ***d'*** distances. This is reflected in the median ***d'*** distances for the pnictogens (0.25 (N) > 0.19 (P) > 0.08 (As) > −0.26 (Sb)), chalcogens (0.23 (O) > 0.16 (S) > −0.03 (Se) > −0.18 (Te)), and halogens (0.16 (F) > 0.11 (Cl) > 0.00 (Br) > −0.12 (I)). Similarly, lighter elements within a family have less van der Waals overlap; pnictogens: 19.3% (N) < 24.8% (P) < 40.8% (As) < 71.8% (Sb); chalcogens: 22.5% (O) < 28.1% (S) < 53.9% (Se) < 65.7% (Te); and halogens: 25.6% (F) < 32.8% (Cl) < 49.5% (Br) < 61.3% (I).

These trends in van der Waals overlap correlate reasonably well with the atomic polarizability, as is shown in Fig. 4.21. The figure reveals correlation coefficients (R^2) of 0.850 for the pnictogens (blue circles), 0.907 for the chalcogens (red squares), and 0.970 for the halogens (pale triangles). This correlation is smaller for the tetrels ($R^2 = 0.555$).

From this analysis, it is concluded that irrespective of the family of elements, the heavier and more polarizable elements act as better σ-hole donor. This observation is in agreement with our computational analysis as outlined in Section 2.1.

3. Concluding remarks

Section 1 of this chapter provides a theoretical perspective on the σ-holes and σ-hole interactions involving p-block elements. It was highlighted that the polarizability of the p-block elements decreases in later groups (14 → 18). The polarizability, as well as the MEP values, of F-saturated elements increases for elements in lower periods (2↓5) with the most significant increase from periods 2 to 3.

The σ-holes are less defined in space for the heavier ChBs and PnBs, thus explaining the lesser directionality of those compared with TrBs and HaBs. This fact is also confirmed by the geometry observed for the acetonitrile complexes where the HaB and TrB present a marked linearity in their complexes, even for the heaviest elements.

The X-ray structures highlighted in Section 2 of this chapter give experimental support to the fact that the elements from groups 14 to 18 can establish directional

TrB, PnB, ChB, HaB, and NgB interactions with Lewis bases or anions. Furthermore, these interactions play a relevant role governing the solid-state architecture of these crystalline organic and inorganic solids. While these interactions are often unnoticed or simply identified as short contacts, we argue that they are actually predictable and general.

The comparative CSD survey (Section 3) reveals that van der Waals overlap within a geometric window opposite an X–Tr/Pn/Ch/Ha bond can occur with all the p-block elements and increases in later groups (14 → 18). Within each group, this observation correlates with the atomic polarizabilities.

The generality and predictability discussed in this chapter illustrate that σ-hole interactions with p-block elements have great potential to be further applied by the scientific community in several fields such as materials, catalysis, supramolecular chemistry, and crystal engineering.

4. Theoretical methods

The atomic polarizabilities, MEP surface calculations, and optimization of the complexes were carried out using Gaussian-16 software [265] and the MP2/def2-TZVP level of theory [266,267]. The BSSE correction [268] was applied during the optimization, thus leading to BSSE-corrected energies and geometries. The NBO calculations [269] focusing on the second-order perturbation analysis were computed using the NBO 3.1 program [270] as implemented in Gaussian-16.

References

[1] A. Pizzi, C. Pigliacelli, G. Bergamaschi, A. Gori, P. Metrangolo, Coord. Chem. Rev. 411 (2020) 213242.
[2] A. Daolio, P. Scilabra, G. Terraneo, G. Resnati, Coord. Chem. Rev. 413 (2020) 213265.
[3] D. von der Heiden, A. Vanderkooy, M. Erdélyi, Coord. Chem. Rev. 407 (2020) 213147.
[4] H. Gill, M.R. Gokel, M. McKeever, S. Negin, M.B. Patel, S. Yin, G.W. Gokel, Coord. Chem. Rev. 412 (2020) 213264.
[5] Y. Xu, P.M.J. Szell, V. Kumar, D.L. Bryce, Coord. Chem. Rev. 411 (2020) 213237.
[6] W. Wang, Y. Zhang, W.J. Jin, Coord. Chem. Rev. 404 (2020) 213107.
[7] S.J. Grabowski, Coord. Chem. Rev. 407 (2020) 213171.
[8] K.M. Fromm, Coord. Chem. Rev. 408 (2020) 213192.
[9] M. Fourmigué, A. Dhaka, Coord. Chem. Rev. 403 (2020) 213084.
[10] S. Scheiner, M. Michalczyk, W. Zierkiewicz, Coord. Chem. Rev. 405 (2020) 213136.
[11] M.S. Taylor, Coord. Chem. Rev. 413 (2020) 213270.
[12] A. Bauzá, A. Frontera, Coord. Chem. Rev. 404 (2020) 213112.
[13] N. Biot, D. Bonifazi, Coord. Chem. Rev. 413 (2020) 213243.
[14] G. Cavallo, P. Metrangolo, T. Pilati, G. Resnati, G. Terraneo, Cryst. Growth Des. 14 (2014) 2697–2702.
[15] G. Terraneo, G. Resnati, Cryst. Growth Des. 17 (2017) 1439–1440.

[16] G.R. Desiraju, P.S. Ho, L. Kloo, A.C. Legon, R. Marquardt, P. Metrangolo, P. Politzer, G. Resnati, K. Rissanen, Pure Appl. Chem. 85 (2013) 1711–1713.

[17] C.B. Aakeroy, D.L. Bryce, G.R. Desiraju, A. Frontera, A.C. Legon, F. Nicotra, K. Rissanen, S. Scheiner, G. Terraneo, P. Metrangolo, G. Resnati, Pure Appl. Chem. 91 (2019) 1889–1892.

[18] S. Zahn, R. Frank, E. Hey-Hawkins, B. Kirchner, Chem. Eur. J. 17 (2011) 6034–6038.

[19] G.S. Girolami, J. Chem. Educ. 86 (2009) 1200–1201.

[20] A. Bauzá, T.J. Mooibroek, A. Frontera, Angew. Chem. Int. Ed. 52 (2013) 12317–12321.

[21] P. Politzer, P. Lane, M.C. Concha, Y. Ma, J.S. Murray, J. Mol. Model. 13 (2007) 305–311.

[22] P. Metrangolo, G. Resnati, Chem. Eur. J. 7 (2001) 2511–2519.

[23] P. Metrangolo, H. Neukirch, T. Pilati, G. Resnati, Acc. Chem. Res. 38 (2005) 386–395.

[24] P. Metrangolo, F. Meyer, T. Pilati, G. Resnati, G. Terraneo, Angew. Chem. Int. Ed. 47 (2008) 6114–6127.

[25] P. Metrangolo, G. Resnati (Eds.), Halogen Bonding. Fundamentals and Applications, Springer-Verlag, Berlin, Heidelberg, 2008.

[26] P. Politzer, J.S. Murray, T. Clark, Phys. Chem. Chem. Phys. 12 (2010) 7748–7757.

[27] M. Erdélyi, Chem. Soc. Rev. 41 (2012) 3547–3557.

[28] a T.M. Beale, M.G. Chudzinski, M.G. Sarwar, M.S. Taylor, Chem. Soc. Rev. 42 (2013) 1667–1680;
b P. Politzer, J.S. Murray, T. Clark, Phys. Chem. Chem. Phys. 15 (2013) 11178–11189.

[29] P. Metrangolo, G. Resnati (Eds.), Halogen Bonding II. Impact on Materials Chemistry and Life Sciences, Springer International Publishing AG, Heidelberg, 2015.

[30] L.C. Gilday, S.W. Robinson, T.A. Barendt, M.J. Langton, B.R. Mullaney, P.D. Beer, Chem. Rev. 115 (2015) 7118–7195.

[31] G. Cavallo, P. Metrangolo, R. Milani, T. Pilati, A. Priimagi, G. Resnati, G. Terraneo, Chem. Rev. 116 (2016) 2478–2601.

[32] K.T. Mahmudov, M.N. Kopylovich, M.F.C. Guedes da Silva, A.J.L. Pombeiro, Dalton Trans. 46 (2017) 10121–10138.

[33] R. Gleiter, G. Haberhauer, D.B. Werz, F. Rominger, C. Bleiholder, Chem. Rev. 118 (2018) 2010–2041.

[34] L. Vogel, P. Wonner, S.M. Huber, Angew. Chem. Int. Ed. 58 (2019) 1880–1891.

[35] S. Scheiner, Acc. Chem. Res. 46 (2013) 280–288.

[36] J.E. del Bene, I. Alkorta, J. Elguero, The pnicogen bond in review: structures, binding energies, bonding properties, and spin–spin coupling constants of complexes stabilized by pnicogen bonds, in: S. Scheiner (Ed.), Noncovalent Forces, Springer, Heidelberg, 2015, pp. 191–264.

[37] L. Brammer, Faraday Discuss 203 (2017) 485–507.

[38] G. Berger, J. Soubhye, F. Meyer, Polym. Chem. 6 (2015) 3559–3580.

[39] A. Caballero, F. Zapata, N.G. White, P. J Costa, V. Félix, P. D Beer, Angew. Chem. Int. Ed. 51 (2012) 1876–1880.

[40] M.J. Langton, S. W Robinson, I. Marques, V. Félix, P. D Beer, Nat. Chem. 6 (2014) 1039–1043.

[41] A. Borissov, J.Y.C. Lim, A. Brown, K.E. Christensen, A. L Thompson, M.D. Smith, P.D. Beer, Chem. Commun. 53 (2017) 2483–2486.

[42] R. Hein, A. Borissov, M.D. Smith, P.D. Beer, J.J. Davis, Chem. Commun. 55 (2019) 4849–4852.
[43] D. Bulfield, S.M. Huber, Chem. Eur. J. 22 (2016) 14434–14450.
[44] A. Dreger, P. Wonner, E. Engelage, S.M. Walter, R. Stoll, S.M. Huberm, Chem. Commun. 55 (2019) 8262–8265.
[45] C. Xu, C.C.J. Loh, J. Am. Chem. Soc. 141 (2019) 5381–5391.
[46] A.R. Voth, F.A. Hays, P.S. Ho, Proc. Natl. Acad. Sci. U. S. A. 104 (2007) 6188–6193.
[47] M.R. Scholfield, C.M.V. Zanden, M. Carter, P.S. Ho, Protein Sci. 22 (2012) 139–152.
[48] A.V. Jentzsch, D. Emery, J. Mareda, S. K Nayak, P. Metrangolo, G. Resnati, S. Matile, Nat. Commun. (2012) 905.
[49] A.V. Jentzsch, S. Matile, J. Am. Chem. Soc. 135 (2013) 5302–5303.
[50] S.K. Nayak, G. Terraneo, Q. Piacevoli, F. Bertolotti, P. Scilabra, J.T. Brown, S.V. Rosokha, G. Resnati, Angew. Chem. Int. Ed. 58 (2019) 12456–12459.
[51] E. Persch, O. Dumele, F. Diederich, Angew. Chem. Int. Ed. Engl. 54 (2015) 3290–3327.
[52] L.A. Hardegger, B. Kuhn, B. Spinnler, L. Anselm, R. Ecabert, M. Stihle, B. Gsell, R. Thoma, J. Diez, J. Benz, J.M. Plancher, G. Hartmann, D.W. Banner, W. Haap, F. Diederich, Angew. Chem. Int. Ed. 50 (2011) 314–318.
[53] Y. Lu, T. Shi, Y. Wang, H. Yang, X. Yan, X. Luo, H. Jiang, W. Zhu, J. Med. Chem. 52 (2009) 2854–2862.
[54] R. Wilcken, M.O. Zimmermann, A. Lange, A.C. Joerger, F.M. Boeckler, J. Med. Chem. 56 (2013) 1363–1388.
[55] C. Heroven, V. Georgi, G.K. Ganotra, P.E. Brennan, F. Wolfreys, R.C. Wade, A.E. Fernández-Montalván, A. Chaikuad, S. Knapp, Angew. Chem. Int. Ed. 57 (2018) 7220–7224.
[56] D.J. Wood, J.D. Lopez-Fernandez, L.E. Knight, I. Al-Khawaldeh, C. Gai, S. Lin, M.P. Martin, D.C. Miller, C. Cano, J.A. Endicott, I.R. Hardcastle, M.E.M. Noble, M.J. Waring, J. Med. Chem. 62 (2019) 3741–3752.
[57] J.Y. C Lim, I. Marques, A. L Thompson, K. E Christensen, V. Félix, P. D Beer, J. Am. Chem. Soc. 139 (2017) 3122–3133.
[58] J.Y.C. Lim, J.Y. Liew, P.D. Beer, Chem. Eur. J. 24 (2018) 14560–14566.
[59] L.M. Lee, M. Tsemperouli, A.I. Poblador-Bahamonde, S. Benz, N. Sakai, K. Sugihara, S. Matile, J. Am. Chem. Soc. 41 (2019) 810–814.
[60] S. Benz, J. López-Andarias, J. Mareda, N. Sakai, S. Matile, Angew. Chem. Int. Ed. 56 (2017) 812–815.
[61] A.P. Gamiz-Hernandez, I.N. Angelova, R. Send, D. Sundholm, V.R.I. Kaila, Angew. Chem. Int. Ed. 54 (2015) 11564–11566.
[62] M. Macchione, A. Goujon, K. Strakova, H.V. Humeniuk, G. Licari, E. Tajkhorshid, N. Sakai, S. Matile, Angew. Chem. Int. Ed. 58 (2019) 15752.
[63] P. Politzer, J.S. Murray, P. Lane, Int. J. Quantum Chem. 107 (2007) 3046–3052.
[64] S. Zhu, C. Xing, W. Xu, G. Jin, Z. Li, Cryst. Growth Des. 4 (2004) 53–56.
[65] A. Priimagi, G. Cavallo, A. Forni, M. Gorynsztejn-Leben, M. Kaivola, P. Metrangolo, R. Milani, A. Shishido, T. Pilati, G. Resnati, G. Terraneo, Adv. Funct. Mater. 22 (2012) 2572–2579.
[66] A. Mukherjee, S. Tothadi, G.R. Desiraju, Acc. Chem. Res. 47 (2014) 2514–2524.
[67] S. Scheiner, Struct. Chem. 30 (2019) 1119–1128.
[68] E. Corradi, S.V. Meille, M.T. Messina, P. Metrangolo, G. Resnati, Angew. Chem. Int. Ed. 39 (2000) 1782–1786.
[69] D.J. Pascoe, K.B. Ling, S.L. Cockroft, J. Am. Chem. Soc. 139 (2017) 15160–15167.

[70] P.L. Bora, M. Novák, J. Novotný, C. Foroutan-Nejad, R. Marek, Chem. Eur. J. 23 (2017) 7315–7323.
[71] S. Bhandary, A. Sirohiwal, R. Kadu, S. Kumar, D. Chopra, Cryst. Growth Des. 18 (2018) 3734–3739.
[72] M.R. Ams, N. Trapp, A. Schwab, J.V. Milić, F. Diederich, Chem. Eur. J. 25 (2019) 323–333.
[73] P. Politzer, J.S. Murray, M.C. Concha, J. Mol. Model. 14 (2008) 659–665.
[74] A.F. Cozzolino, I. Vargas-Baca, S. Mansour, A.H. Mahmoudkhani, J. Am. Chem. Soc. 127 (2005) 3184–3190.
[75] F. De Vleeschouwer, M. Denayer, B. Pinter, P. Geerlings, F. De Proft, J. Comput. Chem. 39 (2018) 557–572.
[76] C. Bleiholder, R. Gleiter, D.B. Werz, H. Köppel, Inorg. Chem. 46 (2007) 2249–2260.
[77] C. Bleiholder, D.B. Werz, H. Köppel, R. Gleiter, J. Am. Chem. Soc. 128 (2006) 2666–2674.
[78] P. Politzer, J.S. Murray, Crystals 9 (2019) 165.
[79] P. Politzer, J.S. Murray, T. Clark, G. Resnati, Phys. Chem. Chem. Phys. 19 (2017) 32166–32178.
[80] A. Frontera, A. Bauzá, Chem. Eur. J. 24 (2018) 7228–7234.
[81] A. Bondi, J. Phys. Chem. 68 (1964) 441–451.
[82] S. Scheiner, J. Phys. Chem. A 122 (2018) 2550–2562.
[83] J.S. Murray, K.E. Riley, P. Politzer, T. Clark, Aust. J. Chem. 63 (2010) 1598–1607.
[84] T. Clark, Wiley Interdiscip. Rev. Comput. Mol. Sci. 3 (2013) 13–20.
[85] A. Bundhun, P. Ramasami, J.S. Murray, P. Politzer, J. Mol. Model. 19 (2013) 2739–2746.
[86] S.J. Grabowski, Phys. Chem. Chem. Phys. 16 (2014) 1824–1834.
[87] S.P. Thomas, M.S. Pavan, T.N.G. Row, Chem. Commun. 50 (2014) 49–51.
[88] A. Bauzá, T.J. Mooibroek, A. Frontera, Chem. Eur. J. 20 (2014) 10245–10248.
[89] A. Bauzá, T.J. Mooibroek, A. Frontera, Phys. Chem. Chem. Phys. 18 (2016) 1693–1698.
[90] E.C. Escudero-Adán, A. Bauzá, A. Frontera, P. Ballester, ChemPhysChem 16 (2015) 2530–2533.
[91] J.J. Roeleveld, S.J. Lekanne Deprez, A. Verhoofstad, A. Frontera, J.I. van der Vlugt, T.J. Mooibroek, Chem. Eur. J. 26 (2020) 10126, https://doi.org/10.1002/chem.202002613.
[92] V.L. Heywood, T.P.J. Alford, J.J. Roeleveld, S.J. Lekanne Deprez, A. Verhoofstad, J.I. van der Vlugt, T.J. Mooibroek, Chem. Sci. 11 (20) (2020) 5289–5293, https://doi.org/10.1039/D0SC01559H.
[93] D. Mani, E. Arunan, Phys. Chem. Chem. Phys. 15 (2013) 14377–14383.
[94] T.J. Mooibroek, Molecules 24 (18) (2019) 3370, https://doi.org/10.3390/molecules24183370.
[95] S. Scheiner, Phys. Chem. Chem. Phys. 23 (10) (2021) 5702–5717, https://doi.org/10.1039/D1CP00242B.
[96] A. Frontera, C 6 (4) (2020) 60, https://doi.org/10.3390/c6040060.
[97] A. Bauzá, A. Frontera, Crystals 6 (2016) 26.
[98] O.M. Huta, I.O. Patsaj, A. Konitz, J. Meszko, J. Blazejowski, Acta Crystallogr. Sect. C Cryst. Struct. Commun. 58 (2002) o295.
[99] E. Haussuhl, J. Schreuer, Z. Kristallogr. - Cryst. Mater. 216 (2001) 616–622.
[100] M. Anioła, A. Katrusiak, CrystEngComm 18 (2016) 3223–3228.

[101] S. Patia, Z. Rappoport (Eds.), The Chemistry of Functional Groups: The Chemistry of Organic Germanium, Tin and Lead Compounds, vol. 19, Wiley, 1995.
[102] J. Parr, in: J.A. McCleverty, T.J. Meyer (Eds.), Comprehensive Coordination Chem. II, vol. 3, Elsevier Pergamon, Oxford, 2004, p. 545.
[103] T. Sato, in: E.W. Abel, F.G.A. Stone, G. Wilkinson (Eds.), Comprehensive Organometallic Chem. II, vol. 11, Pergamon Press, Oxford, 1995, p. 389.
[104] J.T. Pinhey, in: E.W. Abel, F.G.A. Stone, G. Wilkinson (Eds.), Comprehensive Organometallic Chem. II, vol. 11, Pergamon Press, Oxford, 1995, p. 461.
[105] A. Greenberg, G. Wu, Struct. Chem. 1 (1990) 79–85.
[106] P. Hencsei, Struct. Chem. 2 (1991) 21–26.
[107] M.G. Voronkov, V.P. Barishok, L.P. Petukhov, R.G. Rahklin, V.A. Pestunovich, J. Organomet. Chem. 358 (1988) 39–55.
[108] E. Lukevics, V. Dimens, N. Pokrovska, I. Zicmane, J. Popelis, A. Kemme, J. Organomet. Chem. 586 (1999) 200–207.
[109] R.J.P. Corriu, J. Organomet. Chem. 400 (1990) 81–106.
[110] Q. Shen, R.L. Hilderbrandt, J. Mol. Struct. 64 (1980) 257–262.
[111] G. Förgics, M. Kolonits, I. Hargittai, Struct. Chem. 1 (1990) 245–250.
[112] R. Eujen, E. Petrauskas, A. Roth, D.J. Brauer, J. Organomet. Chem. 613 (2000) 86–92.
[113] E. Lukevics, L. Ignatovich, S. Beliakov, J. Organomet. Chem. 588 (1999) 222–230.
[114] P. Livant, J. Northcott, T.R. Webb, J. Organomet. Chem. 620 (2001) 133–138.
[115] S.S. Karlov, P.L. Shutov, A.V. Churakov, J. Lorberth, G.S. Zaitseva, J. Organomet. Chem. 627 (2001) 1–5.
[116] I. Alkorta, I. Rozas, J. Elguero, J. Phys. Chem. A 105 (2001) 743–749.
[117] I. Alkorta, J. Organomet. Chem. 625 (2001) 148–153.
[118] M. Liu, Q. Li, S. Scheiner, Phys. Chem. Chem. Phys. 19 (2017) 5550–5559.
[119] P.G. Taylor, A.R. Bassindale, Y. El Aziz, M. Pourny, R. Stevenson, M.B. Hursthouse, S.J. Coles, Dalton Trans. 41 (2012) 2048–2059.
[120] A.R. Bassindale, M. Pourny, P.G. Taylor, M.B. Hursthouse, M.E. Light, Angew. Chem. Int. Ed. 42 (2003) 3488–3490.
[121] I. Kalikhman, O. Girshberg, L. Lameyer, D. Stalke, D. Kost, J. Am. Chem. Soc. 123 (2001) 4709–4716.
[122] X. Dai, S.-B. Choi, C.W. Braun, P. Vaidya, S. Kilina, A. Ugrinov, D.L. Schulz, P. Boudjouk, Inorg. Chem. 50 (2011) 4047.
[123] J. Tillmann, F. Meyer-Wegner, A. Nadj, J. Becker-Baldus, T. Sinke, M. Bolte, M.C. Holthausen, M. Wagner, H.-W. Lerner, Inorg. Chem. 51 (2012) 8599.
[124] X. Dai, D.L. Schulz, C.W. Braun, A. Ugrinov, P. Boudjouk, Organometallics 29 (2010) 2203–2205.
[125] S.J. Grabowski, Crystals 43 (2017) 43.
[126] P. Scilabra, V. Kumar, M. Ursini, G. Resnati, J. Mol. Model. 24 (2018) 37.
[127] A. Bauza, S.K. Seth, A. Frontera, Coord. Chem. Rev. 384 (2019) 107–125.
[128] L.A. Villaescusa, P. Lightfoot, R.E. Morris, Chem. Commun. (2002) 2220–2221.
[129] E. Lukevics, P. Arsenyan, S. Belyakov, J. Popelis, O. Pudova, Organometallics 18 (1999) 3187–3193.
[130] J.-Q. Wang, D.-Z. Kuang, F.-X. Zhang, Y.-L. Feng, Z.-F. Xu, Chin. J. Inorg. Chem. 19 (2003) 1109–1112.
[131] S. Calogero, P. Ganis, V. Peruzzo, G. Tagliavini, G. Valle, J. Organomet. Chem. 220 (1981) 11–20.

[132] C. Trujillo, G. Sanchez-Sanz, I. Alkorta, J. Elguero, New J. Chem. 39 (2015) 6791–6802.
[133] I. Alkorta, J. Elguero, S.J. Grabowski, Phys. Chem. Chem. Phys. 17 (2015) 3261–3272.
[134] M.D. Esrafili, F. Mohammadian-Sabet, M.M. Baneshi, Int. J. Quantum Chem. 115 (2015) 1580–1586.
[135] U. Adhikari, S. Scheiner, Chem. Phys. Lett. 532 (2012) 31–35.
[136] Q.-Z. Li, R. Li, P. Guo, H. Li, W.-Z. Li, J.-B. Cheng, Comput. Theor. Chem. 980 (2012) 56–61.
[137] A. Bauzá, D. Quiñonero, P.M. Deyà, A. Frontera, CrystEngComm 15 (2013) 3137–3144.
[138] J.S. Murray, P. Lane, P. Politzer, Int. J. Quantum. Chem. 107 (2007) 2286–2292.
[139] I. Alkorta, G. Sánchez-Sanz, J. Elguero, J. Phys. Chem. A 118 (2014) 1527–1537.
[140] R. Shukla, D. Chopra, Phys. Chem. Chem. Phys. 18 (2016) 29946–29954.
[141] S. Sarkar, M.S. Pavan, T.N. Guru Row, Phys. Chem. Chem. Phys. 17 (2015) 2330–2334.
[142] P.R. Joshi, N. Ramanathan, K. Sundararajan, K. Sankaran, J. Phys. Chem. A 119 (2015) 3440–3451.
[143] Lyssenko, ChemPhysChem 16 (2015) 676–681.
[144] P. Scilabra, G. Terraneo, G. Resnati, J. Fluor. Chem. 203 (2017) 62–67.
[145] A. Bauzá, T.J. Mooibroek, A. Frontera, Chem. Commun. 51 (2015) 1491–1493.
[146] A. Bauzá, A. Frontera, T.J. Mooibroek, Cryst. Growth Des. 16 (2016) 5520–5524.
[147] A. Bauzá, A.V. Sharko, G.A. Senchyk, E.B. Rusanov, A. Frontera, K.V. Domasevitch, CrystEngComm 19 (2017) 1933–1937.
[148] A. Bauzá, A. Frontera, T.J. Mooibroek, Chem. Eur. J. 25 (2019) 13436–13443.
[149] J.M. Hoffmann, A.K. Sadhoe, T.J. Mooibroek, Synthesis 52 (2020) 521–528.
[150] A. Bauzá, A. Frontera, T.J. Mooibroek, Nat. Commun. 8 (2017) 14522.
[151] T.J. Mooibroek, CrystEngComm 19 (2017) 4485–4488.
[152] W. Li, L. Spada, N. Tasinato, S. Rampino, L. Evangelisti, A. Gualandi, P.G. Cozzi, S. Melandri, V. Barone, C. Puzzarini, Angew. Chem. Int. Ed. 57 (2018) 13853–13857.
[153] P. Batail, D. Grandjean, F. Dudragne, C. Michaud, Acta Crystallogr. B30 (1974) 2653–2658.
[154] L.T. Eremenko, V.P. Kosilko, G.G. Aleksandrov, I.L. Eremenko, G.V. Lagodzinskaya, Izvestiya Akademii Nauk SSSR, Seriya Khimicheskaya, 2005, pp. 476–478.
[155] G.W. Drake, S. Bolden, J. Dailey, M.J. McQuaid, D. Parrish, Propellants, Explos. Pyrotech. 37 (2012) 40–51.
[156] R.H. Crabtree, The Organometallic Chemistry of the Transition Metals, fourth ed., Yale University, Wiley & Sons. Inc., New Haven, Conneticut, 2005 (section 4.2, page 99).
[157] P.R. Joshi, N. Ramanathan, K. Sundararajan, K. Sankaran, J. Mol. Spectrosc. 331 (2017) 44–52.
[158] N. Ramanathan, K. Sankaran, K. Sundararajan, Phys. Chem. Chem. Phys. 18 (2016) 19350–19358.
[159] T.G. Carter, E.R. Healey, M.A. Pitt, D.W. Johnson, Inorg. Chem. 44 (2005) 9634–9636.
[160] T.G. Carter, W.J. Vickaryous, V.M. Cangelosi, D.W. Johnson, Comments Inorg. Chem. 28 (2007) 97–122.

[161] W.J. Vickaryous, R. Herges, D.W. Johnson, Angew. Chem. Int. Ed. 43 (2004) 5831–5833.
[162] S. Moaven, J. Yu, J. Yasin, D.K. Unruh, A.F. Cozzolino, Inorg. Chem. 56 (2017) 8372–8380.
[163] C. Leroy, R. Johannson, D.L. Bryce, J. Phys. Chem. A 123 (2019) 1030–1043.
[164] V.M. Cangelosi, M.A. Pitt, W.J. Vickaryous, C.A. Allen, L.N. Zakharov, D.W. Johnson, Cryst. Growth Des. 10 (2010) 3531–3536.
[165] S.L. Benjamin, W. Levason, G. Reid, R.P. Warr, Organometallics 31 (2012) 1025–1034.
[166] B.N. Diel, T.L. Huber, W.G. Ambacher, Heteroat. Chem. 10 (1999) 423–429.
[167] K.M. Anderson, C.J. Baylies, A.H.M.M. Jahan, N.C. Norman, A.G. Orpen, J. Starbuck, Dalton Trans. (2003) 3270–3277.
[168] J.P.H. Charmant, A.H.M.M. Jahan, N.C. Norman, A.G. Orpen, T.J. Podesta, CrystEngComm 6 (2004) 29–33.
[169] T. Murafuji, M. Nagasue, Y. Tashiro, Y. Sugihara, N. Azuma, Organometallics 19 (2000) 1003–1007.
[170] D. Heift, R. Mokrai, J. Barrett, D. Apperley, A. Batsanov, Z. Benkö, Chem. Eur. J. 25 (2019) 4017–4024.
[171] C. Silvestru, H.J. Breunig, H. Althaus, Chem. Rev. 99 (1999) 3277–3328.
[172] B. Nekoueishahraki, P.P. Samuel, H.W. Roesky, D. Stern, J. Matussek, D. Stalke, Organometallics 31 (2012) 6697–6703.
[173] G.G. Briand, N. Burford, Adv. Inorg. Chem. 50 (2000) 285–357.
[174] L. Agocs, N. Burford, T.S. Cameron, J.M. Curtis, J.F. Richardson, K.N. Robertson, G.B. Yhard, J. Am. Chem. Soc. 118 (1996) 3225–3232.
[175] S.L. Benjamin, W. Levason, G. Reid, M.C. Rogers, R.P. Warr, J. Organomet. Chem. 708–709 (2012) 106–111.
[176] H.J. Trubenstein, S. Moaven, M. Vega, D.K. Unruh, A.F. Cozzolino, New J. Chem. 43 (2019) 14305–14312.
[177] C. Ganesamoorthy, M.S. Balakrishna, J.T. Mague, H.M. Tuononen, Inorg. Chem. 47 (2008) 7035–7047.
[178] G. Muller, J. Brand, S.E. Jetter, Z. Naturforsch, B Chem. Sci. 56 (2001) 1163–1171.
[179] S. Benz, A.I. Poblador-Bahamonde, N. Low-Ders, S. Matile, Angew. Chem. Int. Ed. 57 (2018) 5408–5412.
[180] J.Y.C. Lim, P.D. Beer, Chem 4 (2018) 731–783.
[181] M. Hirai, J. Cho, F.P. Gabbaï, Chem. Eur. J. 22 (2016) 6537–6541.
[182] J. Qiu, D.K. Unruh, A.F. Cozzolino, J. Phys. Chem. A 120 (2016) 9257–9269.
[183] R. Haiges, A. Vij, J.A. Boatz, S. Schneider, T. Schroer, M. Gerken, K.O. Christe, Chem. Eur. J. 10 (2004) 508–517.
[184] P. Deokar, D. Leitz, T.H. Stein, M. Vasiliu, D.A. Dixon, K.O. Christe, R. Haiges, Chem. Eur. J. 22 (2016) 13251–13257.
[185] W. Levason, M.E. Light, S. Maheshwari, G. Reid, W. Zhang, Dalton Trans. 40 (2011) 5291–5297.
[186] J.S. Murray, P. Lane, T. Clark, P. Politzer, J. Mol. Model. 13 (2007) 1033–1038.
[187] P.R. Varadwaj, A. Varadwaj, H.M. Marques, P.J. MacDougall, Phys. Chem. Chem. Phys. 21 (2019) 19969–19986.
[188] P.R. Varadwaj, Molecules 24 (2019) 3166.
[189] P. Scilabra, G. Terraneo, G. Resnati, Acc. Chem. Res. 52 (2019) 1313–1324.

[190] M.E. Sitzmann, M. Bichay, J.W. Fronabarger, M.D. Williams, W.B. Sanborn, R. Gilardi, J. Heterocycl. Chem. 42 (2005) 1117–1125.
[191] A. Forni, I. Moretti, G. Torre, S. Bruckner, L. Malpezzi, J. Chem. Soc. Perkin Trans. 2 (1987) 699–704.
[192] M. Macchione, A. Goujon, K. Strakova, H.V. Humeniuk, G. Licari, E. Tajkhorshid, N. Sakai, S. Matile, Angew. Chem. Int. Ed. 58 (2019) 15752, https://doi.org/10.1002/anie.201909741.
[193] K. Strakova, L. Assies, A. Goujon, F. Piazzolla, H.V. Humeniuk, S. Matile, Chem. Rev. 119 (19) (2019) 10977–11005, https://doi.org/10.1021/acs.chemrev.9b00279.
[194] S. Benz, M. Macchione, Q. Verolet, J. Mareda, N. Sakai, S. Matile, J. Am. Chem. Soc. 138 (29) (2016) 9093–9096, https://doi.org/10.1021/jacs.6b05779.
[195] S. Benz, J. López-Andarias, J. Mareda, N. Sakai, S. Matile, Angew. Chem. Int. Ed. 56 (2017) 812, https://doi.org/10.1002/anie.201611019.
[196] K. Maartmann-Moe, K.A. Sanderud, J. Songstad, Acta Chem. Scand. 38 (1984) 187–200.
[197] W. Clegg, M.R.J. Elsegood, CSD Communication, 2016.
[198] B.L. Chenard, R.L. Harlow, A.L. Johnson, S.A. Vladuchick, J. Am. Chem. Soc. 107 (1985) 3871–3879.
[199] M. Hedidi, G. Bentabed-Ababsa, A. Derdour, T. Roisnel, V. Dorcet, F. Chevallier, L. Picot, V. Thiéry, F. Mongin, Bioorg. Med. Chem. 22 (2014) 3498–3507.
[200] R.F. Burk (Ed.), Selenium in Biology and Human Health, Springer-Verlag, New York, 1994.
[201] A.L. Braga, J. Rafique, Chemistry of Functional Groups, Organic Selenium and Tellurium (Online Version), John Wiley, 2013 ch. 22.
[202] G.E. Garrett, G.L. Gibson, R.N. Straus, D.S. Seferos, M.S. Taylor, J. Am. Chem. Soc. 137 (2015) 4126–4133.
[203] E. Navarro-García, B. Galmés, M.D. Velasco, A. Frontera, A. Caballero, Chem. Eur. J. (2020), https://doi.org/10.1002/chem.201905786.
[204] A.F. Cozzolino, P.J.W. Elder, I. Vargas-Vaca, Coord. Chem. Rev. 255 (2011) 1426–1438.
[205] S. Scheiner, Int. J. Quantum Chem. 113 (2013) 1609–1620.
[206] V.D.P.N. Nziko, S. Scheiner, J. Phys. Chem. A 118 (2014) 10849–10856.
[207] T. Suzuki, H. Fujii, Y. Yamashita, C. Kabuto, S. Tanaka, M. Harasawa, T. Mukai, T. Miyashi, J. Am. Chem. Soc. 114 (1992) 3034–3043.
[208] S. Fritz, C. Ehm, D. Lentz, Inorg. Chem. 54 (2015) 5220–5231.
[209] I. Subrahmanyan, G. Aravamudan, G.C. Rout, M. Seshasayee, J. Crystallogr. Spectrosc. Res. 14 (1984) 239–248.
[210] R. Neidlein, D. Knecht, A. Gieren, C. Ruiz-Perez, Z. Naturforsch, B Chem. Sci. 42 (1987) 84–90.
[211] D. Chopra, T.N.G. Row, CrystEngComm 13 (2011) 948–951.
[212] C. Laurence, J. Graton, J.F. Gal, J. Chem. Educ. 88 (2011) 1651–1657.
[213] Y.X. Lu, J.W. Zou, Q.S. Yu, Y.J. Jiang, W.N. Zhao, Chem. Phys. Lett. 449 (2007) 6–10.
[214] K. Eskandari, M. Lesani, Chem. Eur. J. 21 (2015) 4739–4746.
[215] J.S. Murray, P. Lane, P. Politzer, J. Mol. Model. 15 (2009) 723–729.
[216] J.S. Murray, P. Politzer, Croat. Chem. Acta 82 (2009) 267–275.
[217] W. Li, Y. Zeng, X. Zhang, S. Zheng, L. Meng, Phys. Chem. Chem. Phys. 16 (2014) 19282–19289.

[218] P. Metrangolo, J.S. Murray, T. Pilati, P. Politzer, G. Resnati, G. Terraneo, CrystEngComm 13 (2011) 6593–6596.
[219] L.A. Hardegger, B. Kuhn, B. Spinnler, L. Anselm, R. Ecabert, M. Stihle, B. Gsell, R. Thoma, J. Diez, J. Benz, J.M. Plancher, G. Hartmann, Y. Isshiki, K. Morikami, N. Shimma, W. Haap, D.W. Banner, F. Diederich, ChemMedChem 6 (2011) 2048–2054.
[220] V. Elakkat, C.-C. Chang, J.-Y. Chen, Y.-C. Fang, C.-R. Shen, L.-K. Liu, N. Lu, Chem. Commun. 55 (2019) 14259–14262.
[221] F. Cecconi, C.A. Ghilardi, S. Midollini, A. Orlandini, A. Vacca, Polyhedron 20 (2001) 2885–2888.
[222] J.L. Sessler, D. An, W.S. Cho, V. Lynch, Angew. Chem. Int. Ed. 42 (2003) 2278–2281.
[223] C. Pettinari, R. Pettinari, F. Marchetti, A. Macchioni, D. Zuccaccia, B.W. Skelton, A.H. White, Inorg. Chem. 46 (2007) 896–906.
[224] A.A.D. Tulloch, A.A. Danopoulos, G.J. Tizzard, S.J. Coles, M.B. Hursthouse, R.S. Hay-Motherwell, W.B. Motherwell, Chem. Commun. (2001) 1270–1271.
[225] W. Oberhauser, C. Bachmann, P. Bruggeller, Polyhedron 14 (1995) 787–792.
[226] A.-J. Chen, C.-C. Su, F.-Y. Tsai, J.-J. Lee, T.-M. Huang, C.-S. Yang, G.-H. Lee, Y. Wang, J.-T. Chen, J. Organomet. Chem. 569 (1998) 39–54.
[227] J. Ruiz, R. Arauz, V. Riera, M. Vivanco, S. Garcia-Granda, E. Pérez-Carreno, J. Chem. Soc., Chem. Commun. 94 (1993) 740–742.
[228] B. Serli, E. Zangrando, E. Iengo, E. Alessio, Inorg. Chim. Acta. 339 (2002) 265–272.
[229] A.F. Hill, A.J.P. White, D.J. Williams, J.D.E.T. Wiltonely, Organometallics 17 (1998) 3152–3154.
[230] E.K. van den Beuken, A. Meetsma, H. Kooijman, A.L. Spek, B.L. Feringa, Inorg. Chim. Acta. 264 (1997) 171–183.
[231] J.J. Stace, K.D. Lambert, J.A. Krause, W.B. Connick, Inorg. Chem. 45 (2006) 9123–9131.
[232] E.P.L. Tay, S.L. Kuan, W.K. Leong, L.Y. Goh, Inorg. Chem. 46 (2007) 1440–1450.
[233] K. Hensen, R. Mayr-Stein, T. Stumpf, P. Pickel, M. Bolte, H. Fleischer, J. Chem. Soc. Dalton Trans. (2000) 473–477.
[234] M. Iwaoka, H. Komatsu, S. Tomoda, J. Organomet. Chem. 611 (2000) 164–171.
[235] K. Hensen, M. Kettner, P. Pickel, M. Bolte, J. Chem. Sci. 54 (1999) 200–208.
[236] T.J. Burchell, D.J. Eisler, M.C. Jennings, R.J. Puddephatt, Chem. Commun. (2003) 2228–2229.
[237] M.C. Etter, R.B. Kress, J. Bernstein, D.J. Cash, J. Am. Chem. Soc. 106 (1984) 6921–6927.
[238] A.R. Cowley, J.R. Dilworth, M. Salichou, Dalton Trans. (2007) 1621–1629.
[239] L.C. Song, G.X. Jin, H.T. Wang, W.X. Zhang, Q.M. Hu, Organometallics 24 (2005) 6464–6471.
[240] R.-F. Song, Y.-B. Xie, J.-R. Li, X.-H. Bu, Dalton Trans. 6 (2003) 4742–4748.
[241] D. Britton, J. Chem. Crystallogr. 42 (2012) 851–855.
[242] M. Mohlen, B. Neumuller, A. Dashti-Mommertz, C. Muller, W. Massa, K. Dehnicke, Z. Anorg. Allg. Chem. 625 (1999) 1631–1637.
[243] L.R.R. Klapp, C. Bruhn, M. Leibold, U. Siemeling, Organometallics 32 (2013) 5862–5872.
[244] G. Cavallo, J.S. Murray, P. Politzer, T. Pilati, M. Ursini, G. Resnati, IUCrJ 4 (2017) 411–419.
[245] A. Bauzá, A. Frontera, Angew. Chem. Int. Ed. 54 (2015) 7340–7343.

[246] J.T. Goettel, K. Matsumoto, H.P.A. Mercier, G.J. Schrobilgen, Angew. Chem. Int. Ed. 55 (2016) 13780−13783.
[247] S.N. Britvin, S.A. Kashtanov, S.V. Krivovichev, N.V. Chukanov, J. Am. Chem. Soc. 138 (2016) 13838−13841.
[248] S.N. Britvin, S. Kashtanov, M.G. Krzhizhanovskaya, A.A. Gurinov, O.V. Glumov, S. Strekopytov, Y.L. Kretser, A.N. Zaitsev, N.V. Chukanov, S.V. Krivovichev, Angew. Chem. Int. Ed. 54 (2015) 14340−14344.
[249] E. Makarewicz, J. Lundell, A.J. Gordon, S. Berski, J. Comput. Chem. 37 (2016) 1876−1886.
[250] J. Miao, Z. Xiong, Y. Gao, J. Phys. Condens. Matter 30 (2018) 44.
[251] S. Borocci, F. Grandinetti, N. Sanna, P. Antoniotti, F. Nunzi, J. Comput. Chem. 40 (2019) 2318−2328.
[252] D.H. Templeton, A. Zalkin, J.D. Forrester, S.M. Williamson, J. Am. Chem. Soc. 85 (1963) 242.
[253] J.H. Burns, R.D. Ellison, H.A. Levy, Acta Crystallogr. 18 (1965) 11−16.
[254] R. Wang, H. Steinfink, Inorg. Chem. 6 (1967) 1685−1692.
[255] J. Haner, G.J. Schrobilgen, Chem. Rev. 115 (2015) 1255−1295.
[256] J.A. Ibers, W.C. Hamilton, Science 139 (1963) 106−107.
[257] K. Lutar, I. Leban, T. Ogrin, B. Zemva, Eur. J. Solid State Inorg. Chem. 129 (1992) 713−727.
[258] K. Matsumoto, J. Haner, H.P.A. Mercier, G.J. Schrobilgen, Angew. Chem. Int. Ed. 54 (2015) 4169−14173.
[259] D.S. Brock, V. Bilir, H.P.A. Mercier, G.J. Schrobilgen, J. Am. Chem. Soc. 129 (2007) 3598−3611.
[260] S. Hoyer, T. Emmler, K. Seppelt, J. Fluorine Chem. 127 (2006) 1415−1422.
[261] A. Bauzá, T.J. Mooibroek, A. Frontera, ChemPhysChem 16 (2015) 2496−2517.
[262] D. Quiñonero, Phys. Chem. Chem. Phys. 19 (2017) 15530−15540.
[263] V.R. Mundlapati, D.K. Sahoo, S. Bhaumik, S. Jena, A. Chandrakar, H.S. Biswal, Angew. Chem. Int. Ed. 57 (2018) 16496−16500.
[264] T.J. Mooibroek, Molecules 24 (2019) 3370.
[265] Gaussian 16, Revision C.01 M.J. Frisch, G.W. Trucks, H.B. Schlegel, G.E. Scuseria, M.A. Robb, J.R. Cheeseman, G. Scalmani, V. Barone, G.A. Petersson, H. Nakatsuji, X. Li, M. Caricato, A.V. Marenich, J. Bloino, B.G. Janesko, R. Gomperts, B. Mennucci, H.P. Hratchian, J.V. Ortiz, A.F. Izmaylov, J.L. Sonnenberg, D. Williams-Young, F. Ding, F. Lipparini, F. Egidi, J. Goings, B. Peng, A. Petrone, T. Henderson, D. Ranasinghe, V.G. Zakrzewski, J. Gao, N. Rega, G. Zheng, W. Liang, M. Hada, M. Ehara, K. Toyota, R. Fukuda, J. Hasegawa, M. Ishida, T. Nakajima, Y. Honda, O. Kitao, H. Nakai, T. Vreven, K. Throssell, J.A. Montgomery Jr., J.E. Peralta, F. Ogliaro, M.J. Bearpark, J.J. Heyd, E.N. Brothers, K.N. Kudin, V.N. Staroverov, T.A. Keith, R. Kobayashi, J. Normand, K. Raghavachari, A.P. Rendell, J.C. Burant, S.S. Iyengar, J. Tomasi, M. Cossi, J.M. Millam, M. Klene, C. Adamo, R. Cammi, J.W. Ochterski, R.L. Martin, K. Morokuma, O. Farkas, J.B. Foresman, D.J. Fox, Gaussian, Inc., Wallingford CT, 2016.
[266] C. Møller, M.S. Plesset, Phys. Rev. 46 (1934) 618−622.
[267] F. Weigend, R. Ahlrichs, Phys. Chem. Chem. Phys. 7 (2005) 3297−3305.
[268] S.F. Boys, F. Bernardi, Mol. Phys. 19 (1970) 553−566.
[269] A.E. Reed, L.A. Curtiss, F. Weinhold, Chem. Rev. 88 (1988) 899−926.
[270] E.D. Glendening, A.E. Reed, J.E. Carpenter, F. Weinhold, NBO Version 3.1.

CHAPTER

Crystal engineering and pharmaceutical crystallization

5

Geetha Bolla[1], Bipul Sarma[2], Ashwini K. Nangia[3,4]

[1]*Department of Chemistry, National University of Singapore, Singapore;* [2]*Department of Chemical Sciences, Tezpur University, Tezpur, Assam, India;* [3]*School of Chemistry, University of Hyderabad, Hyderabad, Telangana, India;* [4]*CSIR-National Chemical Laboratory, Pune, Maharashtra, India*

1. Introduction to supramolecular chemistry, crystal engineering, pharmaceutical cocrystals, and methods of their preparation

1.1 Supramolecular chemistry

The basic principles of chemistry were developed over a century ago to make molecules from atoms linked by covalent bonds and coordination bonds. Beyond molecular chemistry, supramolecular chemistry is about the assembly of molecules via noncovalent intermolecular interactions. The term supramolecular chemistry [1–4] was introduced by Lehn (1987): "Just as there is a field of molecular chemistry based on the covalent bond, there is a field of supramolecular chemistry, the chemistry of molecular assemblies and of the intermolecular bond." Supramolecular chemistry is *chemistry beyond the molecule*. This subject deals with the structures and properties of supermolecules, which are not just an additive but at times cooperative, i.e., the whole is more than the sum of its parts. This field has grown in the 1990s into two distinct branches: the study of supramolecular assembly in solution and the solid state. The concepts, principles of recognition, and the nature of interactions that mediate supramolecular construction are nearly the same in solution and solid state. Supramolecular chemistry is one of the most active and fast growing fields even though molecular chemistry was developed several decades ago and has grown dramatically after the 1970s with the advent of nuclear magnetic resonance (NMR) spectroscopy. Studies of supramolecular aggregates in solution were motivated to understand and mimic the events of mutual molecular recognition that take place in biological systems, for example, enzyme–substrate and protein–ligand interactions. Inspired by the accidental discovery of crown ethers by Pedersen in 1967, the research groups of Lehn and Cram started to explore the chemistry of synthetic receptors for small charged and neutral molecules. Noting the importance of molecular recognition, three pioneering

Hot Topics in Crystal Engineering. https://doi.org/10.1016/B978-0-12-818192-8.00004-4

scientists—Charles J. Pederson, Donald J. Cram, and Jean-Marie Lehn—were awarded the Nobel Prize in 1987 for their outstanding contributions in this area described as "*chemistry of molecular assemblies and of the intermolecular bond.*" It was defined in terms of the noncovalent interactions between host and guest molecules to give a host–guest complex or a supermolecule [5–8]. Fig. 5.1 depicts various examples of synthetic host–guest systems. The host is generally a molecular entity possessing convergent binding sites, whereas the guest has divergent binding arms. The study of supramolecules in solution is referred to as molecular recognition and that in the solid state as crystal engineering [9]. The different noncovalent interactions to construct supramolecular complexes are hydrogen bonds, halogen bonds, π–π, electrostatic, and charge-transfer interactions. The main objective of synthetic supramolecular chemistry and crystal engineering is to design molecular architectures by the control of weak and strong noncovalent interactions [10]. A fundamental difference between supermolecules in solution and crystal structures is that the former are a bound, organized assembly of molecules, whereas the latter have additional conditions of periodicity, infinite repetitive arrangement, and symmetry in three dimensions.

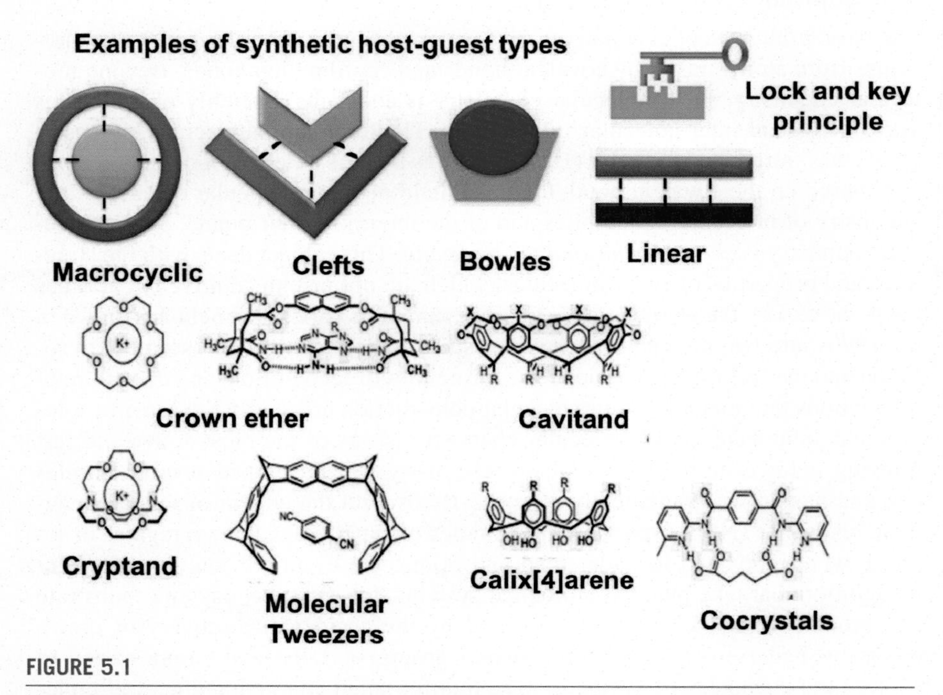

FIGURE 5.1

Host: Organic molecules containing convergent binding sites. Guest: Molecules or ions containing divergent binding sites. The lock-and-key principle is illustrated in a few examples of host–guest complexes.

1.2 Mechanically interlocked molecules

In the 1960s, several chemists reported the formation of molecules consisting of subcomponents connected together without being covalently linked. Compounds comprising these molecules were isolated and described as mechanically interlocked molecules (MIMs) [11,12]. The two archetypal examples of MIMs are the catenanes and the rotaxanes (Fig. 5.2), which were published in 1971 and knots by Gottfried Schill. The word catenane is derived from the Latin word catena, meaning chain, is a molecule with two or more topologically linked macrocyclic component parts, while rotaxane is a molecule comprising at least one macrocyclic component part, i.e., ring(s) (rota) with at least one linear component part, i.e., axle(s) threaded through the ring(s) and terminated by bulky end groups (stoppers), which are large enough to prevent dethreading of the dumbbell(s) resulting from the existence/formation of chemical (covalent and/or coordinative) bonds between the stoppers and the axle(s). Catenanes and rotaxanes are molecules and even when they are not supramolecular entities or supermolecules, they most likely will harbor intramolecular noncovalent bonds. While catenanes assume the nontrivial topologies of links, rotaxanes are topologically trivial for the simple reason that their component parts may be separated by continuous deformation. Over the past few years, MIM investigations have expanded from the domain of organic chemistry to

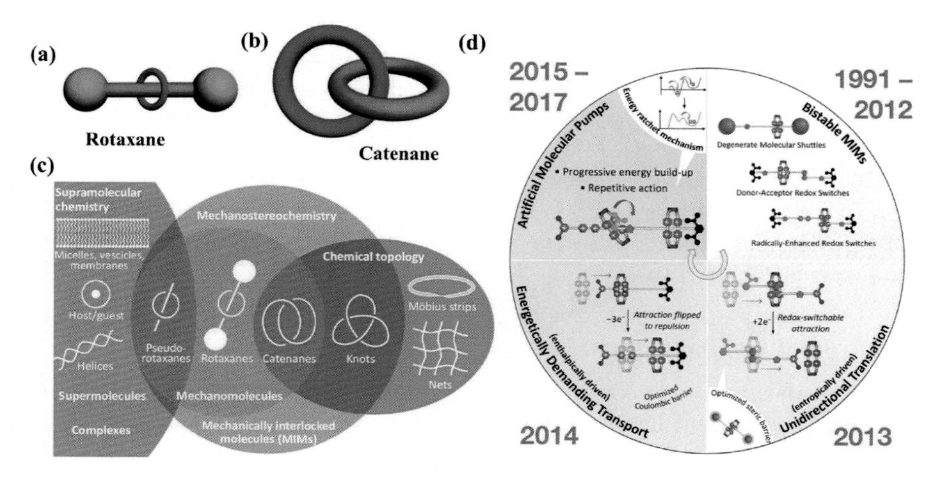

FIGURE 5.2

(A, B) Graphical representations of a catenane and rotaxane, which comprise the two main classes of mechanomolecules. (C) Mechanically interlocked molecules (MIMs), gathering molecules from different fields. A Venn diagram showing the grouping of MIMs (rotaxanes, catenanes, and knots) from supramolecular chemistry, mechanostereochemistry, and chemical topology [11]. (D) A timeline for the design and synthesis of bistable MIMs (1991–2012), leading to unidirectional transport (2013) and energetically demanding transport (2014), and artificial molecular pumps (2015–17) [12].

create a much larger interest in the scientific community. The construction of increasingly more sophisticated architectures continues to fascinate (and sometimes frustrate) chemists from many different backgrounds. Supramolecular chemists have been joined by materials scientists, physicists, and theoreticians in understanding the influence of the mechanical bond on the molecular conformations (Fig. 5.2C and D) of catenanes and rotaxanes, their physicochemical and mechanical properties, and emerging potential applications stemming from their unique properties. In 2016, the Royal Swedish Academy of Sciences awarded the Nobel Prize in Chemistry conjointly to Jean-Pierre Sauvage, Sir J. Fraser Stoddart, and Bernard L. Feringa for their "design and synthesis of molecular machines." A large part of the work performed by Sauvage and Stoddart is connected to the design and development of MIMs. Some may claim that the golden age of MIMs is yet to come, reachable only with valuable and tangible applications, as contrasted with proof-of-concept examples. One may say that the golden age of MIMs has arrived with the flourishing of molecular machines in several domains of chemistry. Numerous examples of the use of MIMs have been described that exploit their unique properties arising from the presence of mechanical bonds [11].

MIMs were synthesized using various synthetic strategies that are constantly being optimized to reach higher yields, which will lead to novel architectures. The rational design of these architectures integrating multiple subcomponents that are not connected by covalent bonds is the key to advancing the field of MIMs. The design of atomically precise molecules implies reaching a high level of understanding of their behavior and working processes. Recent analytical characterization techniques such as spectroscopic, ensemble, and single molecule are being used to obtain precise descriptions of MIMs at the nanoscale. MIM-based proof-of-concept research is now flourishing, especially in the artificial molecular machine community. The next evolutionary step is to address applications that employ unprecedented molecular functionalities to relocate MIMs from "basic chemistry" to "applied nanoengineering."

1.3 Crystal engineering

Crystal engineering is the construction of crystalline materials from molecules or ions using noncovalent interactions [13,14]. The word "crystal engineering" was first introduced by the physicist Pepinsky [15] at the meeting of American Physical Society in Mexico City in August 1955 in an abstract entitled *Crystal Engineering: A New Concept in Crystallography.* This idea was elaborated by Schmidt [16,17] during the period 1960 to 1970 in the context of engineering crystal structures having intermolecular contact geometry appropriate for photodimerization of cinnamic acids in the solid state. He noted that the introduction of a dichlorophenyl group in unsaturated molecules steers crystallization at 4 Å, a distance that is optimal for photodimerization of alkenes. Desiraju [18,19] generalized the definition of crystal engineering in 1989 as "*the understanding of intermolecular interactions in the context of crystal packing and the utilization of such understanding in the design of new solids with desired physical and chemical properties*." The subject started with the study of organic solids and their crystal structures. Crystal engineering today is a mainline interdisciplinary research activity, dealing with the self-assembly of molecular crystals, metal–organic architectures, nanostructures, coordination polymers using hydrogen and halogen bonding, electrostatic, and van der Waals interactions, and metal coordination bonding. The Royal Society of Chemistry and the American Chemical Society have journals dedicated to advances in crystal engineering, namely *CrystEngComm* and *Crystal Growth & Design*. In addition, the recent *IUCrJ* journal launched in 2014 during the International Year of Crystallography (IYCr2014) to celebrate 100 years after the first Nobel Prize related to crystallography, awarded in 1914 to Max von Laue. *IUCrJ* is an open-access journal with the objective of attracting broad scientific significance diffraction methods.

The lifelong odyssey (Fig. 5.3) of the X-ray, crystals, small molecule, proteins, and biomaterials started with the first Nobel Prize to Röntgen in 1901 for X-rays, followed by the second to Max von Laue in 1914. The subject of crystal engineering is at the intersection of chemistry and X-ray diffraction (XRD) and deals with chemical crystallography. It provides an understanding of the structures of molecular solids. The relationship between molecules and crystals was first addressed by W.H. Bragg in 1912. He recognized certain structural units of organic molecules,

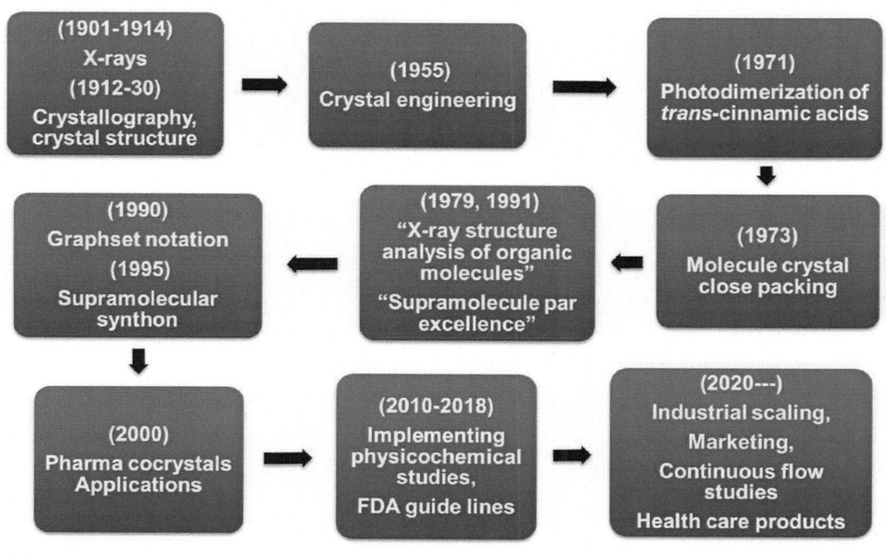

FIGURE 5.3

Evolution of crystallography in the past 100 years.

e.g., benzene and naphthalene going from one crystal structure to another and interestingly diamond along with a few inorganic crystals such as NaCl [20]. William Lawrence Bragg shared the Nobel Prize with his father William Henry Bragg in 1915 as the duo established the basis of X-ray crystallography. The third Nobel Prize related to crystallography was awarded to Charles Glover Barkla in 1917, who established that the X-rays have the properties of light, and that interference and diffraction processes should occur, as for light from a grating. The first direct evidence of a structure was reported in 1923, the first complete and accurate crystal structures of hexamethylenetetramine (HMT) of cubic symmetry, two molecules of HMT per unit cell, which meant that the positions of the carbon and nitrogen atoms could be determined from just two parameters. Soon a very interesting and the next real breakthrough was the remarkable work of Kathleen Lonsdale, who used X-ray crystallography to solve one of the biggest mysteries of chemistry: the structure of the planar benzene ring perhaps most notably continuing the studies of William Henry Bragg on naphthalene and anthracene with accurate reflections data [21]. Although Bragg's findings allowed the width of a benzene ring to be calculated, they did not settle the issue of whether the benzene ring is flat or puckered. Lonsdale chose to study hexamethylbenzene (HMB), not a difficult crystal of low symmetry and triclinic setting (Fig. 5.4) [22–27]. Both crystal structures HMT and HMB

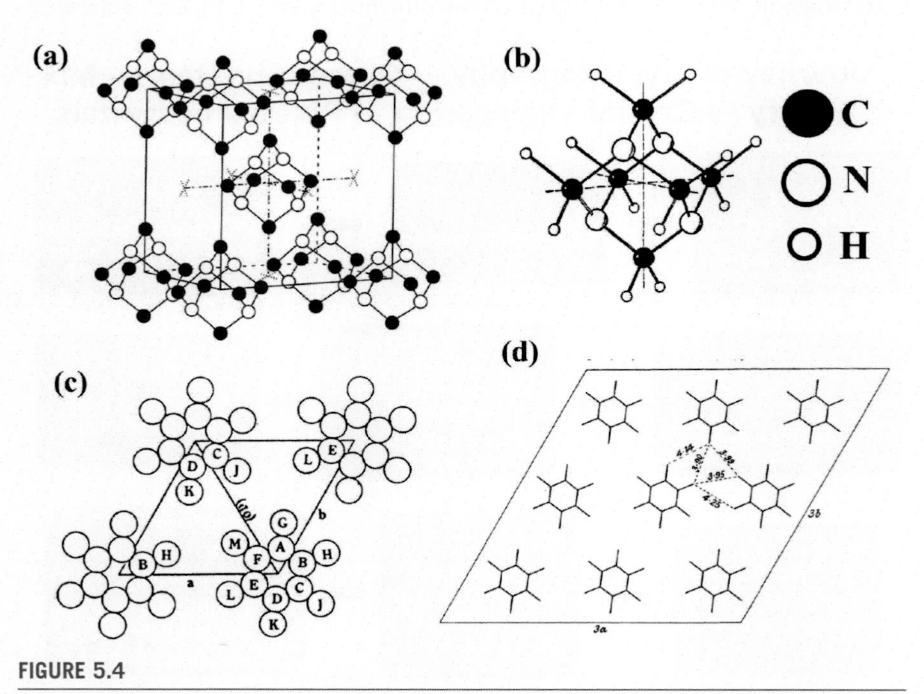

FIGURE 5.4

(A, B) Crystal lattice of HMT. (C, D) HMB crystal packing diagram (001) plane, circles represent carbon atoms [22–27]. *HMB*, hexamethylbenzene; *HMT*, hexamethylenetetramine.

opened the subject to XRD of organic crystal structures (Fig. 5.4), and till date, the research is continuing [22,25,26]. J.M. Robertson reviewed organic chemical crystallography during the period 1920–70s, a period during which many of the developments were made in X-ray crystallography to bring this powerful technique into the broader aspects of structural science and chemical crystallography [27]. Subsequently, J.D. Bernal [28] proposed molecular structures of aromatic hydrocarbons related to phenanthrene from crystal unit cell parameters.

The term "crystal engineering" was first used by Pepinsky [15] in 1955 and then significantly expanded by Schmidt [17,18] in the study of photodimerization reactions of *trans*-cinnamic acids in the 1970s (Fig. 5.5). He formulated the topochemical principle based on the 4 Å rule for dimerization of aromatic conjugated acids and built the solid-state reaction chemistry rules. Photoirradiation of α and β forms of *trans*-cinnamic acid with intermolecular distance of 4.2 Å in the crystal structure gave photodimerized product α/β-truxillic/truxinic acid, whereas the γ form with intermolecular distance of >4.7 Å was photostable. This was the birth of structure–reactivity relationship in the solid state.

Over the following years, there were a number of developments, which brought to bear upon the central question of molecule → crystal. During the 1970s and 1980s, several solid-state reactions were investigated, and the course and outcome of these reactions were correlated on the basis of the topochemical principle

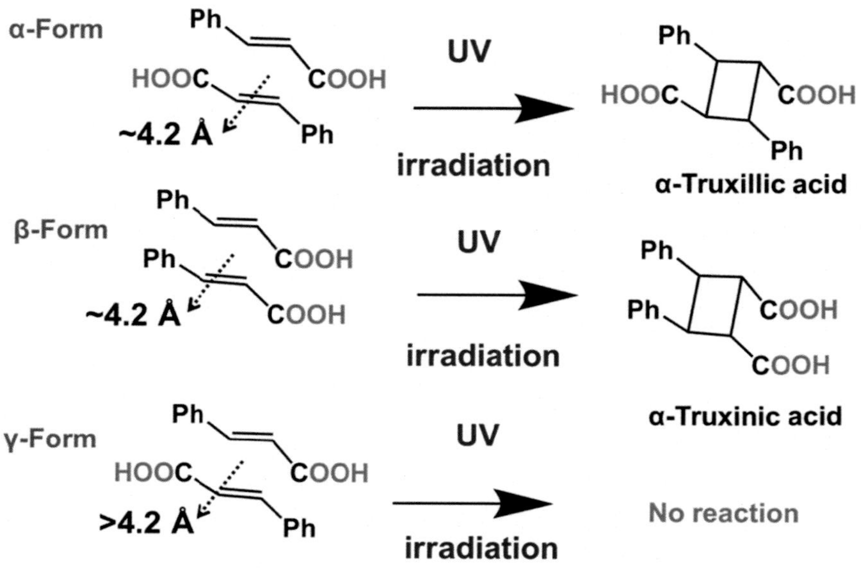

FIGURE 5.5

The first rule on the crystal engineering of aromatic acids to adopt a crystal structure of the β-form of cinnamic acid. This empirical chloro rule was used to make alkenes that would undergo photodimerization to give mirror symmetry cyclobutane derivatives.

[29–32]. In parallel to these developments, a distinctly different thought stream that addressed the molecule → crystal question was led by Kitaigorodskii, who stated that the packing of molecular solids was largely governed by the size and shape of molecules, also called as the principle of close packing [29,30]. His hypothesis was that a molecular crystal is made up of molecules, and within a molecular crystal, it is possible to identify groups of atoms such that for every atom in a group, at least one interatomic distance within this group is significantly shorter than the smallest interatomic distance to an atom in another group. In 1986, Desiraju attempted to rationalize Schmidt's observations on the unit cell parameters of Cl-substituted aromatic compounds on the basis of short Cl···Cl interactions, what one would today call as halogen bond [33,34]. The present-day phase of modern crystal engineering started with the monograph by Desiraju (in 1989) titled *Crystal Engineering: The Design of Organic Solids*. A few years later, Desiraju introduced the supramolecular synthon concept (1995) to design multicomponent molecular crystals. He further focused on the realm of weak interactions and their importance in crystal packing and how they play a role in the assembly of crystal structures. Gavezzotti [35] classified crystal structures of polynuclear aromatic hydrocarbons. Etter [36,37] (1990) postulated hydrogen bonding rules and termed hydrogen bonds as strong, directional, and important in determining crystal structures. A method of classifying and labeling hydrogen bond networks as graph sets was proposed. Supramolecular synthons and graph sets facilitate in the understanding and classification of crystal structures as supramolecular architectures. Dunitz [31,32] aptly stated that a crystal is a supermolecule par excellence of macroscopic dimensions; millions of molecules held together in a long and periodic arrangement by noncovalent interactions. Therefore, crystallization is an impressive model of supramolecular self-assembly in the solid state. Schmidt emphasized that the physical and chemical properties of crystalline solids are critically dependent upon the distribution of molecular components within the crystal lattice. Crystal engineering has implications in materials science, synthetic architectures, and pharmaceutical development. It is interesting that the first description of a molecular crystal as a network (nodes and node connectors) is the organic crystal structure of hydroquinone by determined by Powell [38] in 1948 and subsequently that of 1,3,5,7-tetracarboxylic acid by Ermer [39] in 1989. The journey of chemical crystallography with advances in XRD is depicted in Fig. 5.3.

1.4 Hydrogen bonding and crystal engineering

The hydrogen bond is a well-known intermolecular interaction in molecular recognition, supramolecular chemistry, crystal engineering, and materials science [40–44]. Hydrogen bonding is the most versatile and widely applied interaction in supramolecular chemistry due to its high directionality and tunable strength. The linear directionality of the hydrogen bonds allows the molecules to aggregate in an organized fashion for the design of complex multimolecular structures. Supramolecular synthons lead to supramolecular synthesis and designed molecular self-assembly in crystal engineering. The original ideas about hydrogen bonding from the time of Pauling (1939) [45] defined this electrostatic interaction as being formed by two rather electronegative atoms D and A. The hydrogen bond is designated as D—H···A, where the acceptor A and donor D are electronegative atoms. A hydrogen bond, $D^{(\delta-)}$—$H^{(\delta+)}$···$A^{(\delta-)}$, is an attractive interaction in which the electropositive hydrogen comes in between two electronegative atoms so that it brings D and A closer than their van der Waals sum [46,47]. The interaction is strong and directional enough to hold D—H and A atoms together at room temperature. As the D and A atoms become less electronegative, the interaction becomes weak enough that its directionality is not so linear, behaving more like a weak hydrogen bond and eventually as a hydrophobic interaction. This gradual change in the property of hydrogen bond makes it versatile for a prominent role in chemistry, biology, and materials science. The energy of a hydrogen bond usually lies in the range 0.5—40 kcal/mol and depends on the nature of D and A. It has four chemical characteristics: electrostatics (acid/base); polarization (hard/soft); van der Waals (dispersion/repulsion); and covalency (charge transfer). A hydrogen bond is neither a strong van der Waals interaction nor a weak covalent bond. It is not even a strong directional dipole—dipole interaction. Schematic representation of the distance profile with potential energy and typical intermolecular interaction/nonbonded contact between two atoms is displayed in Fig. 5.6. The energy is the lowest at the equilibrium distance d_0, and it is negative for all distances $d > d_0$ and for a small distance shorter

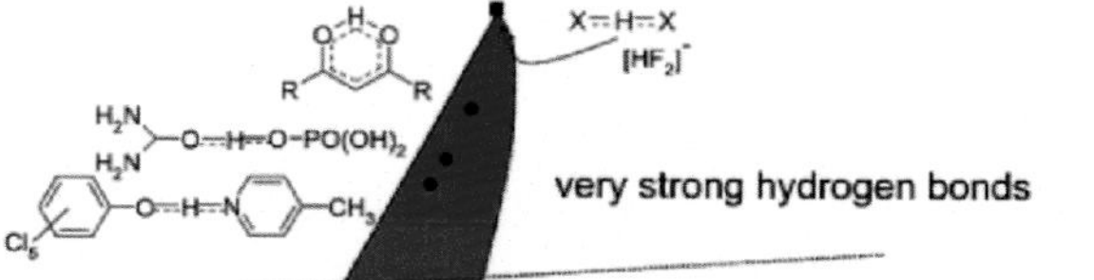

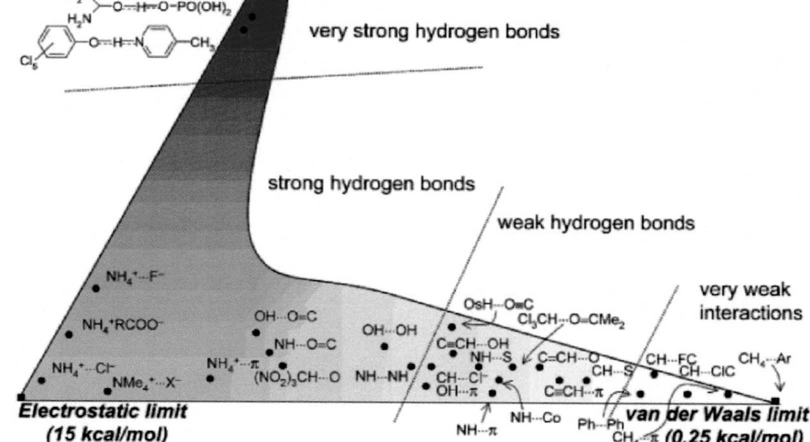

FIGURE 5.6

The hydrogen bridge depicting various types of hydrogen bonds with widely differing energies [47].

than d_0; the energy becomes positive for very short distances. The zero energy line separates the region stabilizing (E < 0) from destabilizing (E > 0) regions. For all distances $d > d_0$, the force is attractive, and for all distances $d < d_0$, it is repulsive [4]. In short, in a small distance region of $d < d_0$, the hydrogen bond is repulsive and yet attractive.

Dunitz postulated that a crystal is an ideal paradigm of a supermolecule, a *supermolecule par excellence* [31,32]. It is an assembly of literally millions of molecules self-crafted by mutual recognition at an amazing level of precision. The high degree of order in a 3D crystal lattice is the final outcome of the complementary dispositions of shape feature and functional groups in the interacting near-neighbor molecules. Once nucleation happens, the growth stage places the molecules in the crystal structure. The close packing principle of Kitaigorodskii [29] postulates that molecules in a crystal pack such that the projections of one molecule dovetail into the hollows of the neighbor, i.e., bumps fit into hollows, just like lock and key, but more smooth in geometry. These intermolecular interactions or noncovalent interactions in organic compounds are of two types: isotropic medium range forces (C···C, C···H, H···H interactions), which are defined by the shape, size, and close packing; and anisotropic long-range forces, which are electrostatic and include hydrogen bonds and heteroatom interactions (O—H···O, N—H···O, C—H···O, C—H···N, O—H···π, halogen···halogen, nitrogen···halogen). The observed three-dimensional architecture in a crystal is the free energy minimum that results from the interplay of the demands of isotropic van der Waals forces whose magnitude is proportional to the size of molecules, and anisotropic hydrogen bond interactions whose strengths are related to the donor atom acidity and acceptor basicity. Therefore, a proper understanding of the strength and directional preferences of hydrogen bonds and intermolecular interactions lies at the heart of controlling the target crystal structure.

The strength and directionality of the hydrogen bond compared with other noncovalent interactions such as van der Waals and π—π stacking account for their significance as a structure-directing force in molecular crystals. Hydrogen bonds are classified as very strong, strong, and weak (Table 5.1) [47]. Very strong hydrogen

Table 5.1 Classification of hydrogen bonds.

Hydrogen bonds, D—H···A				
Strength	**Examples**	**D—A (in Å)**	**H···A (d, Å)**	**D—H···A (angle θ, °)**
Very strong X—H ~ H···Y	[F—H—F]$^-$	2.2—2.5	1.2—1.5	175—180
Strong X—H < H···Y	O—H···O	2.6—3.0	1.6—2.2	145—180
	O—H···N	2.6—3.0	1.7—2.3	140—180
	N—H···O	2.7—3.1	1.9—2.3	150—180
	N—H···N	2.8—3.1	2.0—2.5	135—180
Weak X—H << H···Y	C—H···O	3.0—4.0	2.0—3.0	110—180

bonds are usually ionic or charged O—H···O$^-$ and O$^+$—H•••O with energy of 15—40 kcal/mol; strong O—H···O and N—H···O hydrogen bonds have energy of 4—15 kcal/mol; and weak C—H···O interactions are worth 0.2—4 kcal/mol (Fig. 5.6). The very weakest of hydrogen bonds (C—H···O/N) [11,12] are barely distinguishable from van der Waals interactions, while the very strongest ones (O—H···O$^-$ and F—H···F$^-$) [47] are almost close to weak covalent bonds. Strong hydrogen bonds such as O—H···O, O—H···N, and N—H···O have energies of 5—10 kcal/mol. Although single noncovalent interactions are very weak in terms of energy when compared with conventional covalent bonds, many such interactions collectively add up and play an important role in chemical and biological systems. Furthermore, they can be made and broken easily, facilitating rapid molecular recognition and chemical reaction, resulting in dynamic systems. Hydrogen bonding is therefore the master key for supramolecular recognition, making it the most significant interaction in crystal engineering.

To give a few examples of hydrogen bonds, in hydrogen bifluoride ion $[HF_2]^-$, the hydrogen atom lies equidistant between two electronegative fluorine atoms, and here the hydrogen bond is termed a quasi-covalent bond. With X-ray crystallography, it is possible to know the bond length and angle properties of hydrogen bonds. A typical hydrogen bond like N—H···O=C is present in carboxamides, sulfonamides, amino acids, peptides, and proteins. The tendency of all hydrogen bonds is toward linearity, and the D—H···A angle θ tends toward 180 degrees (strong directional interaction). Any crystal structure is a balance between several intermolecular interactions of varying strengths, directionality, and distance dependence characteristics.

2. Cocrystal, supramolecular synthon, preparation, examples, applications, pharmaceutical cocrystals, spring and parachute model, ternary cocrystals, cocrystal polymoprhs, and drug—drug cocrystals

2.1 Supramolecular synthon

The term synthon was introduced by Corey in 1967 in an article entitled *General Methods for the Construction of Complex Molecules* [48,49]. Desiraju [10] extended the molecular retrosynthesis ideas to supramolecular chemistry in the solid state and defined that *supramolecular synthons are structural units within supermolecules, which can be formed and/or assembled by known or conceivable synthetic operations involving intermolecular interactions*. The supramolecular synthon concept (1995) has since then become an accepted tool in crystal engineering for the design of crystal structures and supramolecular architectures. A detailed understanding of the hydrogen bond donor and acceptor groups in a molecule and the selection of complementary compounds (cocrystals formers) can result in the predictable synthesis of cocrystals. Several systematic examples of supramolecular synthons are

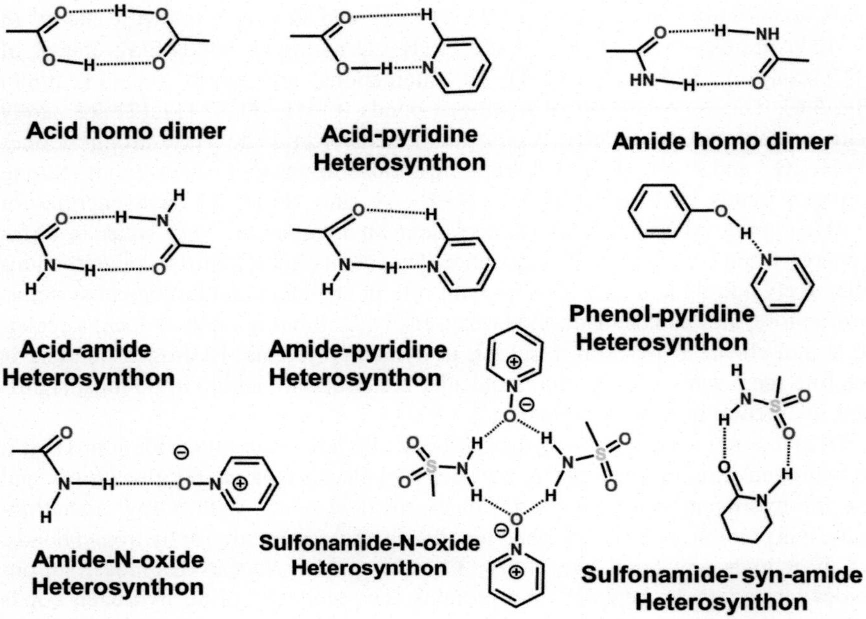

Acid homo dimer
Acid-pyridine Heterosynthon
Amide homo dimer
Acid-amide Heterosynthon
Amide-pyridine Heterosynthon
Phenol-pyridine Heterosynthon
Amide-N-oxide Heterosynthon
Sulfonamide-N-oxide Heterosynthon
Sulfonamide-syn-amide Heterosynthon

FIGURE 5.7

A few homo- and heterosynthons in crystal engineering.

reported in the literature, e.g., acid–pyridine [50–52], acid–amide [53,54,55], hydroxyl–pyridine [56,57], amino-pyridinium–carboxylate [58,59], iodo–nitro [60], amide–*N*-oxide [61,62], sulfonamide–*N*-oxide [63], OH–or NH–pyridine-*N*-oxides [64], sulfonamide–lactam/syn-amides [65–68], and halogen bonds [69–71] (Fig. 5.7). Synthons between similar or like functional groups are supramolecular homosynthons (acid–acid or amide–amide) and those between dissimilar or unlike groups (acid–amide, acid–pyridine) are termed heterosynthons [50–72]. Apart from the known functional groups in crystal engineering in the period 1995–2000, new synthons were added such as for the sulfonamide group and pyridine-N-oxide for drugs and pharmaceuticals.

2.2 Cocrystals

Cocrystals are solids that are crystalline single phase materials composed of two or more different molecular and/or ionic compounds generally in a stoichiometric ratio which are neither solvates nor simple salts.

Aitipamula et al. [73].

A cocrystal is a stoichiometric multicomponent molecular crystal wherein the different components are assembled by heteromolecular interactions (Fig. 5.8), such as hydrogen bonds, halogen bonds, $\pi-\pi$ stacking, and van der Waals

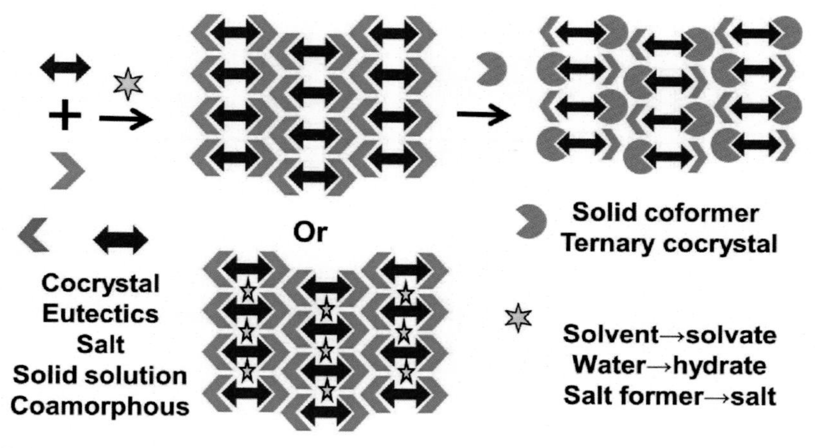

FIGURE 5.8

Schematic representation of solvates, hydrates, cocrystals, and salts. If one of the components is a drug, the product is a pharmaceutical cocrystal, and for ionic complexes, it is salt.

interactions. The supramolecular synthon [10,48,49], and in case of cocrystals the heterosynthon, i.e., hydrogen bonding between unlike functional groups [72], plays a fundamental part in the designed assembly of molecular cocrystals.

2.3 Preparation of cocrystals

A major advance in the preparation of multicomponent cocrystals was reported from the group of Jones [74,75] who showed that mechanochemical grinding and adding a few drops of solvent during grinding/kneading, referred to as solvent-assisted grinding, gave product cocrystals reproducibly. Another common method is first dissolution of the components in a suitable molar ratio by heating or ultrasonication and then slow evaporation of the solvent at ambient conditions for crystals to grow from solution. Both methods are empirical and, with a bit of trial-and-error, yield successful results. The third one is Kofler contact method [76–78]. Here, higher melting point component (A) melted and recrystallized before molten component (B) is brought into contact with it and creates a zone of mixing (Fig. 5.9). The melting crystallization method offers the advantage that solvation/hydration is not observed as concomitant product of crystallization.

2.4 Examples and applications of the cocrystals

The first two molecule cocrystals (molecular complex is the classical term) were reported by Wöhler [79] in 1844. During his studies on quinone, he observed that by mixing solutions of quinone (colorless) and hydroquinone (yellow), a crystalline

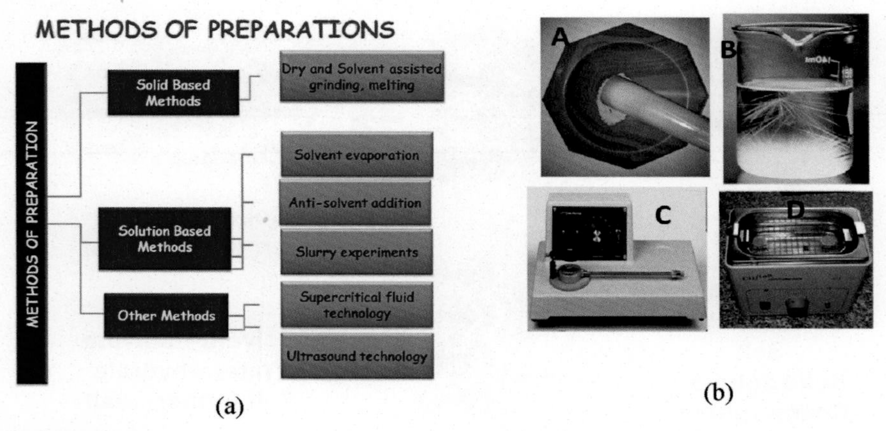

FIGURE 5.9

Classification of cocrystal preparation methods: (A) mechanochemical grinding, (B) solution crystallization, (C) Kofler contact melt method, and (D) ultrasonication.

substance was formed, which he labeled as green hydroquinone. Following the initial publication [79], Ling and Baker published several related cocrystals made from halogenated quinones and green hydroquinone, referred to as quinhydrone [80]. The complete structure of a monoclinic quinhydrone crystal was reported in 1958 [81], which showed quinone and hydroquinone molecules in alternate zigzag chains of O–H···O hydrogen bonds (Fig. 5.10A). This unusual observation on donor–acceptor hydrogen-bonded cocrystal is today the mature field of pharmaceutical cocrystals (Fig. 5.10B).

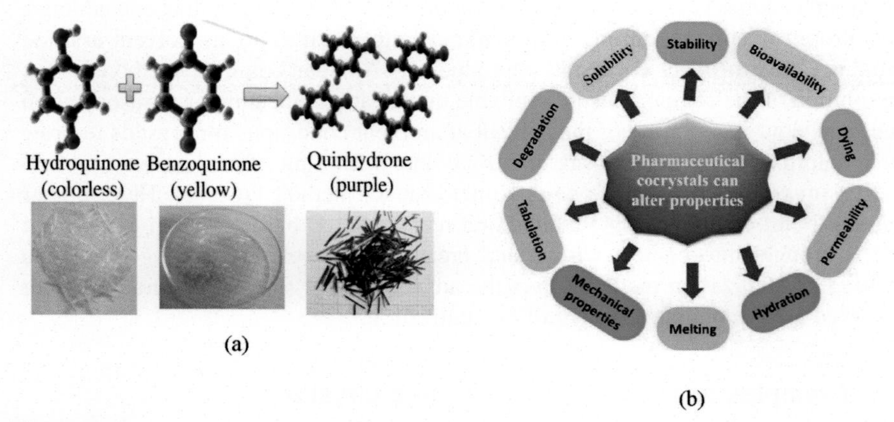

FIGURE 5.10

(A) Molecular and packing diagram of the crystal structure of quinhydrone 1:1 cocrystal of quinone and quinol shows the color change in the product. (B) Various applications of the cocrystals in pharmaceuticals materials for improving their properties.

2.5 Pharmaceutical cocrystals

Pharmaceutical cocrystals belong to one class of cocrystals in which one component must be an active pharmaceutical ingredient (API), and the coformer used to make the cocrystal must be a generally recognized as safe (GRAS) substance [82]. A cocrystal incorporates pharmaceutically acceptable guest molecules and the active drug molecule into the crystal lattice. The past decade and a half was a period of intense research on the preparation of cocrystal systems to improve the physicochemical properties of drugs. When the known traditional approaches (e.g., salt formation, micronization, solid dispersion) fail to increase the dissolution rate, or improve the stability and bioavailability of drugs [83–85], then the advantages of cocrystals are realized, such as APIs with nonionizable functional groups can be processed to improve their solubility, dissolution rate, moisture uptake, and bioavailability, mechanical strength, and compressibility. Remenar and coworkers [86] reported the first dramatic example of the cocrystal of itraconazole with aliphatic dicarboxylic acids, which exhibited higher solubility and faster dissolution rate in comparison with the free base. Moreover, the dissolution profile of the cocrystal with L-malic acid matched with that of the commercial product (Sporanox) containing amorphous itraconazole (Fig. 5.11A and B). The cocrystals approach was recently reported to improve solubility and bioavailability of meloxicam, a drug closely related to tenoxicam in the Biopharmaceutics Classification System (BCS) class II category of low aqueous solubility [87–91]. It is difficult to predict which particular strategy will provide the best solution for a given physicochemical property improvement among

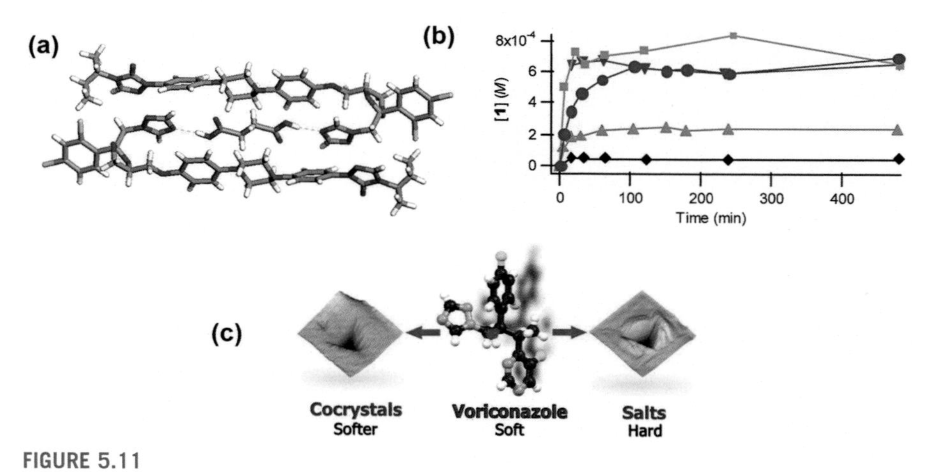

FIGURE 5.11

(A) Itraconazole cocrystals with succinic acid. (B) Dissolution profile for Sporanox beads (*green*) and *cis*-itraconazole cocrystal with L-malic acid (*red*), L-tartaric acid (*blue*), succinic acid (*orange*), and the drug (*black*) in 0.1 N HCl at 25°C. (C) Voriconazole cocrystal for superior tableting property [100].

the known methods such as crystalline salts, amorphous dispersion in polymer matrix, inorganic matrix as a carrier, nanoparticles, coground mixtures, prodrugs of higher solubility, and pharmaceutical cocrystals [91–93].

A few well-known examples of pharmaceutical cocrystals from the past decade are discussed. Indomethacin–saccharin cocrystal exhibits higher solubility than indomethacin at all pH values. The dissolution rate of cocrystals was higher relative to that of other simpler forms, and even, bioavailability was higher in dogs [94]. Carbamazepine–saccharin cocrystal was compared with the marketed drug Tegretol [95]. The cocrystal was further tested for physical and chemical stability, and it is comparable with the marketed product (Tegretol). Finally, comparison of oral bioavailability in dogs showed that the cocrystal is a promising and viable alternative to the anhydrous polymorph in the formulated oral product. Cocrystal, salt, and solvates have been observed in the anticonvulsant drug lamotrigine [96–98]. Structural studies of two polymorphic forms of the 1:1 lamotrigine/methylparaben cocrystals, 1:1 lamotrigine–nicotinamide cocrystal and its monohydrate, 1:1 saccharin salt of lamotrigine, 2:1 adipate and malate salts and dimethanol solvate of nicotinate salt, as well as dimethanol and ethanol monohydrate solvatomorphs of lamotrigine, were studied [95]. Some of the more promising crystal forms were further characterized for their dissolution rate, solubility, and pharmacokinetic behavior. Thirteen new solid forms of etravirine were crystallized in a cocrystal/salt screen to improve the solubility of this anti-HIV drug [99]. One anhydrous form, five salts (HCl, mesylate, sulfate, besylate, tosylate), two cocrystals (adipic acid; 1,3,5-benzenetricarboxylic acid), and five solvates (formic, acetic acid, acetonitrile, and 2:1,1:1 methanolates) were obtained. These results show that compared with salts, cocrystals are stable high solubility forms of drugs. Voriconazole [100] is an antifungal drug, which has a soft drug tablet due to which it is difficult to compact the pellet when pressure is applied on the solid. Cocrystals of voriconazole with fumaric acid, 4-hydroxybenzoic acid, and 4-aminobenzoic acid and salts with hydrochloric acid and oxalic acid were studied by nanoindentation applied on single crystals of the salts and cocrystals. Interestingly, the salts exhibited better hardness compared with the drug and cocrystals in the order salts $\gg$ drug $>$ cocrystals (Fig. 5.11C).

2.6 Spring and parachute model

Recently, Nangia [101] elaborated on the "spring and parachute" mechanism of solubility enhancement for amorphous drugs and pharmaceutical cocrystals. The steps are as follows: (1) the cocrystal dissociates into amorphous or nanocrystalline drug clusters after separating from the cocrystal in solution, (2) this transient metastable form dissolves rapidly compared with the crystalline state (the spring) and slowly reaches the stable crystalline modification in the dissolution medium following Ostwald's Law of Stages, resulting in (3) high apparent solubility of the drug for cocrystals over an extended period of time in the aqueous medium (the parachute). The sequence of mechanistic pathway for solubility improvement of pharmaceutical cocrystals is outlined in Fig. 5.12. The reason for the peak solubility exhibited by

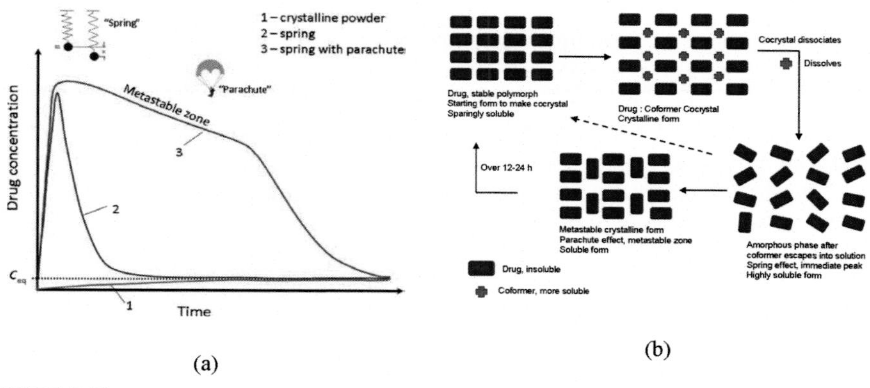

FIGURE 5.12

(A) The spring and parachute concept to achieve high apparent solubility for insoluble drugs. (1) The crystalline (stable) form has low solubility (2) A short-lived metastable species (i.e., amorphous phase) shows peak solubility but quickly drops (within minutes to an hour) to the low solubility of the crystalline form. (3) Highly soluble drug forms are maintained for a long enough time (usually hours) in the metastable zone. (B) Possible mechanisms of pharmaceutical cocrystals in dissolution medium.

Adapted from N.J. Babu, A. Nangia, Solubility advantage of amorphous drugs and pharmaceutical cocrystals. Cryst. Growth Des. 11 (2011) 2662–2679.

amorphous drugs is reasonably well understood: high free energy and micronized particle size. Dissociation of the hydrogen bonded drug···coformer cocrystal in aqueous medium liberates the more soluble coformer into the solution. The less soluble API aggregates in an amorphous phase because of the sudden "crashing out" of the coformer. This phenomenon results in drug aggregates which are amorphous-like giving peak drug solubility for a short period (spring effect). Instead of immediate precipitation of the amorphous drug to the stable crystalline form (the spring, curve 2 in Fig. 5.12A), the fact that the amorphous drug clusters are generated by dissociation of the cocrystal in stages (Fig. 5.12B), the crashing out or transformation of the metastable phase to the stable form occurs in a finite metastable zone width (MSWZ; the parachute, red curve in Fig. 5.10A). Thus the "spring" effect is achieved by dissociation of the cocrystals to an amorphous-like drug form. If the amorphous phase directly transforms to the stable crystalline form without the intermediacy of metastable polymorphs (dash arrow), then the drug will exhibit spring effect only. The maintenance of peak solubility profile, or parachute, happens for a long enough time (120–300 min) to give high-dose solubility because transformation of these amorphous clusters to stable crystalline phases and/or crystal growth is a slow process in the dissolution medium. Finally, the drug will transform to the lowest energy phase (usually the crystalline form with the lowest solubility).

The zwitterionic drug tenoxicam (TNX) belongs to the oxicam family of nonsteroidal antiinflammatory drugs and exhibits good efficacy comparable with piroxicam. Its therapeutic action as a drug is limited by poor aqueous solubility (14 mg/L). Because of poor aqueous solubility and absorption limitation, its utility as an oral drug pill is less impressive. To overcome this limitation, cocrystals of TNX with benzoic, salicylic acid, catechol, resorcinol, and pyrogallol and salts with piperazine (PIP), HCl, and methanesulfonic acid were studied (Fig. 5.13) [102]. Weak carboxylic acids and phenolic coformers resulted in tenoxicam dimer (conformer A) in the crystal structure. Strong acids, such as HCl and methanesulfonic acid (MSA), as well as bases, e.g., piperazine, gave salts with no TNX dimer and a different molecular packing (conformer B). An interesting feature of tenoxicam like molecules is that being neutral—zwitterionic in nature, they can form salts with both acids and bases. This is advantageous in solid form and salt screen to expand the available supramolecular space. Solubility and intrinsic dissolution rate (IDR) of TNX cocrystals/salts were carried out to optimize the solid form with the best physicochemical properties. There is good correlation between coformer solubility and dissolution enhancement observed with TNX cocrystal. TNX—RES (10×) and TNX—PPZ (5×) exhibited the highest improvement and good stability for 24 h in the slurry medium (Table 5.2).

2.7 Ternary cocrystals

Recently, ternary cocrystals have acquired significance due to the relevance of cocrystals with applications in fields ranging from pharmaceuticals to material science to explosives. Ternary cocrystals contain three different compounds in a definite stoichiometric ratio assembled via hydrogen and/or halogen bonds [103]. The design

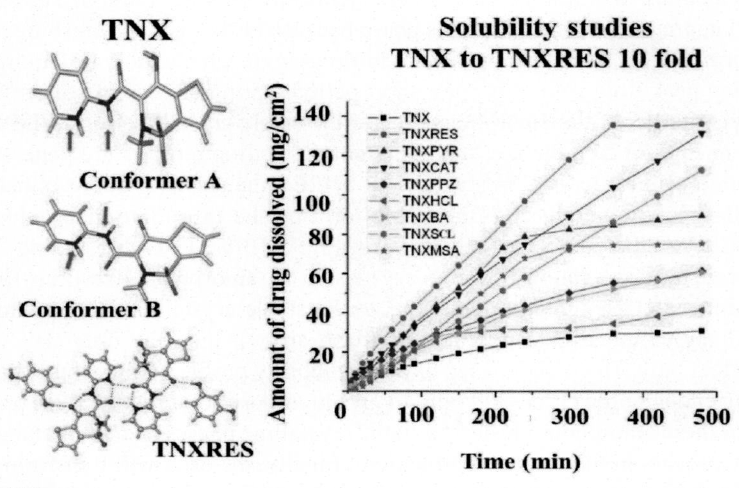

FIGURE 5.13

Tenoxicam (TNX) structures with coformers. Intrinsic dissolution rate of TNX cocrystals, salts in pH 7 buffer medium at 37°C.a

Table 5.2 Dissolution profile of the tenoxicam cocrystals and salts.

	Absorption coefficient (mM^{-1}/cm)	Solubility[a] (mg/L) after 24 h slurry in pH 7 buffer	Intrinsic Dissolution Rate[a] ($mg\ cm^{-2}\ min$)	Solubility (g/L) of the coformer	Residue after 24 h slurry in pH 7 buffer	Residue after 24 h slurry in pH 1.2 medium
TNX	19.48	450	0.16	–	TNX	TNX
TNX–BA (1:1)	11.86	196 (×0.4)	0.32 (×2.0)	2.9	TNX, TNX–BA	TNX–BA
TNX–SCL (1:1)	15.23	130 (×0.3)	0.20 (×1.3)	2.0	TNX, TNX–SCL	TNX–SCL
TNX–CAT (1:1)	13.82	2634 (5.8)	0.50 (×3.1)	430.0	TNX, TNX–CAT	TNX, TNX–CAT
TNX–RES (1:1)	11.86	4556 (×10.1)	0.67 (×4.2)	1100.0	TNX–RES	TNX–RES
TNX–PYR (1:1)	15.13	3394 (×7.5)	0.40 (×2.4)	580.0	TNX, TNX–PYR	TNX, TNX–PYR
TNX–PPZ (1:0.5)	14.79	2484 (×5.5)	0.40 (×2.5)	Freely soluble	TNX–PPZ	TNX, TNX–PPZ
TNX–HCl (1:1)	11.22	261 (×0.6)	0.21 (×1.3)	Freely soluble	TNX	TNX
TNX–MSA (1:1)	16.62	186 (×0.4)	0.19 (×1.2)	Freely soluble	TNX	TNX

PYR, *pyrazinamide;* TNX, *tenoxicam;* BA, *benzoic acid; CAT, catechol; PPZ, piperazine; RES, resorcinol; IDR, intrinsic dissolution rate; SCL, salicylic acid; MSA, methanesulfonic acid.*

[a] *The number in parentheses indicates the number of times higher solubility of cocrystal/salt compared with reference TNX.*

strategies for binary cocrystals are numerous, whereas in the case of ternary cocrystals in which three neutral solid compounds are present in a single crystal structure, the reports are fewer and the exercise is challenging. The design and directed assembly of three different molecules with complementary functional groups during cocrystallization to make ternary crystals is particularly difficult because it involves subtle discrimination and selective recognition between sites of molecules with a consideration of their size, shape, conformation, and solubility [103–108]. There are multiple possibilities for molecules to interact in a three-component assembly: (1) Will they assemble as a binary complex? (2) Will they form ternary through an intermediate binary formation? (3) Will there be solid product after cocrystallization or a solvate/hydrate? (4) How congruent are the functional groups and the solubility of molecules in diverse solvent systems? (5) Will they have congruency in nucleation, a preevent to crystallization? The components must interact congruently to form cocrystals; otherwise, there will be a binary complex and unreacted third component. (6) Lastly, the hydrogen bonding between the molecules must be stronger and specific compared to that with solvent/water, else multiple products will be formed. In the numerous crystallization trials under various conditions in different solvents, multiple nucleation possibilities can occur, such as solvates/hydrates, polymorphs of single components, binary combinations and their polymorphs/solvates/hydrates, or just starting materials (Fig. 5.14). To overcome all these possible obstacles, more effort and time with different experiments are necessary and also sufficient quantity of starting materials for the numerous experiments. The effort and explorations are justified because such alternative drug formulations are required for APIs.

There are a few reports on ternary systems starting from the beginning of this century. The systematic synthesis of ternary cocrystals was first reported by Aakeroy [104,105] (2001, 2002) exploiting Etter's hydrogen bonding rules [36,37,106,107]: the stronger acid···pyridine interaction (the strongest hydrogen bond) and the weaker acid···amide interaction (the second-best hydrogen bond in terms of strength) (Fig. 5.14A) result in ternary assembly of carboxylic acids, pyridine, and amide functional groups. A decade later, Desiraju [108] introduced another new concept to design ternary cocrystals by shape and size mimicry. Thus, 2- or 5-methylresorcinol (component 1), 4,4′-bipyridine (component 2), and a similar shape/size anthracene, pyrene, phenazine, 2,2′-bithiophene (component 3) were cocrystallized (Fig. 5.14C). Kuroda [105] reported crystal structure of three components BN (bis-β-naphthol), BQ (p-benzoquinone), and AN (anthracene) to study color changes in different systems. Aitipamula [109] reported isostructural ternary cocrystals of anti-TB (tuberculosis) drug isoniazid with nicotinamide and fumaric acid/succinic acid (SA). There was thus a surge of interest in ternary systems in the past decade. These studies in different ternary systems suggest that functional group selection, synthon combination, optimum crystallization conditions, and mechanochemistry are the key to success.

Desiraju postulated that crystallization of ternary cocrystals with molecules M1, M2, and M3 is difficult because there are many possibilities: single components

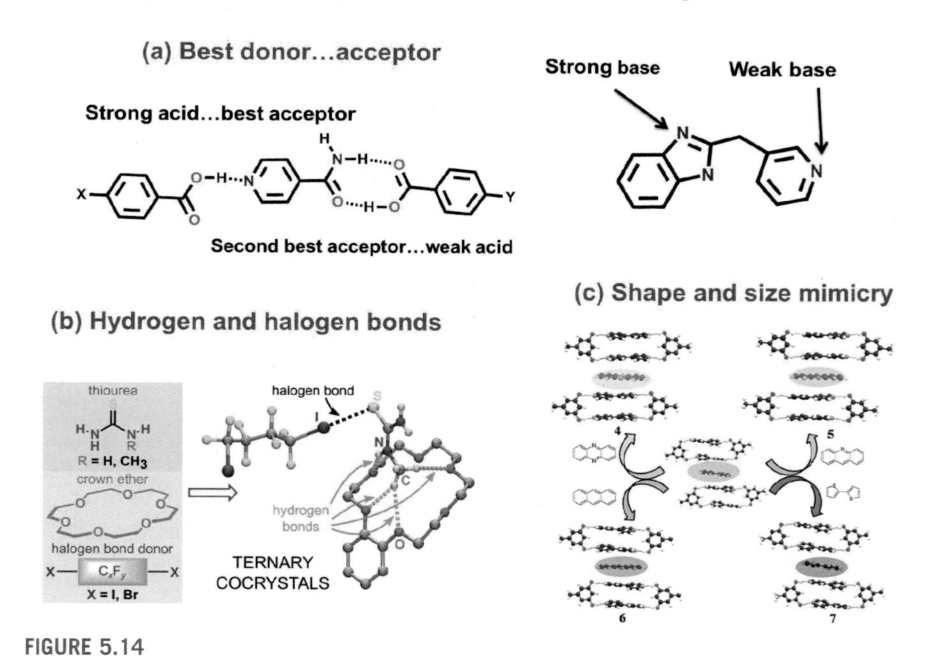

FIGURE 5.14

Important examples of ternary cocrystals where the strong donor and strong acceptor interact first and size/shape match assist in ternary directed assembly. (A) Best donor—acceptor. (B) Hydrogen and halogen bonds. (C) Shape and size mimicry.

(M1, M2, M3), binary cocrystals (M1···M2, M1···M3, M2···M3), solvated cocrystals, or polymorphs of single-component crystals. If synthon between M1···M2 binary adduct is stronger than interactions M2···M3 or M1···M3, then one may expect that initial formation of M1···M2 in solution can direct crystallization toward the association of M3 to give an M1···M2···M3 ternary cocrystal. If the M1···M2 is too strong or the binary cocrystal is too insoluble, it may be preferentially separated. A fine-tuning and balance of synthons and solubility are major factors role to result the ternary cocrystals [110,111]. A new designed strategy of ternary cocrystals was reported recently with prominent success, based on the orthogonality of two heterosynthon strategies like hydrogen bonding between crown ethers and thioureas and the halogen bonding between thioureas and perfluorohalocarbons. The strategy resulted in a high 75% success rate (Fig. 5.14B) [1110].

There were no reports on sulfonamides, an important and ubiquitous drug class, to make ternary systems. Ternary cocrystals of sulfonamide with pyridine carboxamides and lactams/syn-amides were attempted by Nangia [68]. Thus, *para*-substituted benzene sulfonamide, such as 4-carboxybenzene sulfonamide (SMBA = A) with coformers pyridine carboxamides (B=NAM, INA, PAM), could form acid-amide or acid-pyridine heterosynthons, and then lactams/2-hydroxypyridone derivatives (C=VLM, CPR/2HP, MeHP, OmeHP) will assemble ternary cocrystals of sulfonamides. Five ternary systems (SMBA−NAM−HP, SMBA−INA−HP, SMBA−NAM−MeHP, SMBA−NAM−OMeHP, SMBA−PAM−MeHP) were produced (Fig. 5.15A and B) using this design strategy. This model study was extended to the drug acetazolamide (ACZ) [113]. Ternary adduct

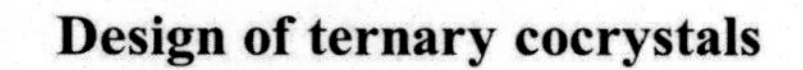

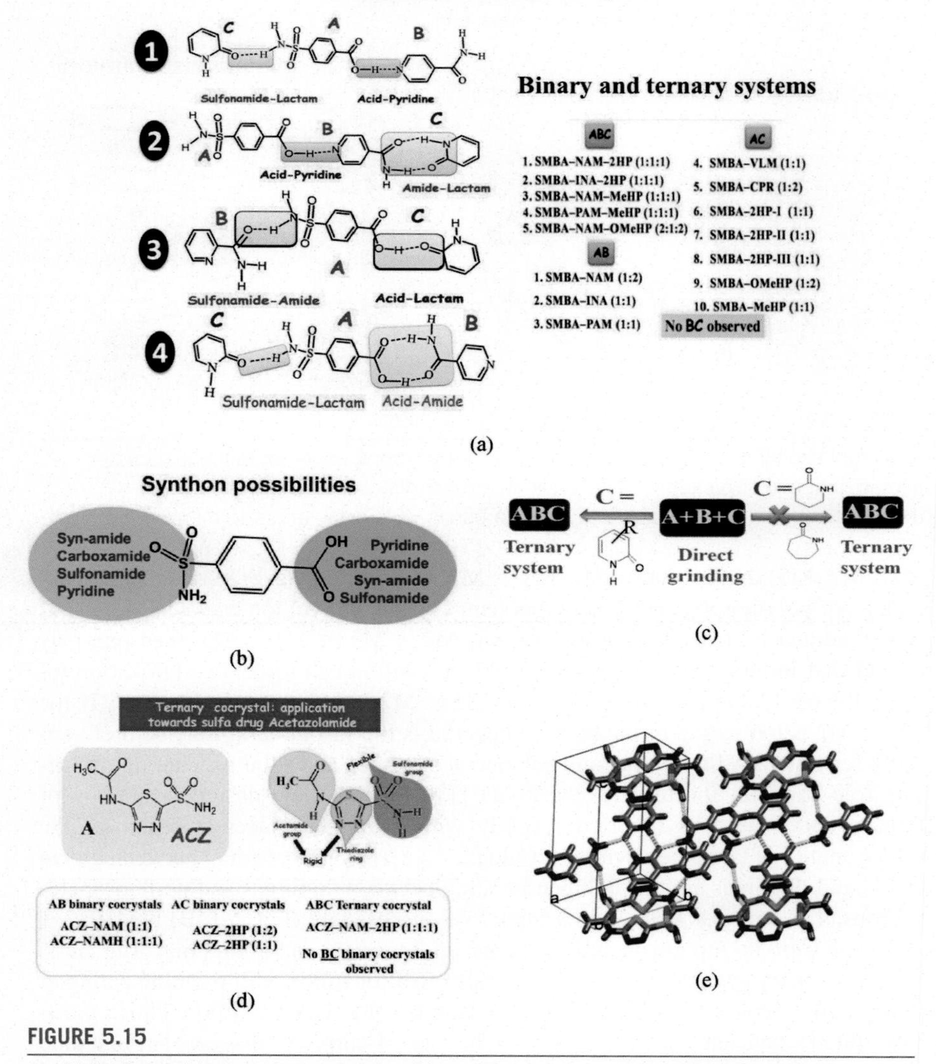

FIGURE 5.15

(A, B) Sulfonamide ternary cocrystals with pyridine carboxamides and syn-amides. (C, D) Acetazolamide ternary cocrystals. (E) Ternary cocrystals of sulfonamides.

of ACZ with lactams (valerolactam [VLM] and caprolactam [CPR]) and cyclic amides (2-pyridone, labeled as 2HP and its derivatives MeHP, OMeHP) together with pyridine amides (nicotinamide and picolinamide, NAM, PAM) were successfully formed. This crystal engineering study resulted in the first ternary cocrystal of ACZ with amide coformers (GRAS like), ACZ–NAM–2HP (1:1:1) (Fig. 5.15C–E).

2.8 Cocrystal polymorphs

Similar to single component polymorphic systems, multicomponent cocrystals and salts can also result in polymorphs under different conditions [76,114–117]. Especially for pharmaceutical cocrystals and salts, polymorphism takes on a commercial and manufacturing significance, similar to Ranitidine and Ritonavir in the mid- to late 1990s. In general, cocrystal polymorphs can result from the diversity in hydrogen bonding (synthon polymorphs), changes in conformation, and packing motifs between the components. There are a few examples of cocrystal polymorphs among APIs, for example, cabamazepine–nicotinamide, cabamazepine–saccharin (dimorphs), ethenzamide–ethylmalonic acid (dimorphs), ethenzamide–gentisic acid (trimorphs) [117], ethenzamide–24DHBA (dimorphs) [118], furosemide–nicotinamide (tetramorphs), and temozolomide (TMZ)–4-hydroxy benzamide (HBZA) [120] Ethenzamide resulted in different cocrystal polymorphs [109,116–122].

Five cocrystals polymorphs (forms I–V) and one hydrate of furosemide–nicotinamide (FS–NCT) have been reported [123]. Crystal structures of tetramorphs were determined from X-ray powder diffraction data (Fig. 5.16A–E). The supramolecular synthons in cocrystals polymorphs are different as FS has sulfonamide and COOH groups and NCT has Py-N and amide groups. This combination and variety of donor–acceptor groups results in diversity of synthons and crystal structure polymorphs. The presence of a large number of polymorphs in cocrystals suggests the importance of polymorph screening and characterization in drug cocrystals. The second example is trimorphic cocrystals of two APIs: ethenzamide and gentisic acid (Fig. 5.16F–I). Preferential crystallization of a particular polymorph from a specific solvent was observed in this case, so the effect of solvent selection on polymorphic outcome of crystallization becomes important. Cyclic syn-carboxamides (lactams) cocrystals of celecoxib with different synthons and cocrystals with five to eight member ring lactams were obtained [53]. Among them, VLM gave trimorphic cocrystals with synthon variety in celecoxib sulfonamide functional group (dimer and catemer) and others gave only one form exclusively (Fig. 5.16J–M). The trimorphic cocrystals of celecoxib with δ-VLM contain sulfonamide dimer and catemer together with cocrystals dimer, whereas the CPR cocrystal contains $SO_2N{-}H\cdots O{=}C$ heterosynthon [53]. The alternation of synthons with even–odd ring coformers offers a crystal engineering approach to sulfonamide cocrystals. Single-crystal X-ray structures of celecoxib cocrystals with GRAS lactams and new synthons between the sulfonamide and lactam functional group are explained as even–odd ring size changes and provide a starting point for synthon-based crystal engineering of sulfonamide drugs.

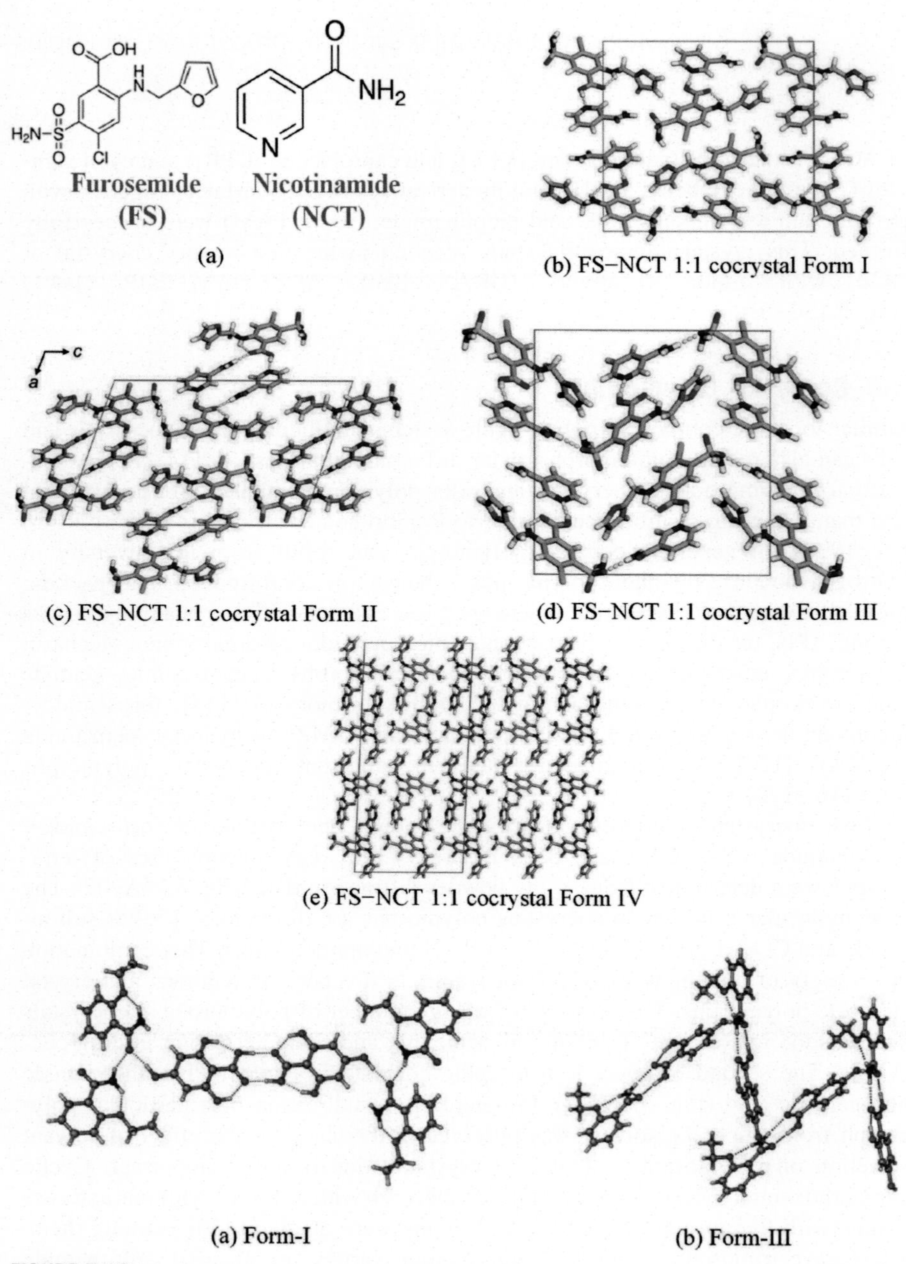

FIGURE 5.16

(A) Molecular structures of the furosemide, nicotinamide. (B–D) 2D packing of cocrystals polymorphs of FS–NCT (1:1). (E) FS–NCT 1:1 cocrystal form IV. (F) Molecular structures of the ethenzamide (EA) and gentisic acid (GA). (G–I) 2D packing of the cocrystal polymorphs for EA–GA (1:1) polymorphs. (J–M) Sulfonamide cocrystals form homodimer synthons in CEL–VLM, cocrystal form I. Hydrogen bonding of the CEL–VLM-II (2:2) cocrystals. Two CEL and VLM asymmetric molecules (capped stick and ball and stick model) are connected N–H···O catemer chains. Crystal structure packing of CEL–VLM-III (1:1) ratio. Two CEL molecules are connected by N–H···O catemer chain to VLM dimer in cocrystal structure. *FS–NCT*, furosemide–nicotinamide; *VLM*, valerolactam.

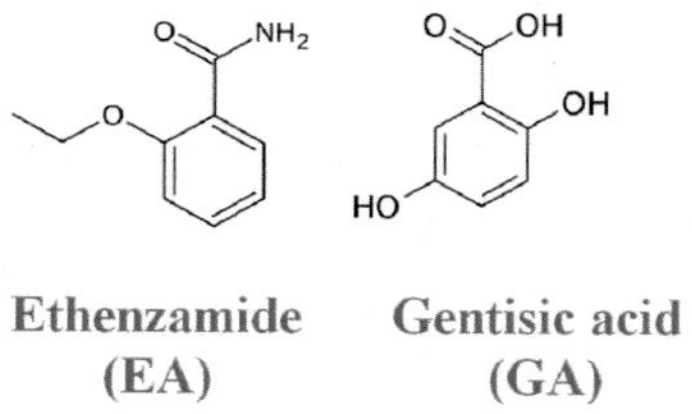

O
NH2
O
OH
OH
HO
Ethenzamide
(EA)
Gentisic acid
(GA)

(g)

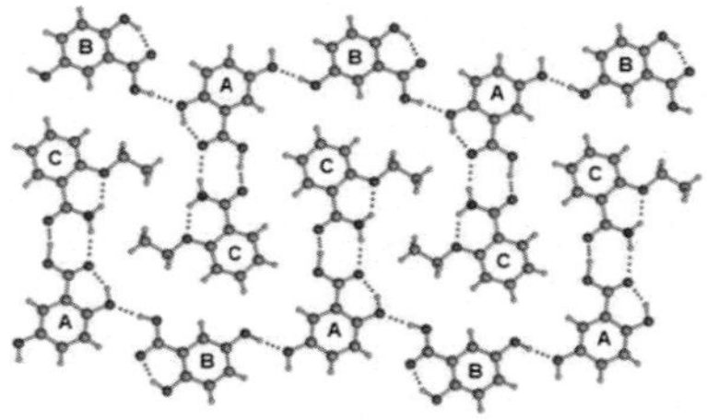

(h) EA–GA 1:1 cocrystal Form I

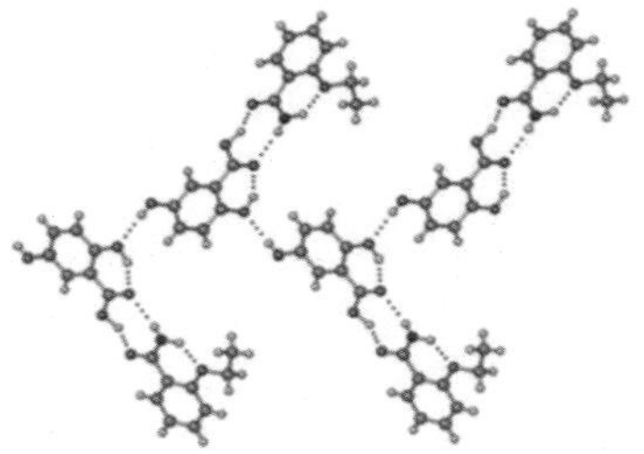

(i) EA–GA 1:1 cocrystal Form II

CF3
H2N
CEL
p-Tol
PYR
VLM
CPR
AZL

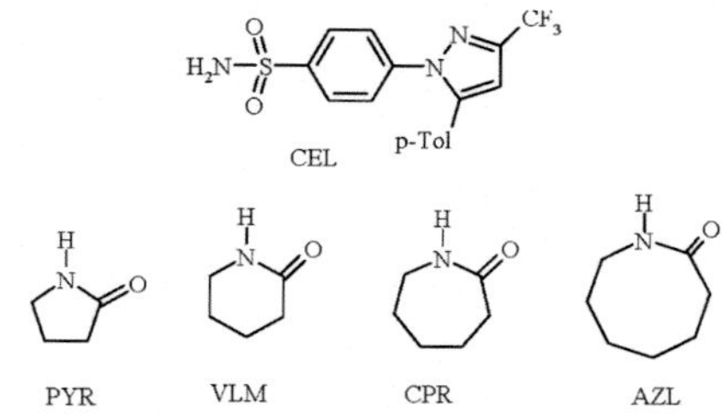

(j) Celecocxib and coformers structures

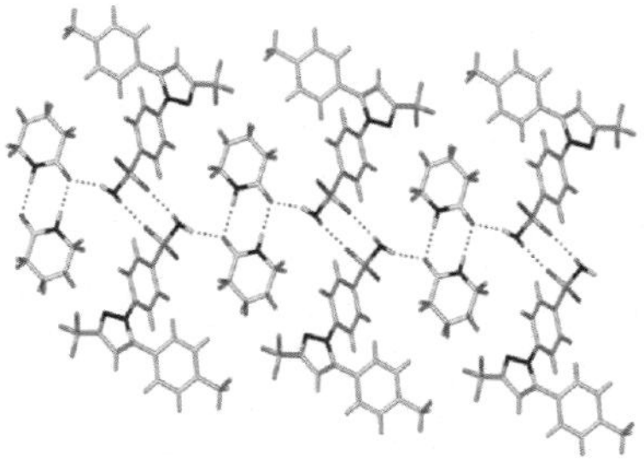

(k) CEL–VLM Form I

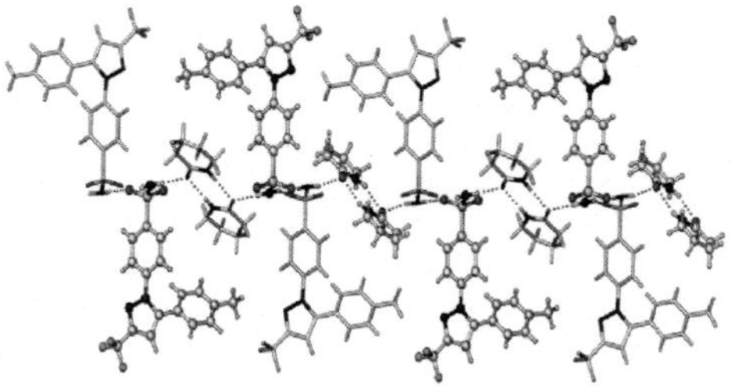

(l) CEL–VLM Form II

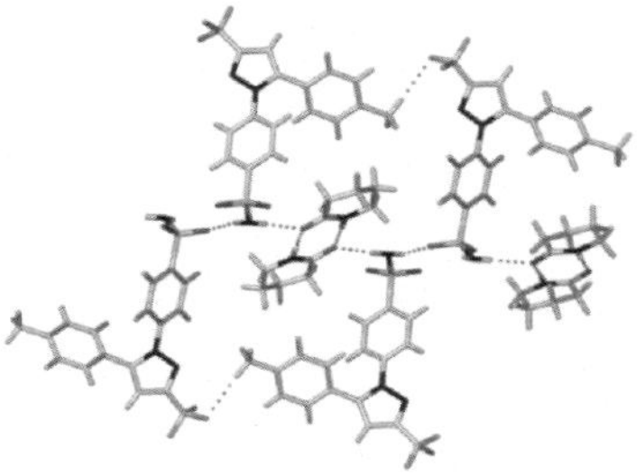

(m) CEL–VLM Form III

FIGURE 5.16 cont'd.

2.9 Cocrystals of different stoichiometry

The design of cocrystals generally relies on the interaction sites available in the parent molecule, and accordingly, stoichiometry of the starting materials is planned. However, the experimental conditions also dictate the formation of a specific stoichiometry cocrystal. There are examples of different stoichiometry cocrystals composing of the same components having different properties. Jones [124] et al. demonstrated crystal structure prediction (CSP)-based calculations to predict the stoichiometry of urea (U) with acetic acid (A) cocrystal system (Fig. 5.17). Relying on global minimization of lattice energy, they performed CSPs considering 1:1, 1:2, and 1:3 initial stoichiometries. They generated the experimental stoichiometry 1:2 and rationalized it in experimental condition based on the comparison of global minimum. In another work, they have also used CSP method to predict possible crystal structures and stoichiometry of theobromine with acetic acid [124].

Cocrystallization of nicotinamide and fumaric acid resulted 1:1 and 2:1 different stoichiometry having amide—acid heterosynthon (1:1 stoichiometry) and amide—amide homosynthon (2:1 stoichiometry) and different conformations of fumaric acid (Fig. 5.18) [125].

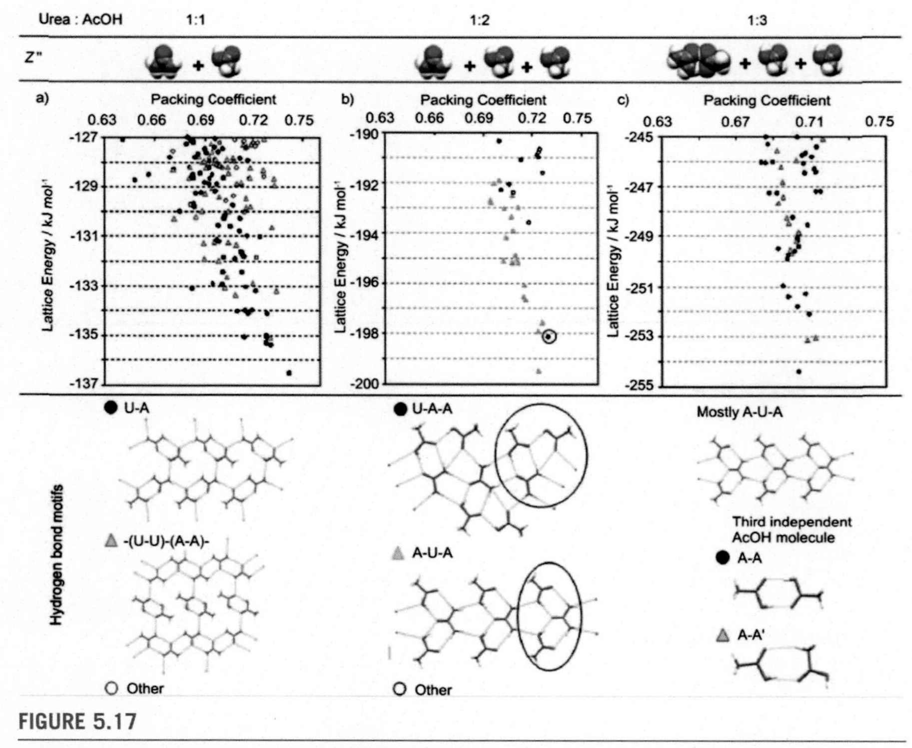

FIGURE 5.17

Lattice energy versus packing coefficients of the generated urea: acetic acid crystal structures in stoichiometry 1:1, 1:2, and 1:3.

Four cocrystals with different stoichiometry of BCS class I drug theophylline with o-aminobenzoic acid were studied by Sarma et al. [126] The *o*-isomer of amino benzoic acid participates in intramolecular H-bonding, leading to variable number of symmetry independent entities in the cocrystal lattice (CC-I to CC-IV), but in *m*- and *p*-isomer, there is no chance of intramolecular H-bonding, resulting in the formation of only 1:1 stoichiometry cocrystal (CC-V and CC-VI, respectively) (Figs. 5.19 and 5.20). These stoichiometric cocrystals show different physicochemical properties such as aqueous solubility and cell membrane permeation, which essentially determine the bioavailability of the drug cocrystal (Fig. 5.20) [126].

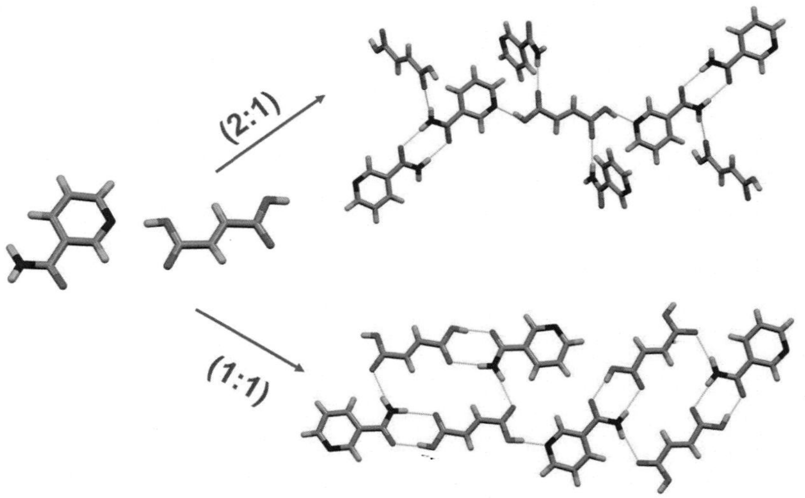

FIGURE 5.18

Amide—acid heterosynthon and acid—acid homosynthon observed in 1:1 and 2:1 cocrystal of nicotinamide and fumaric acid.

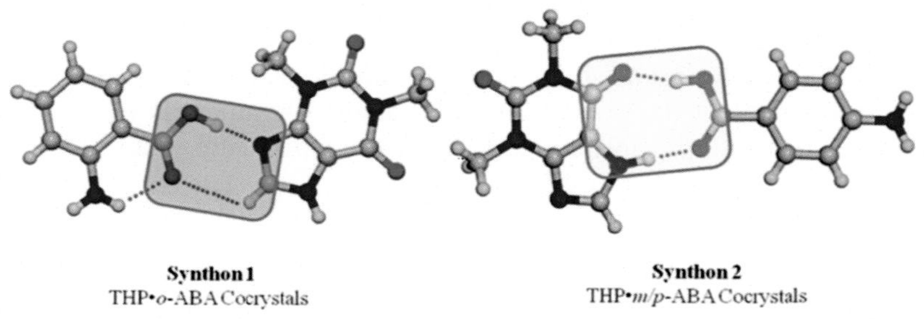

FIGURE 5.19

The hydrogen bond synthon observed in different stoichiometry cocrystal of THP with *o*-ABA (synthon 1) and *m/p*-ABA (synthon 2). *ABA*, aminobenzoic acid; *THP*, theophylline.

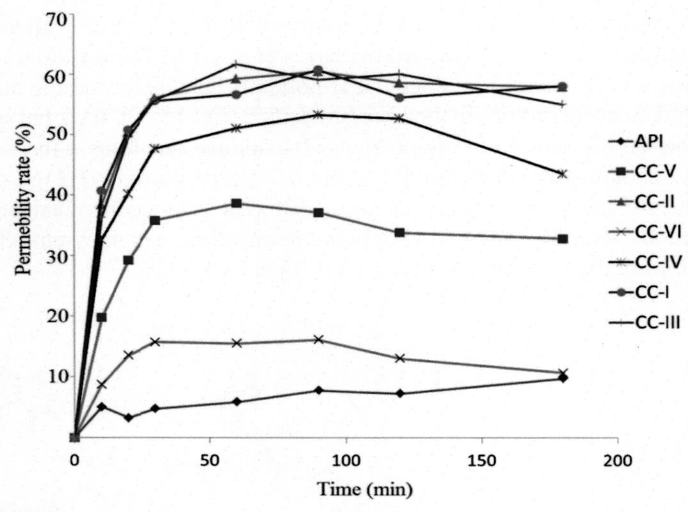

FIGURE 5.20

Comparison of membrane permeability of drug theophylline and the different stoichiometry cocrystal with *o*-aminobenzoic acid (CC-I to CC-IV) and *m*- and *p*-aminobenzoic acid (CC-V and CC-VI).

2.10 Drug–drug cocrystals

The concept of modifying the physicochemical properties of a drug molecule by forming a pharmaceutical cocrystal has generated immense interest due to their multidrug therapy [127]. These multicomponent crystals have potential in drug–drug cocrystals. Coadministration of a combination of theophylline and phenobarbital as a 2:1 cocrystal [128], ethenzamide–gentisic acid [117], and meloxicam–aspirin cocrystal [129] decreased the time required to reach the effective concentration in human plasma compared with the parent drugs. ACZ–theophylline (1:1) cocrystal had a faster dissolution rate than that of the physical mixture [130]. Pharmaceutical composition comprising therapeutically effective amount of lamivudine–zidovudine cocrystal for HIV treatment is reported [131]. A single-step codeposition of albuterol sulfate–ipratropium bromide [132], isoniazid–4-aminosalicylic acid, and pyrazinamide–4-aminosalicylic acid was shown for tuberculosis therapy [133]. Celecoxib–venlafaxine [134] and tramadol–celecoxib (will be discussed in Section 7) cocrystals for pain management are reported [134]. Amoxicillin–clavulanate cocrystal improved the antibiotic activity against non–beta-lactamase bacterial *Sarcina lutea* [135]. Sulfamethazine–theophylline [136] and pyrazinamide–diflunisal [137] are well known drug–drug cocrystal combinations. Antituberculosis drugs isoniazid (INH) and pyrazinamide (PYR) with 4-aminosalicylic acid (PAS) are reported

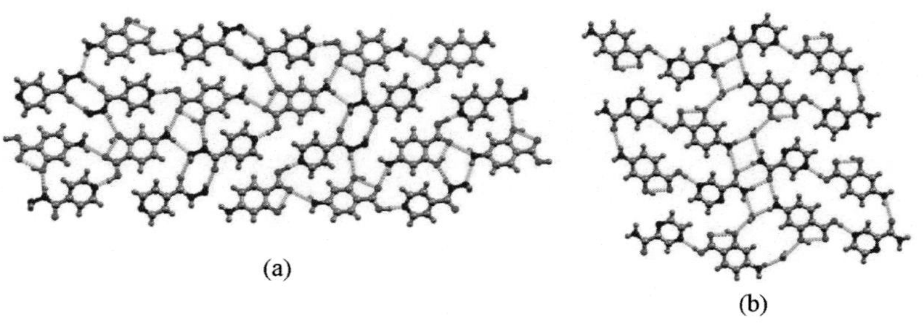

FIGURE 5.21

(A) INH—PAS crystal packing motif. (B) PYR—PAS, 2D crystal packing of hydrogen bonds. Both structures contain the acid—pyridine synthon. *INH*, isoniazid; *PAS*, pyrazinamide with 4-aminosalicylic acid; *PYR*, pyrazinamide.

(Fig. 5.21) [138]. INH—PAS is a 1:1 cocrystal, whereas PYR—PAS as 1:1 monohydrate and its crystal structure contains the acid···pyridine heterosynthon. In INH—PAS, the H-atom is located on the carboxylic acid and is indicative of a cocrystal, whereas there is partial proton transfer in PYR—PAS. The extent of proton transfer depends on the temperature indicative of salt—cocrystal intermediate states. Drug—drug cocrystals are likely to be more significant for TB and cancer because of multidrug therapy (Table 5.3).

2.11 Supramolecular gels

Supramolecular gels are colloidal in nature in which a small amount of solid material immobilizes the flow of a large amount of solvent; gels are preliminarily confirmed by the simple tube inversion method. Gels are prepared from low-molecular-weight (<3000) organic compounds capable of immobilizing solvents, which are known as gelators [139,140]. Gelator molecules aggregate through noncovalent interactions to give different morphologies such as strands, tapes, and fibers. Unexpectedly gel formation resulted in two of the salts, namely niflumic acid (NFA)—PIP and NFA—BZA II, after dissolving the salts in nitrobenzene (NB), methyl salicylate (MS), menthol, and mesitylene. Solubility and dissolution rate of the novel solid forms NFA—CPR, NFA—PIP, NFA—benzenesulfonic acid (BSA), and NFA—tyramine (TYA) showed the best behavior for NFA—TYA salt with improvement of 42 and 54 times in solubility and dissolution rate compared with the reference drug [141]. The remaining solids such as NFA—PIP, NFA—BSA salts, and NFA—CPR cocrystal exhibited improvement of 39, 18, 1.4 times in solubility and 7.8, 10, 2 times for dissolution rate. Supramolecular gelators of antiinflammatory drugs for topical application complement the oral route (Fig. 5.22). These gelators have a propensity to form self-assembled fibrillary networks by noncovalent interactions within the immobilized solvent molecules as supramolecular gels.

Table 5.3 Drug–drug cocrystals.

Aspirin cocrystals	Sildenafil–acetylsalicylic acid (1:1), 1:1:1 cocrystal salt involving sildenafil cation, acetylsalicylate and salicylic acid, meloxicam–aspirin (1:1) cocrystals
Antituberculosis drug cocrystals	Antituberculosis drugs isoniazid (INH), pyrazinamide (Pyz), and 4-aminosalicylic acid (PAS). Ternary cocrystals as drug-bridged drug with fumaric acid to connect a Pyz and INH
Bicalutamide cocrystals	Bicalutamide (Bic)-salicylamide and Bic-benzamide.
Caffeine, theophylline cocrystals	Caffeine with various substituted hydroxybenzoic acid derivatives, theophylline (polymorphs) and with epalrestat, sulfacetamide with caffeine, and theophylline with hydroxybenzoic and dihydroxybenzoic acids with phenolic coformers
Carbamazepine	Carbamazepine–indomethacin, 1:1 cocrystal of carbamazepine, and 4-aminosalicylic acid
Pyrimethamine cocrystals	With carbamazepine, theophylline, aspirin, α-ketoglutaric acid, saccharin, p-coumaric acid, succinimide, and L-isoleucine
Drug–nutraceutical cocrystals	With nicotinic acid, pyridoxine, vanillic acid, resveratrol, pterostilbene, quercetin, etc.
Diflunisal and diclofenac cocrystals	With theophylline, furosemide–caffeine cocrystals
Few other examples	Gefitinib–furosemide salt hydrate, gliclazide–metformin salt hydrochlorothiazide–isoniazid, lamivudine–zidovudine, myricetin–piracetam, norfloxacin–sulfathiazole salt hydrate, oxaprozin–salbutamol salt, pyrazinamide–diflunisal, vitamin D2–vitamin D3, and temozolomide–theophylline. Dapsone–sulfanilamide, flavone, luteolin, caffeine next flufenamic acid cocrystals with 2-chloro-4-nitrobenzoic acid and ethenzamide

3. Heterogeneous nucleation, self-assembled monolayers, surface-mounted metal–organic frameworks

3.1 Heterogeneous nucleation and crystallization up on predesigned surfaces

A crystal cannot grow instantaneously; nucleation occurs at first followed by growth, and each crystal has its own growth history. Hence, understanding the crystal morphology of small organic molecules has become the grand challenge for solid-state research groups. A fundamental understanding of molecular faces and the ability to grow crystals with desired properties has fascinating possibilities. Consequently, numerous methods have been developed to understand crystal growth; among them, tailor-made additive (a molecule very similar in structure) induced crystal mechanism due to surface adsorption of a particular crystal face via weak supramolecular interaction in solution. Results show that such additives have a dramatic effect on crystal growth and habit due to the preferential adsorption of the

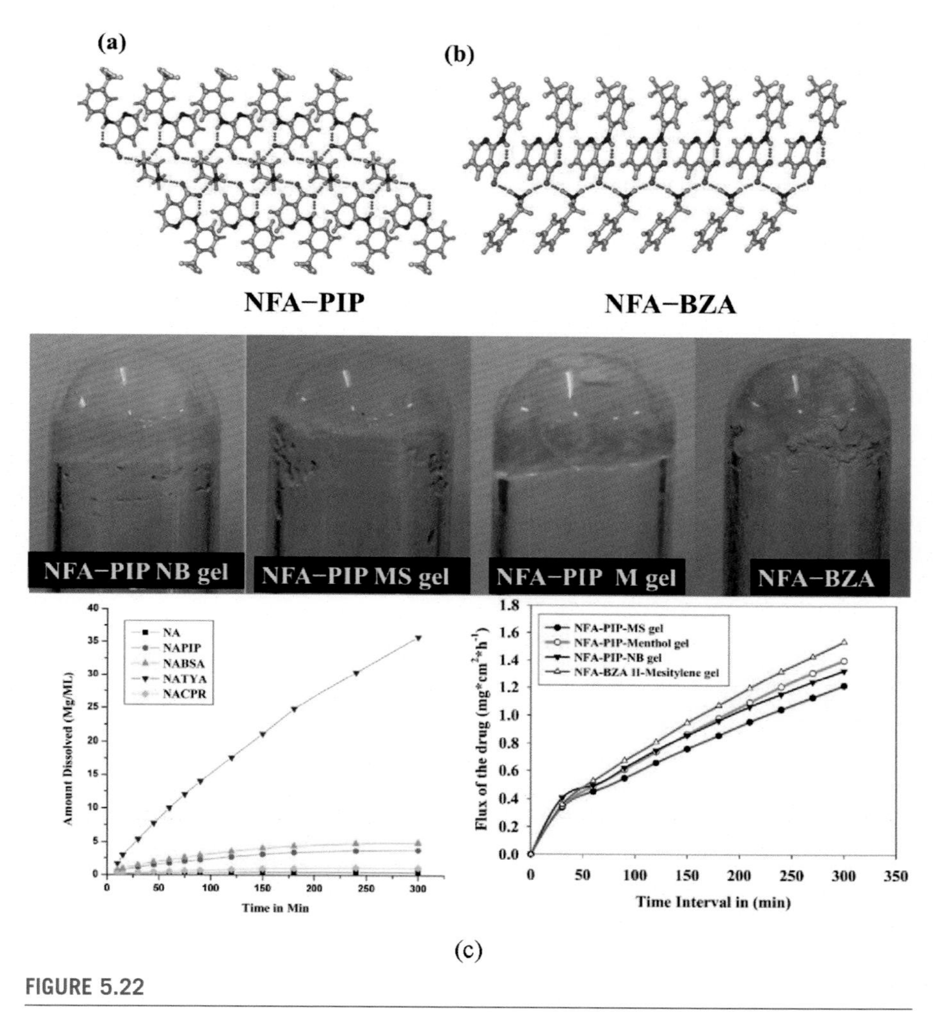

FIGURE 5.22

(A, B) Crystal packing of the NFA–PIP, NFA–BZA. (C) Formation of gel and their dissolution, permeability profile. *BZA*, benzamide.

additive on specific crystal faces called active sites for nucleation and growth. A bound additive generally perturbs the regular growth of upcoming layers based on attachment energies and allows different directional growth, which finally leads to changes in crystal morphology. An efficient method for kinetic resolution of racemic conglomerates by crystallization in the presence of tailored additives was described and explained as stereoselective adsorption of the additive on the surface of the growing crystals of the enantiomer resulting in a drastic decrease in the rate of growth, which thus leads to preferential crystallization of the opposite enantiomer (rule of reversal). This was proven by the crystallization of the conglomerates

(R,S)-glutamic acid hydrochloride, (R,S)-threonine, (R,S)-*p*-hydroxyphenyl) glycine *p*-toluenesulfonate, and (R,S)-asparagine hydrate in the presence of specific amino acids as additives. Another example is the dramatic change in morphology in the presence of carboxylic acid additives on amide crystals, $RCONH_2/RCOOH$, e.g., in asparagine/aspartic acid system, the carboxyl O atom is a much weaker proton acceptor than the corresponding amide O atom. Hence, there is a large energy loss in replacing an amide—amide hydrogen bond by an amide—acid hydrogen bond at the site of the additive [142—145].

When we examine the basics of nucleation theory (Fig. 5.23A), growth mechanisms of crystals are still unclear, and many theories have been postulated. Classical nucleation theory and crystal growth explanation suggest that crystallization occurs mainly by nuclei that resemble mature crystal phases. However, a critical size where the nuclei volumes free energy begins to offset the unfavorable surface free energy arising from the interface with the growth medium. Hence, crystallization under heterosurface confinement offers an opportunity to examine nucleation, growth control, morphology control, polymorph control, and sometimes both polymorph and morphology and growth direction and phase transformations at which kinetics and thermodynamics of nucleation and growth mainly depend on the contact angle between them. Collectively, these investigations have increased our understanding of crystallization, and they suggest strategies for controlling crystal growth. The concept of classical nucleation theory is not applicable in many cases, since it is not an appropriate framework for understanding crystal nucleation from solution despite its popularity and the theory fail to account for all the essential degrees of freedom to describe the nucleation of crystals. Hence, heterogeneous nucleation (Fig. 5.23B) is attractive for many researchers to understand crystal nucleation and growth.

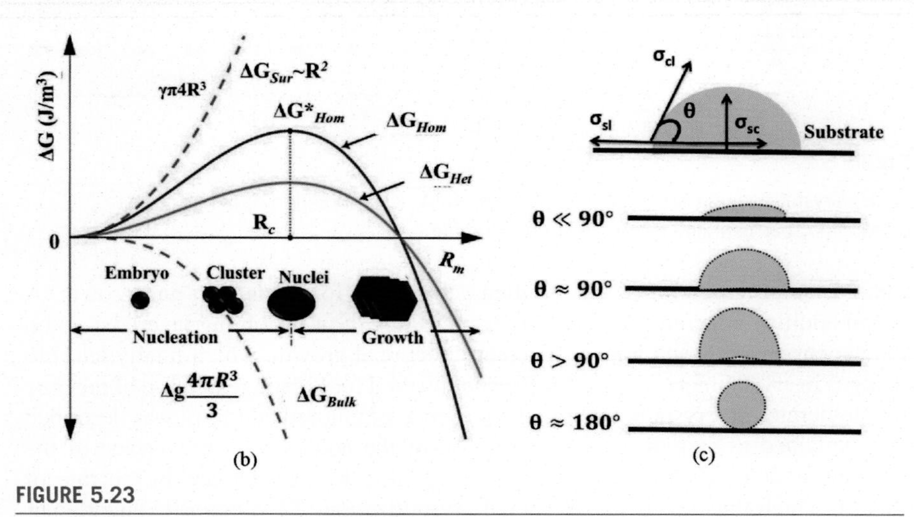

FIGURE 5.23

(A) Homogeneous nucleation. (B) Heterogeneous nucleation.

An off-shoot of crystal nucleation is polymorphic screening of pharmaceutical drugs and small organic molecules. Polymorphic screening and crystal morphology engineering of organic molecules have been studied in the past few decades through multiple methods. However, a fundamental understanding of heterogeneous nucleation on molecular surfaces has been challenging and has increased interest from the pharmaceutical industry. Different approaches to heterogeneous nucleation are heterosurface design such as polymer, self-assembled monolayers (SAMs), surface-mounted metal–organic frameworks (SURMOFs), polymers, nanoconfinements, supramolecular gels, substrates, crystal up on crystal (epitaxial), low-molecular-weight gels, ionic liquids, etc. [142–147]. Despite the fact that SAMs [142] are a well-established methods used to form less stable polymorphs or crystalline shapes (different faces), results are often not practical due to the limitation in contribution to the total surface, applicable at room temperature, cost-effective, and less stable. Polymer and gel-induced polymorphic studies were reported recently and showed advantages over solution crystallization; however, they are often limited in either morphology or polymorphic yield, and finally, a detailed explanation of the underlying mechanisms is not fully clear. On that account, a new one-pot method with multiple new design of heterosurface is important to obtain metastable polymorphs.

3.2 Self-assembled monolayers

Recently, heterogeneous surface medication methods were developed through SAMs for organic devices, understanding the calcite nucleation and growth, and biomedical devices to give completely new types of behavior that open unprecedented applications, such as ultrasensitive label-free biosensors and SAM/organic transistors, polymorph screening, and morphological crystal engineering because the presence of a foreign surface in a crystallizing system can influence the pathway of crystallization for the simple reason that a nucleation barrier can be lowered, as a reduction in the interfacial free energy occurs. The ability to functionalize a surface with desired properties and their highly ordered structure makes SAMs attractive templates for nucleation and crystal growth due to the SAM surface chemistry that leads to specific intermolecular interactions with the nucleating plane of crystals, thereby controlling their morphology and crystal form. In principle, cleaned gold-plated substrates are dipped into the corresponding solution over 14–24 h for the monolayer to assemble. It was noted that OH and COOH head functional group thiols resulted in better direction compared to other functional groups. However, depending on the need or objective, one may incorporate hydrophilic or hydrophobic nature of substrate by selecting the head group.

SAMs of 4-hydroxy-4-mercaptobiphenyl, 4-(4-mercaptophenyl)pyridine, and their mixed SAMs with 4-methyl-4-mercaptobiphenyl were prepared on gold (111) surfaces where SAMs can serve as nucleation planes/templates and able to modify the morphology and growth of glycine crystals [143]. After an induction period, it was observed that nucleation on 100% of OH surfaces, the glycine crystallographic plane is {011}, whereas for 0% and 50% OH surfaces, it is {h0l} face. Furthermore, at 25%, 75%, and 100% surface pyridine concentrations, the crystallographic planes are {010}, {121}, and {1105}, respectively. These differences are attributed to differences in H-bonding between glycine molecules in the nucleating layer and SAM surface and mean that due to interfacial H-bonding increase, the dipoles of glycine molecules within the crystal become more perpendicular to the SAM surface. The direction of dipoles of glycine molecules, which are nucleated on a pyridine surface, is not as close to the surface normal as those of molecules that nucleated on a hydroxyl surface. This implies that the overall H-bonding interactions between the COO^- and NH_3^+ groups of glycine (Fig. 5.24), and the hydroxyl groups of the SAMs surface are stronger than those between the NH_3^+ and the pyridine group. Furthermore, 4-mercaptobenzoic acid SAMs on gold induce glycine to nucleate and grow as rod-shaped particles. However, the crystal habit is considerably different than the crystals grown from aqueous solution (bipyramidal) with the {011}, {110}, {010}, and {120} as major morphological importance faces. Inspection of the {12−1} crystallographic plane shows hydrogen bonds to be formed between SAMs and glycine growing crystal faces.

The glycine study was extended by modifying the SAM substrate through patterning due to an increased interest in the development of new drug delivery methods and controlling crystal size and solid form purity. Patterned SAMs of lateral dimensions ranging from 25 to 725 μm functionalized metallic islands were used to screen polymorphs under different conditions, as they acted as nucleation sites to control the size of the crystals. Gold islands are formed by evaporation of titanium

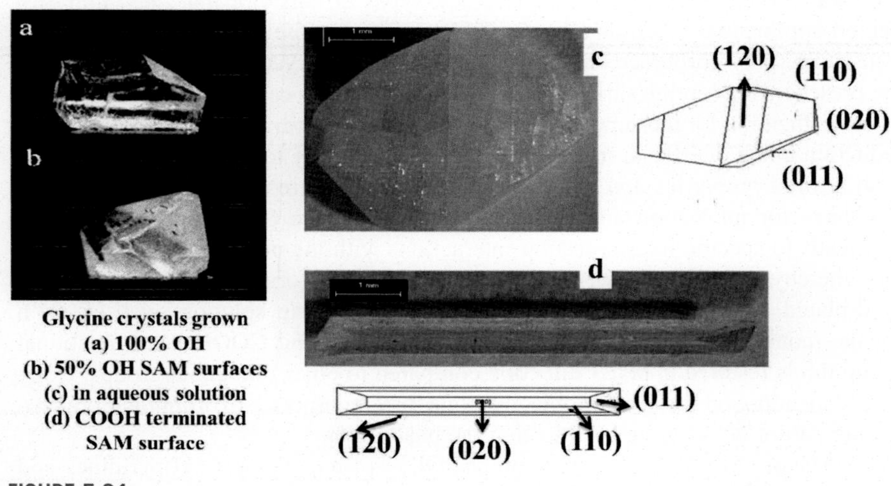

FIGURE 5.24

Glycine growing crystal faces with SAMs. *SAMs*, self-assembled monolayers.

Taken from A. Ulman, Formation and structure of self-assembled monolayers. Chem. Rev. 96 (1996) 1533–1554.

through a mesh onto glass slides, and next evaporation of gold and 4-mercaptobenzoic acid was selected for SAMs [144]. Octadecyltrichlorosilane (OTS) was used to backfill the glass-exposed substrate. Fig. 5.24 shows the patterned SAMs and crystallization of aqueous glycine solutions on different sizes of islands; each island resulted in glycine crystal. Very interestingly, crystal habits are not like bipyramidal, needle, or triangle prisms and maintain a uniform array of nucleation at hydrophilic regions, which confirms the consistent deposition is reproducible on each island and homogeneous. Raman microscopy was employed to identify the polymorphic forms (CH_2 symmetric stretch of polymorphs at 2972, 2953, and 2964 cm^{-1}) at hydrophilic square islands. Patterned SAMs can direct the polymorph selectivity as a new method for particle engineering. This bottom-up approach involves the nucleation and growth of crystals, polymorphic on hydrophilic square islands surrounded by hydrophobic regions [145].

In previous studies [142], patterned SAMs (Fig. 5.25) were used to produce micrometer-sized crystals, and the study is extended with circular-shaped islands with diameters of 500 nm using photolithography and SAMs (Fig. 5.26). The circular islands are covered with a hydrophilic SAM (3-aminopropyl-triethoxysilane), while the rest of the surface is functionalized with a hydrophobic SAM as OTS. Glycine solution was immersed into a vessel filled with hexane and water (hexane is immiscible with water and the solubility of glycine in hexane is extremely low) and the two solutions separated, and the patterned substrate was dipped into glycine solution and slowly retracted into hexane without exposing it to air. The glycine solution wetted the hydrophilic circular islands, generating arrays of 500-nm

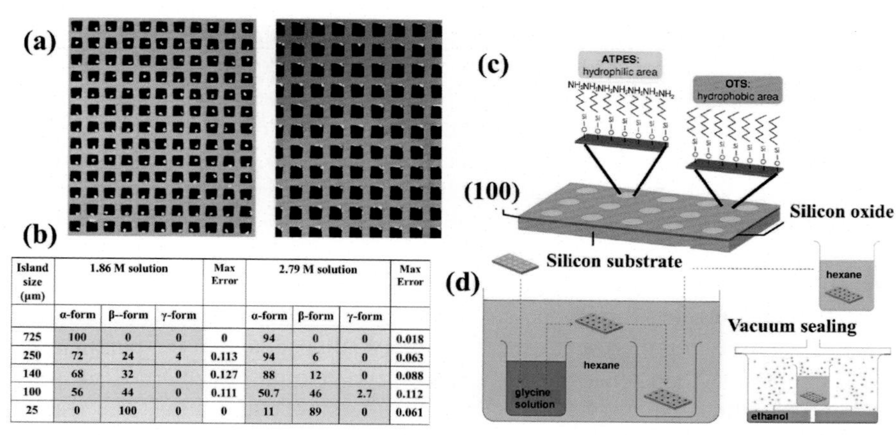

Island size (μm)	1.86 M solution			Max Error	2.79 M solution			Max Error
	α-form	β--form	γ-form		α-form	β-form	γ-form	
725	100	0	0	0	94	0	0	0.018
250	72	24	4	0.113	94	6	0	0.063
140	68	32	0	0.127	88	12	0	0.088
100	56	44	0	0.111	50.7	46	2.7	0.112
25	0	100	0	0	11	89	0	0.061

FIGURE 5.25

Crystallization on confined engineered surfaces: a method to control crystal size and generate different polymorphs.

Modified from A.Y. Lee, I.S. Lee, S.S. Dette, J. Boerner, A.S. Myerson. Crystallization on confined engineered surfaces: a method to control crystal size and generate different polymorphs. J. Am. Chem. Soc. 127 (2005) 14982–14983.

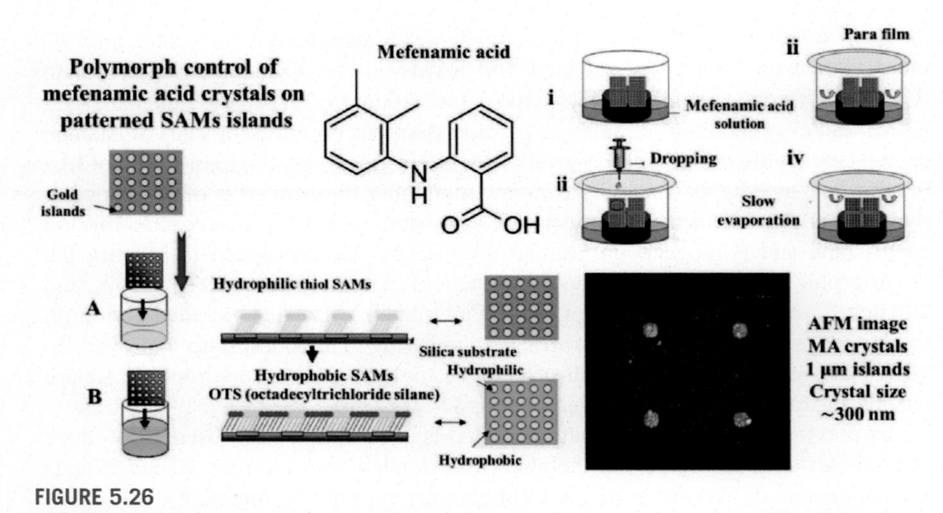

FIGURE 5.26

Mefenamic acid crystallization on surface of patterned SAMs. *SAMs*, self-assembled monolayers.

Taken from P. Bora, B. Saikia, B. Sarma, Oriented crystallization on organic monolayers to control concomitant polymorphism. Chem. Eur. J. 26 (2020) 699–710.

hemispherical solution droplets on the substrate surface covered by hexane. Glycine solution droplets remained undersaturated; next the supersaturation level droplets were controlled by diffusion of antisolvent (EtOH). After 70 h, glycine crystals of 200–500 nm were formed [146]. As the solubility of crystals was measured as a function of crystal size according to Ostwald–Freundlich equation to predict solubility enhancement in small crystal size, it has immediate need for the production of the nanosized pharmaceutical crystals to aid the formulation of poorly soluble drugs.

This approach is extended to pharmaceuticals (Fig. 5.26). Seven different SAMs were employed to study the nucleation behavior of the nonsteroidal antiinflammatory drug mefenamic acid (MA). The observations selectively proved that SAMs forming a strong interaction with the –COOH group of MA molecules preferably resulted in form II of MA. The ability to prepare crystalline MA as small as ~300 nm while controlling the polymorphic form was demonstrated. As control of polymorphism during crystallization is a practical challenge in the pharmaceutical industry, the SAM approach for a specific polymorph on patterned surfaces would be an alternative approach for drug polymorph crystallization. The results suggested that the SAMs forming strong interactions with the MA –COOH molecules depleted their polarity and pushed the system to form $\pi\cdots\pi$ and C–H$\cdots\pi$ interactions, which lowered the free energy barrier of nucleation and promoted the growth of form II.

The antimicrobial sulfa drug sulfathiazole [147] (reported five polymorphs, Fig. 5.27) is known for its concomitant crystallization due to conformational

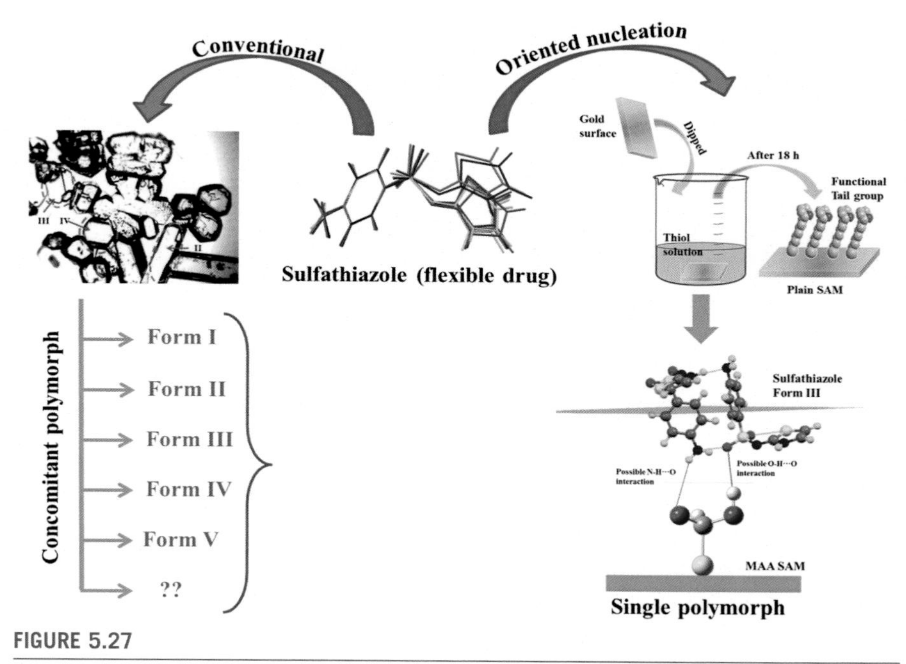

FIGURE 5.27

Control concomitant polymorphism of sulfathiazole by oriented crystallization on the surface of self-assembled monolayers.

Taken from P. Bora, B. Saikia, B. Sarma, Oriented crystallization on organic monolayers to control concomitant polymorphism. Chem. Eur. J. 26 (2020) 699–710.

flexibility. Hence, during industrial development stage, practical problems of concomitant crystallization need to be addressed for marketing. Functionalized SAMs were employed to control concomitant polymorphs and pure form III crystals grown on mercaptoacetic acid (MAA) and pure form II on mercaptosuccinic acid (MSA) provided selective crystallization through designed SAMs. These observations suggest that the functional groups in the SAM play an important role for the nucleating crystal, control the crystallization process, and give a clue for biological templates as well as open up opportunities toward polymorphs selection [147].

3.3 Surface-mounted metal–organic frameworks

SAMs are a well-known method and used to form reproduce various less stable polymorphs or crystalline morphologies (different faces), but the results are often not practical due to the limitation in contribution to the total surface. On account to that, recently, SAMs are extended as SURMOFs, which are principally deposition of metal–organic frameworks (MOFs) on SAM heterosurfaces. Apart from SAMs and SURMOFs, few more additional directions are reported by different groups toward pharmaceutical drugs. SURMOFs (Fig. 5.28) have been studied recently with

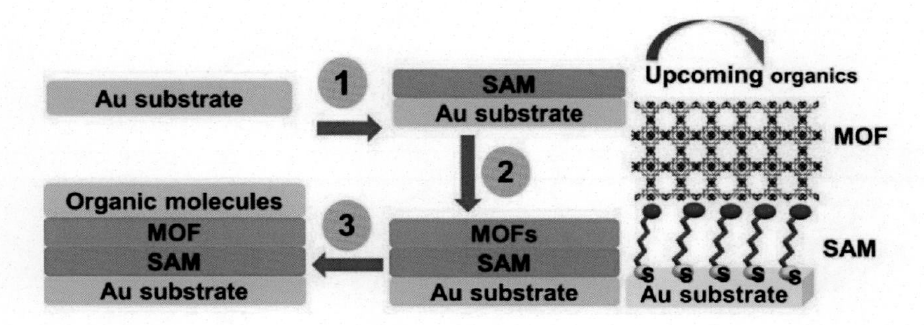

FIGURE 5.28

SURMOF substrates preparation. *SURMOF*, surface-mounted metal–organic framework.

Taken from G. Bolla, A.S. Myerson, SURMOF induced morphological crystal engineering of substituted benzamides. Cryst. Growth Des. 18 (2018) 7048–7058; G. Bolla, A.S. Myerson, SURMOF induced polymorphism and crystal morphological engineering of acetaminophen polymorphs: advantage of heterogeneous nucleation. CrystEngComm 20 (2018) 2084–2088.

different applications such as photovoltaics, CO_2 reduction, memory devices, supercapacitors, and batteries. The advantage of controlled orientation of the MOFs on the basis of the ground SAM functional group would allow preferential growth of the target functional group of small organic molecules, but this young branch of chemistry has not yet been explored in the direction of morphological engineering. MOFs being highly porous crystalline materials, their impact, and contribution on the surface as a heterogeneous layer are quite effective when compared with the usual SAM surface. In addition, they can lead to different nucleation and growth directions than the usual crystallization path.

The template SURMOF crystallization [148–150] method involves three steps: (1) SAMs are prepared using gold substrates and thiol solutions; (2) MOF film preparation from solutions using layer-by-layer dipping; and (3) crystallization of the organic solid. These SURMOF substrates were designed to investigate how the template functionalization of highly porous surfaces (SURMOFs) can influence the nucleation of functional organic molecules and morphological crystal engineering of benzamides (BZAs), Fig. 5.29. The selection of BZAs was justified due to their lack of complex formation during crystallization with selected Hong Kong University of Science and Technology (HKUST) MOFs.

In order to unnderstand the viability of SURMOF approach in the crystallization of small organic molexules, BZA, 4-amino benzamide (ABZA), and HBZA were studied No complexation was observed between BZAs and SURMOF films.

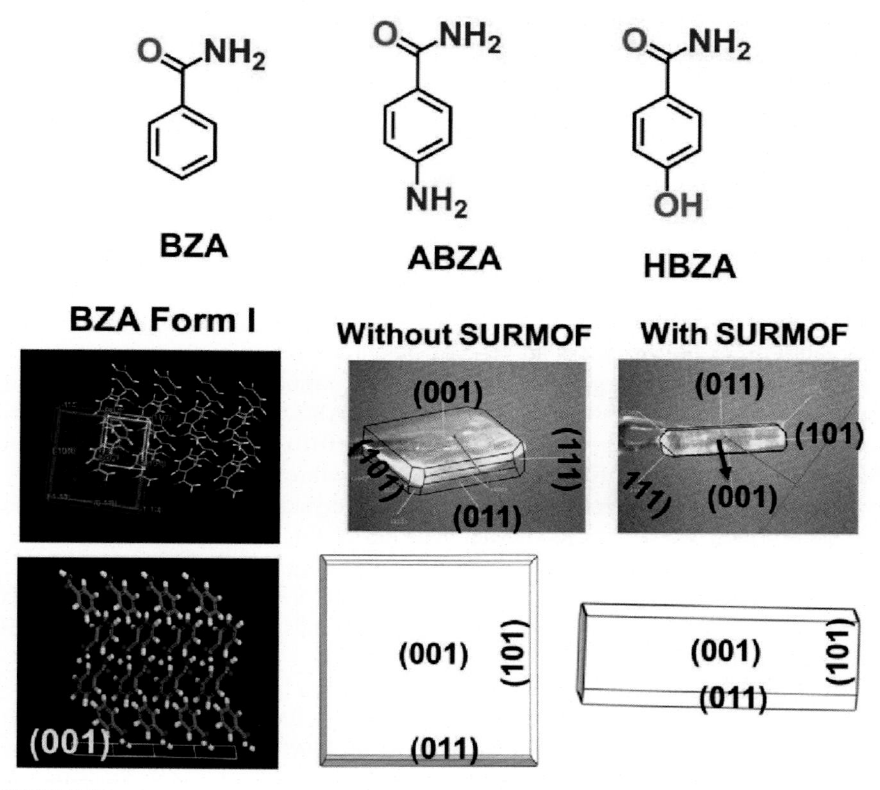

FIGURE 5.29

Benzamide crystals grown on SURMOFs. *SURMOFs*, surface-mounted metal–organic frameworks.

Taken from G. Bolla, A.S. Myerson, SURMOF induced morphological crystal engineering of substituted benzamides. Cryst. Growth Des. 18 (2018) 7048–7058.

Solution crystallization of BZA crystals resulted in plate morphology of (001), (011), and (101) as major faces, whereas SURMOF crystallization suggested needle morphology with (001) and (011) morphologically important faces, and the major (101) face was retarded. Packing of BZA clearly showed the (001) face along the *b*-axis as N–H···O and along the *a*-axis as N–H···O catemers. SURMOF-induced crystals inhibited the (001) face growth via weak interactions between the HKUST and amide. The (001) face of BZA is exposed to amide as the next (011) face and has no hydrogen bond groups, whereas the (101) crystal face showed alternate amide groups with attachment energies to (001), (011), and (101) as −14.9, −44.50, and −47.89 kcal/mol, which showed that the growth rate between (011) and

(101) is not high (~3 kcal/mol). Hence, controlling one face with modified SURMOF surface crystallization is an alternative to the additive-induced crystallization as the result matches to attain plate to needle crystals of BZA. In a second example, ABZA solution showed crystallization without substrate in EtOH as block morphology with (100) and (110) as major faces, with the primary amide chains N—H···O, whereas ABZA crystals on the surface of the SURMOF substrate showed needles on the surface and block morphology at the edges, (111) as the major face and (001), (110), (100), (011), and (010) as minor faces, which confirmed that nucleation of ABZA on MOF surface is different from normal solvent crystallization. Indeed, edge-grown crystals showed needles of (100) face as the primary surface and also further showed (111) face of minor morphological importance. Hence, MOF pores can block the surface, which allows other faces to grow and further leads to different growth compared with normal experiments. The third example of HBZA was grown similar to BZA and ABZA. Well-grown single crystals showed rod morphology with (001), (011), and (010) faces, and next crystals on the surface of the SURMOF heterogeneous substrate through solution phase epitaxy crystallization method were studied with multiple crystals at various places of the designed surface. These crystals showed plate morphology as (001) of major importance and (011), (010), (100), and (101) of minor morphological importance. The MOF-induced crystals grew along the *c*-axis through N—H···O on the surface and block morphology at the edges, whereas edge crystals were a balance of MOF and solution crystallization.

To show the advantage of heterogeneous nucleation by SURMOFs, an example of drug acetaminophen (N-acetyl-para-aminophenol, APAP) is presented (Fig. 5.30) in which formation of the less stable polymorph of APAP and morphology changes using SURMOFs were observed. APAP is a well-known API (used as an analgesic and antipyretic drug; also referred to as paracetamol), and three polymorphs are reported wherein form I is less soluble and exhibits poor tableting compaction compared with form II and another metastable form III. Forms I and II crystalize concomitantly in solution crystallization, form II was reported as needles, whereas form I has blocks and prisms. From formulation and tableting point of view, form II is better due to its layered packing. However, the reproducibility of form II during solution crystallization still remains a great challenge, but melt crystallization or the use of an additive can produce form II, but these methods are not suitable for industrial processing for bulk-scale formulations. Hence, studies continued to find suitable methods for the production of the form II. Crystal nucleation on the porous surface of SURMOF contributed significantly to unique nucleation kinetics, allowed for metastable phase stabilization, and engineered the SURMOF with block morphology providing a dual advantage. This successful demonstration of MOF-induced heterogeneous nucleation offers a new approach and opens challenges for metastable polymorph discovery along with inherent morphology differences based on complementary interactions at the SURMOF interface.

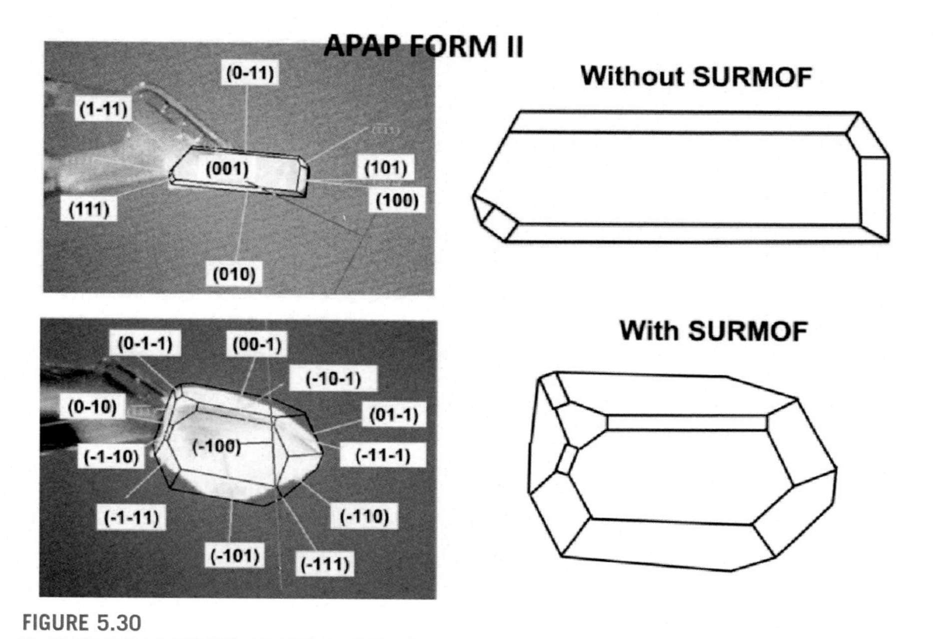

FIGURE 5.30

APAP crystal on SURMOFs resulted in form II with block morphology. First time report of APAP form II crystals of block morphology. *APAP*, *N*-acetyl-*para*-aminophenol; *SURMOFs*, surface-mounted metal–organic frameworks.

Taken from G. Bolla, A.S. Myerson, SURMOF induced polymorphism and crystal morphological engineering of acetaminophen polymorphs: advantage of heterogeneous nucleation. CrystEngComm 20 (2018) 2084–2088.

4. Solid phases as ionic cocrystals, solid solutions, and eutectics, coamorphous

4.1 Ionic cocrystals

Ionic cocrystals (ICCs) [151–156] are solid-state materials obtained by cocrystallization of organic molecules, including APIs, with inorganic salts. Hydrogen bonding between common donor and acceptor groups (OH, NH, O, N, etc.) between HB donors and anions (OH···Cl^-, NH···Br^-, etc.) can alter the physicochemical properties of drugs. Cocrystallization with inorganic salts also offers a viable route to the resolution of racemic mixtures by using enantiopure coformers. ICCs are a subset of cocrystal family, which combines the characteristics of molecular crystals with those of ionic salts to improve the physicochemical properties of pharmaceuticals in improved formulations. For example, when an alkali or alkaline earth halide is used as coformer and APIs interact directly with the metal cations, the halide ions interact with other molecules (and water). Recently, this type of cocrystal has been used for chiral resolution.A robust crystallization process in water for the ICC, piracetam–CaCl2–2H2O shows the Ca2+ cation in an octahedral coordination with the oxygens of four different molecules (Fig. 5.31) and importantly shows

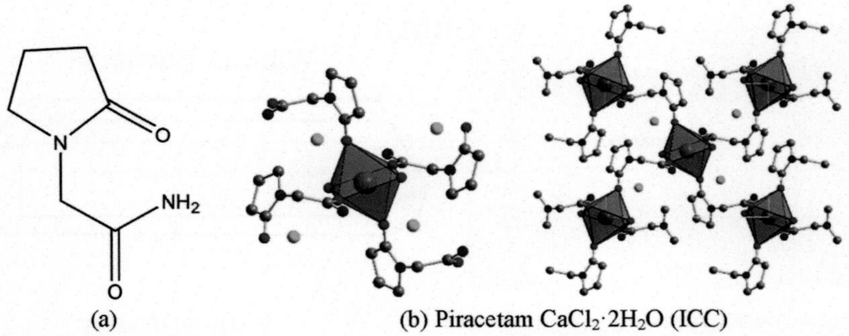

(a) (b) Piracetam $CaCl_2 \cdot 2H_2O$ (ICC)

FIGURE 5.31

(A) Piracetam. (B) Ca^{2+} cation in the crystal structure of piracetam $CaCl_2 \cdot 2H_2O$. Ref. [151].

the significance of the common ion effect in the crystallization of ICCs. Other ICCs are reported with barbituric acid, diacetamide, malonamide, and nicotinamide with the inorganic salt coformer $CaCl_2$.

4.2 Eutectics and solid solutions

Eutectics are multicomponent solids with diverse applications but are less studied in terms of their molecular structure organization and bonding interactions [157–162]. A eutectic is defined based on its low melting point compared with the individual components, and these are closer to the individual species in that their crystalline arrangement is similar to the parent components, but they are different with respect to the structural order and properties. A solid solution possesses structural homogeneity throughout the structure (single phase), but a eutectic is a heterogeneous ensemble of individual components whose crystal structures are like discontinuous solid solutions (phase separated). Structural analysis of cocrystals, solid solutions, and eutectics has led to an understanding that materials with strong adhesive (hetero) interactions between the unlike components will lead to cocrystals, whereas those having stronger cohesive (homo/self) interactions will more often give rise to solid solutions (for similar structures of components) and eutectics (for different structures of components). In effect, crystal engineering ideas from cocrystal design were extended to eutectics and solid solutions as novel composite materials [157]. The preparation of binary cocrystals and ternary and quaternary solid solutions of 1,4-diazabicyclo-[2.2.2]octane (DABCO) and 4-X-phenols ($X = Cl, CH_3, Br$) is reported (Fig. 5.32) [162]. If they are different, a ternary cocrystal BAC or a ternary solid solution $A(B_nC_{1-n})_2$ $(0 < n < 1)$ can be formed, depending on the similarity in the size and shape of the molecules. Adding another isosteric component (D) can lead to a quaternary solid solution $A(B_nC_mD_p)_2$ $(n + m + P = 1, 0 < n, m, P < 1)$. DABCO was used as the central molecule and 4-X-phenols ($X = Cl$, CH_3, Br) as the peripheral molecules. The Cl, Me, and Br substituents of the

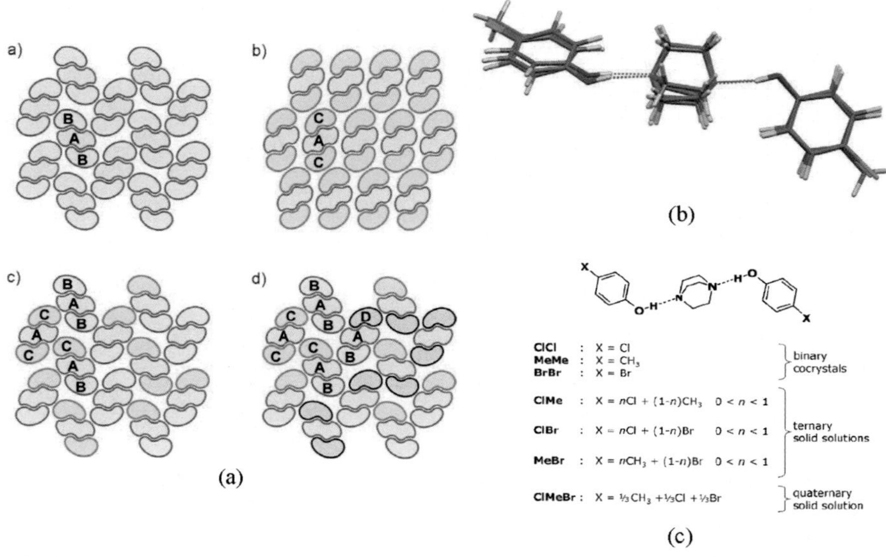

FIGURE 5.32

(A) Cocrystals and solid solutions of molecule A with molecules B, C, and/or D with similar shape and size. Binary cocrystals AB_2 and AC_2 adopt different crystal structures. Ternary solid solution of A, B, and C The trimolecular assemblies may have BAB, CAC, or BAC configurations. (B) Superposition of the trimolecular assemblies in the binary cocrystals ClCl, MeMe, and BrBr. Dashed lines show the O–H···N hydrogen bonds. C, gray; H, white; N, blue; O, red; Cl, green; Br pink. (C) General scheme where molecule structures showed the design of eutectics.

4-X-phenols have similar shape and comparable size and fit in with the recurring O–H···N hydrogen bond in trimolecular assembly. The size, shape, and chemical similarity of 4-X-phenols enable the formation of solid solutions upon crystallization of DABCO with two or three of the phenols.

Solid solutions and eutectics exhibit close similarity to the XRD line patterns of the pure constituents; the structural boundary between cocrystal and eutectic through two systems has been shown. Benzoic acid combines with structural analogs 4-fluorobenzoic acid, pentafluorobenzoic acid, and BZA to form a solid solution, a cocrystal, and a eutectic. 4-Fluorobenzoic acid and benzoic acid form continuous isomorphous solid solutions, whereas pentafluorobenzoic acid resulted in cocrystal when treated with BZA (Fig. 5.33) [155,159].

Another example is curcumin, which combines with isomeric dihydroxy benzenes to give different products (Fig. 5.34) [163–166]. It forms a cocrystal with resorcinol but eutectic with hydroquinone. The strong O–H adhesive interactions replace the C–H···O and O–H···O of the parent crystal structures and form

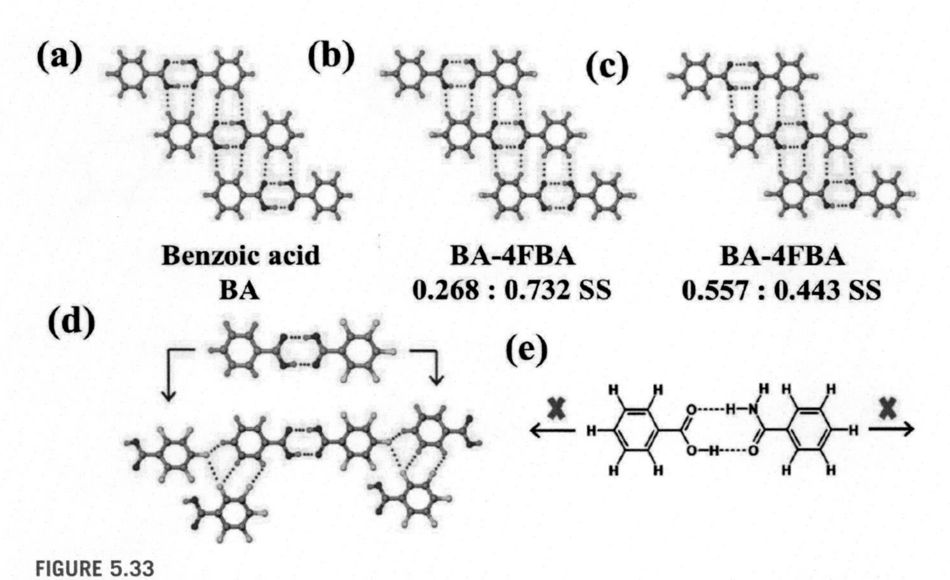

FIGURE 5.33

(A) Benzoic acid—4-fluorobenzoic acid solid solutions. (B) Growth of the benzoic acid—pentafluorobenzoic acid dimer in the cocrystal structure. (C) The absence of such auxiliary interactions in benzoic acid—benzamide system means that instead of a cocrystal, the product is a eutectic.

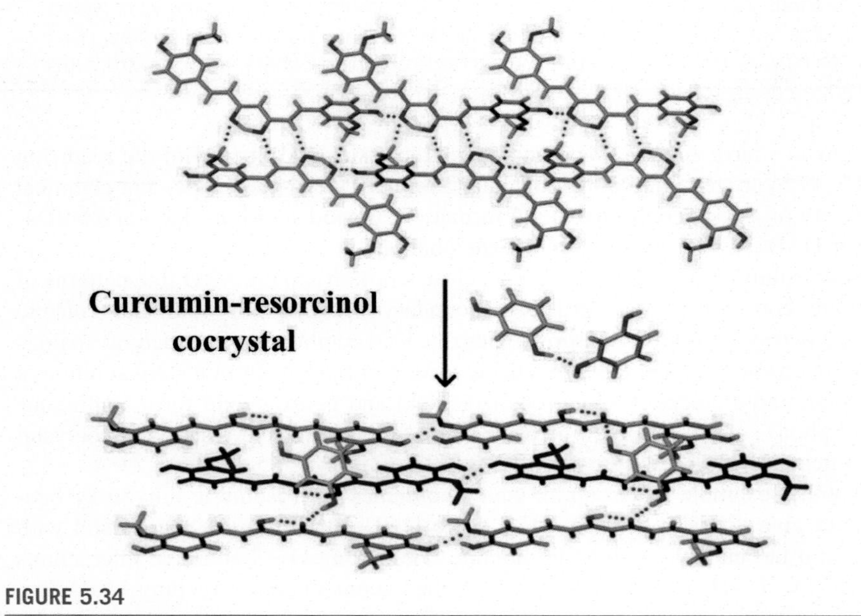

FIGURE 5.34

Curcumin—resorcinol cocrystal structure is sustained by the dominance of strong O—H···O adhesive interactions.

cohesive bonds, consistent with strong hydrogen bond donor—best acceptor pairing. The strong heteromolecular interactions of resorcinol OH groups give good molecular packing in the cocrystal. However, the high symmetry of p-substituted hydroquinone together with lower symmetry of the keto-enol structure in curcumin gives shape mismatch for efficient packing as compared with resorcinol, and hence the product is a eutectic.

Two isomorphous cocrystals of nitazoxanide (NTZ) with p-aminosalicylic acid (PASA) and p-aminobenzoic acid (PABA) as well as their alloys were prepared (Fig. 5.35A and B) [167]. The cocrystals exhibit faster dissolution rates and higher pharmacokinetic properties compared with the reference drug. The cocrystal alloy NTZ—PABA:NTZ—PASA (0.75:0.25) exhibited fourfold higher bioavailability in Sprague Dawley (SD) rats.

4.3 Coamorphous and ionic liquids

Coamorphous drug complexes [168—174] are a novel approach in which the drug and coformer attain high thermodynamic functions of the amorphous phase to result in an improved pharmaceutical product. A coamorphous system is a multicomponent single-phase amorphous solid, which lacks periodic arrangement in the crystal lattice and is associated by weak but discrete intermolecular interactions between the components. A coamorphous solid is contrasted with a cocrystal, salt, or eutectic primarily by its amorphous nature when subjected to powder XRD. Compared with salts, eutectics, and cocrystals, coamorphous solids are a new entry to pharmaceutical crystal forms. Curcumin—artemisinin coamorphous solid (1:1) (Fig. 5.36i) exhibited a dramatic increase in the pharmacokinetic profile of curcumin (AUC_{0-12} 2.6 mg h/mL, C_{max} 1 mg/mL) in SD rats [173].

Another interesting system is indapamide (Fig. 5.36ii) [174]. The stronger sulfonamide—pyridine ($SO_2NH_2 \cdots N\text{-}Py$) and sulfonamide—carboxamide ($SO_2NH_2 \cdots O{=}C{-}NH$) hydrogen bonds direct the formation of cocrystals, while the weaker sulfonamide—amine ($SO_2NH_2 \cdots N{-}H$) hydrogen bond results toward coamorphous product. IDP—PIP and IDP—ARG coamorphous solids exhibit remarkable stability under accelerated ICH conditions. The stronger sulfonamide—pyridine and sulfonamide—carboxamide heterosynthons with rigid/aromatic coformers provide cocrystal products, while the weaker sulfonamide—amine synthon with flexible molecules give coamorphous products of IDP. This observation is consistent with the synthon strength and molecular likeness/dissimilarity guidance to give cocrystal, solid solution, or eutectic product. The higher aqueous solubility and better diffusion membrane permeability of IDP-ARG mean that amino acids can enhance not only powder dissolution but also diffusion kinetics. The change in the crystalline to coamorphous state of the APIs with suitable coformers has proven to be a simple and efficient methodology for improving drug bioavailability. These findings lead to novel coamorphous forms in solid-state pharmaceuticals.

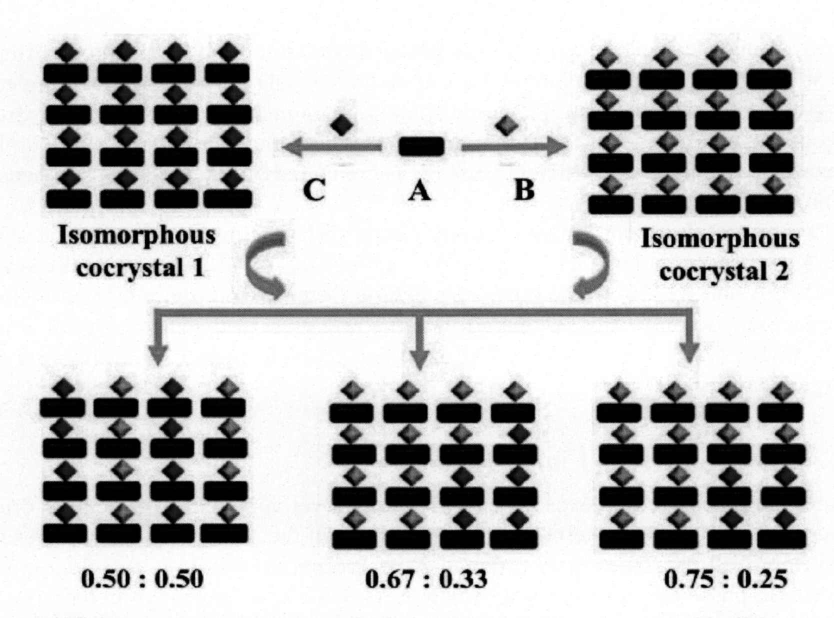

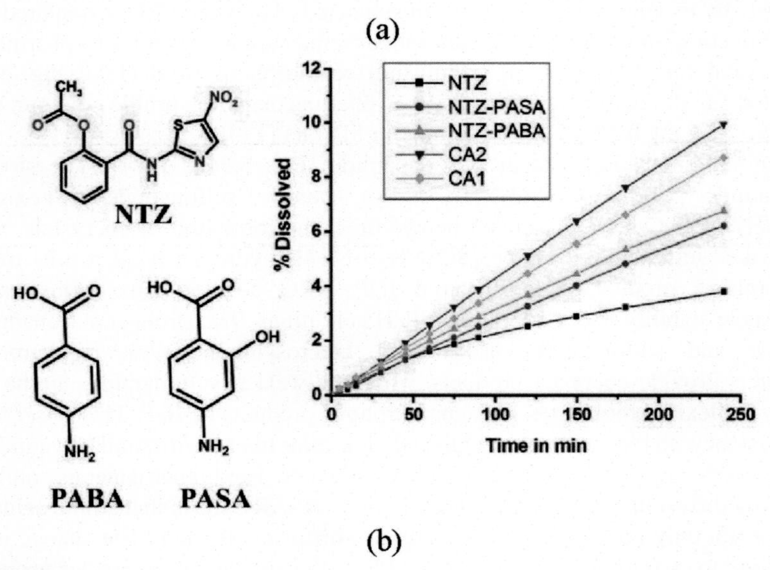

FIGURE 5.35

(A) Schematic representation of isomorphous cocrystals. (B) IDR of cocrystals and cocrystal alloys of NTZ in 3% cetyltrimethylammonium bromide (CTAB) (pH 7) buffer. *NTZ*, nitazoxanide.

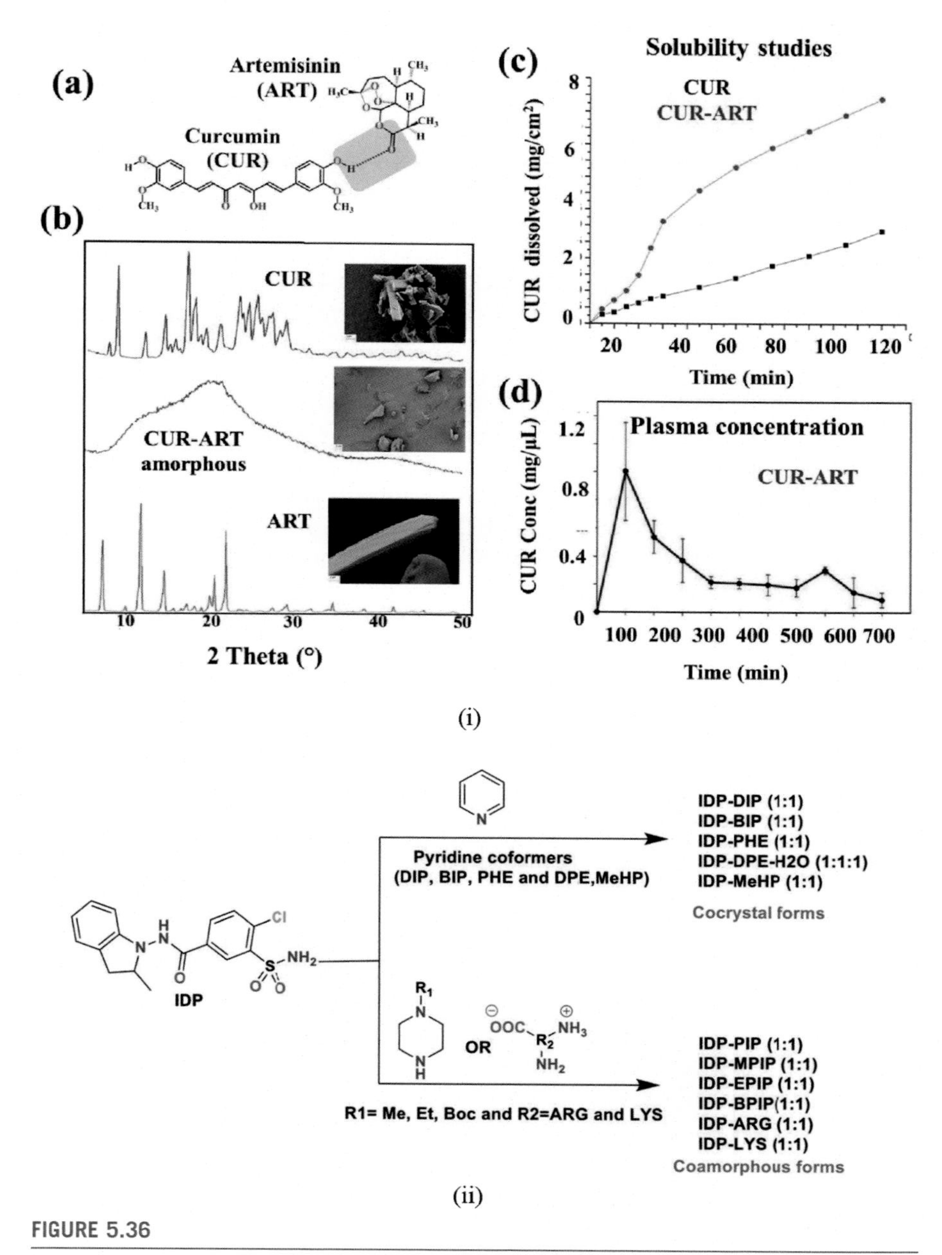

FIGURE 5.36

(i) Curcumin–artemisinin coamorphous solid (1:1) prepared and showed pharmacokinetic profile. (ii) Indapamide coamorphous solids.

5. Applications of pharmaceutical cocrystals

TMZ is a frontline prodrug for the treatment of glioblastoma multiforme (brain cancer) approved by the US Food and Drug Administration (US FDA) in 1999. A limitation with this otherwise potent and selective DNA alkylating agent is degradation of the prodrug to the inactive product 5-aminoimidazole-4-carboxamide by incipient hydrolysis during storage (Fig. 5.37). This transformation not only makes the drug less effective and makes patients anxious. In order to solve the stability problem of TMZ, a cocrystal of SA was prepared which showed that TMZ−SA is stable for over 6 months and has comparable dissolution rate; there were no significant differences in the pharmacokinetic profile of the cocrystal compard to TMZ reference drug in SD rats. Significantly, the active drug released from the cocrystal has detected the brain tissue of rats. Hence, these results suggest that TMZ−SA cocrystal will be a potential lead for an improved TMZ formulation [175−177].

The need to improve solubility and permeability of drugs is labeled as BCS and the Biopharmaceutics Drug Disposition Classification System (BDDCS) (Fig. 5.38). The major advances associated with the BCS is to focus on improvement of physiologically based pharmacokinetics such as dissolution and permeability of poorly absorbed BCS class II drugs [178].

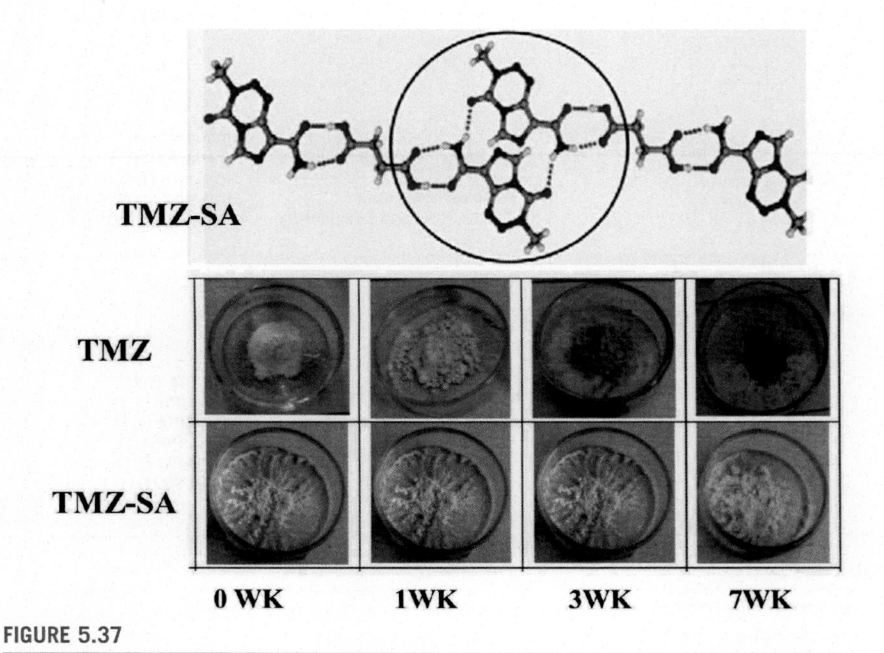

FIGURE 5.37

TMZ−SA improved stability in succinic acid cocrystal. *TMZ*, temozolomide−succinic acid.

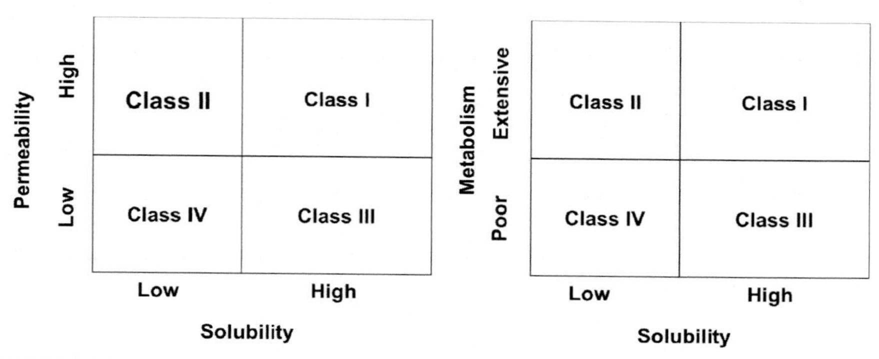

FIGURE 5.38

BCS (left) and BDDCS (right) presented in a Cartesian spatial representation. *BCS*, Biopharmaceutics Classification System; *BDDCS*, Biopharmaceutics Drug Disposition Classification System.

Apart from solubility and permeability, there are several other physicochemical properties of drugs where improvement is required. Cocrystals can be used to optimize physicochemical properties such as solubility, stability, hydration [179], melting point, high energy, conductivity, drying and mechanical properties such as flowability, compressibility, and pharmacokinetic properties of bioavailability/permeability of APIs without changing the chemical composition and pharmacological action (Fig. 5.39, Table 5.4).

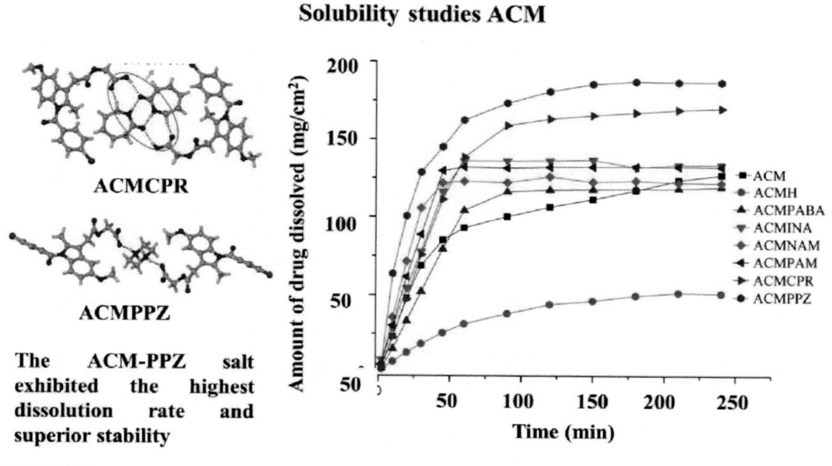

FIGURE 5.39

Acemetacin cocrystals exhibit improved solubility and stability compared with the parent drug which is prone to hydratation.

Table 5.4 Reports on cocrystals with multiple functions.

Improved property	Parent compound: Cocrystal former/compound name
Solubility and dissolution rate	Fluoxetine hydrochloride: benzoic, succinic, fumaric acid; piroxicam: saccharin; carbamazapine: nicotinamide, saccharin; indomethacin: saccharin itraconazole: malic, tartaric, succinic acid; curcumin: resorcinol; tenoxicam: resorcinol; bumetanide cocrystals; curcumin cocrystals; furosemide cocrystals, tenoxicam cocrystals, piroxicam cocrystals, meloxicam cocrystals, celecoxib
Mechanical property	Multicomponent crystals: voriconazole as a case study; theophylline, methyl gallate; paracetamol cocrystal; piroxicam–saccharin
High energy materials	Diacetone diperoxide, triacetone triperoxide cocrystals; cocrystals of CL-20; 2,4,6-trinitrotoluene cocrystal
Dying applications	Pyridine-2,4-dione based heterocyclic dye-pyridine N–Me and N–Et substituents
Conductivity improvement	Phenazine: deuterated chloranilic acid; phenazine–chloranilic acid; anilic acid: phenazine etc.
Melting	Hexamethylene bisacetamide cocrystals; liquid propofol–isonicotinamide; alkane dicarboxylic acids–isonicotinamide
Physical stability/ chemical degradation	Temozolomide cocrystals; femotidine cocrystals, angrographolide cocrystals
Hydration stability	Caffeine-oxalic/glutaric acid; acetacin cocrystals; niclosamide cocrystals; theophylline cocrystals, nitofurantoin cocrystals with PABA; acemetacin cocrystals; etoricoxib cocrystals
Bioavailability	Sildenafil–dicarboxylic acids; lithium ionic cocrystal with glucose Green tea epigallocatechin-3-gallate cocrystals; cocrystals of quercetin; meloxicam cocrystals
Permeability	Acyclovir cocrystals, salts and cocrystal of etodolac: advantage of solubility, dissolution and permeability; drug–drug cocrystal of febuxostat and piroxicam for the treatment of gout; dabrafenib cocrystals with fumaric acid, succinic acid, and adipic acid; novel pharmaceutical cocrystals of bumetanide; entacapone: improving aqueous solubility, diffusion permeability, and cocrystal stability with theophylline; salts and salt cocrystals of the antibacterial drug pefloxacin

6. Manufacturing of cocrystals and salts. Batch versus continuous cocrystallization and applications of the flow chemistry

Crystallization is the key separation and purification process in the bulk and fine chemicals and food and pharmaceutical industries to produce a broad diversity of materials. The process can influence the performance of downstream process operations such as filtration, drying, milling, and formulation. The physicochemical properties of end product essentially depend on crystal characteristics, i.e., particles of controlled size, morphology, crystal structure, and phase purity, which are the major outcomes of a crystallization process. Therefore, crystallization events, viz., nucleation, crystal growth, and polymorphic transformation are vital steps to ensure high quality of the end product. Crystallization processes can be run in two methods: (1) batch operation and (2) continuous operation. The batch method is the most common method followed in pharmaceutical industry traditionally, whereas continuous processing affords multiple advantages and enhanced reproducibility of results. This is because the continuous crystallization occurs under identical and controlled conditions, compared with batch process where conditions can change with time as well as deviations of conditions from batch to batch [180]. One important advantage of batch method is that the crystallizer can be cleaned thoroughly at the end of each batch to prevent contamination (seeding) that may induce secondary nucleation of undesired phases. In the batch crystallization process, cooling, antisolvent, or reactive crystallization has been the norm for the vast majority of industrial-scale pharmaceutical crystallization processes with well-defined parameters [181]. However, operating costs of a batch system can be significantly higher than those of a comparable continuous unit with built-in flexibility for control of temperature, supersaturation, nucleation, crystal growth, and all the other process parameters that influence crystal size distribution [182]. Thus, continuous crystallization processes that result in higher yield and uniform purity of the end product have been a standard method for the crystallization of bulk commodity chemicals and large-volume specialty chemicals [183–185].

Two highly desirable crystal characteristics are (1) mean particle size and (2) particle size distribution. They essentially affect the transport process of the API through the circulatory system to the target organs. The correct polymorphic crystal form is another crucial constraint, as one polymorph may have the desired therapeutic effect while the other may be inactive (low bioavailability) or have harmful properties due to high supersaturation. Based on crystallizer type, different continuous crystallization approaches are demonstrated. There are two main types of continuous crystallizer: the mixed-suspension mixed-product removal (MSMPR) crystallizer and the continuous tubular crystallizer or plug flow continuous crystallizer (PFR) (Figs. 5.40 and 5.41) [186,187]. The MSMPR crystallizer is easier to convert from batch and also cost-effective over the tubular crystallizer, which has the advantages of narrow residence time. The choice of the use of MSMPR or PFR system depends on the kinetics of the process. Generally, MSMPR process is preferred for low conversions and long residence times and the PFR for higher conversion with short residence times. Typical multistage continuous crystallization experiments are performed in three-stage continuous crystallizer. Each stage is operated as an MSMPR. Each stage is a jacketed interconnected glass reactor with independent temperature control and magnetic stirring. The feeder is filled with API solution and continuously added to the first reactor using low-flow pumps followed by the removal of the slurry from each stage and transferred to the next stage using

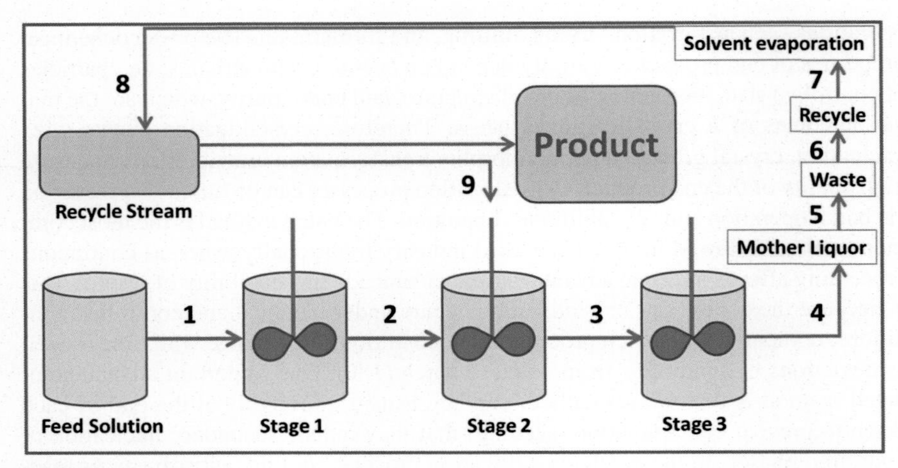

FIGURE 5.40

Schematic diagram of a typical continuous crystallization process with recycle process attachment, the mixed-suspension mixed-product removal (MSMPR).

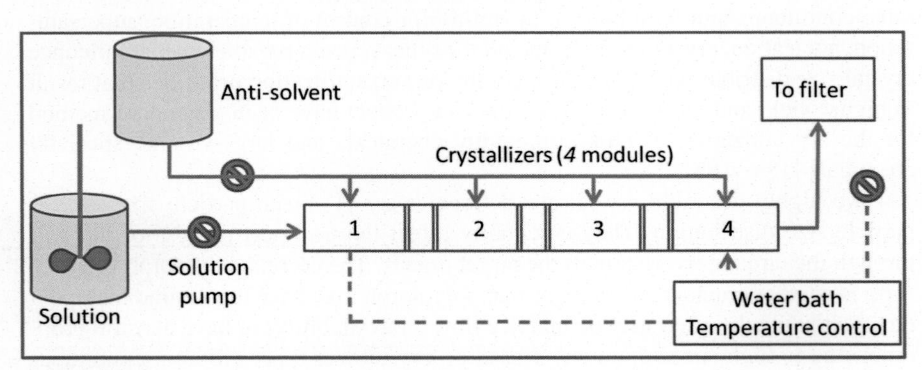

FIGURE 5.41

Schematic process flow diagram of the continuous crystallization system with multistage antisolvent addition in continuous tubular crystallizer or plug flow continuous crystallizer (PFR).

medium-flow peristaltic pumps. A timer setup controls the alternating operation of these pumps. All stages in continuous process can be combined with simulated recycle stage in later stages and get control over crystal size distribution and phase purity, essentially enabling several downstream corrective actions. Risk factors such as phase transformations and chemical degradation continue to pose challenges in continuous processing. Nevertheless the process guarantees for better-quality powder mixtures and the ability to control key physical properties such as drug bioavailability.

Pharmaceutical cocrystals produced by any operations are expected to provide high-quality products that qualify to the drug regulatory guidelines. Usual formulation methods such as neat grinding/solvent drop grinding and spray-drying may result in tiny crystals that compromise purification of the product due to inclusion of excipient/solvent into the API crystal lattice. Such processes thwart isolation and characterization. Kim and coworkers have developed an effective crystallization protocol for a proprietary substance that is prone to "oil out" by seeding at low supersaturation [188]. The controlled protocol results in polymorphic phases with uniform crystal habit. Ultrasound was employed as a tool for particle size control with plate-like or needle-like habit. The role of temperature cycling was further demonstrated as highly operative for particle uniformity and crystal habit modification for blend homogeneity and bioavailability considerations. Sonication was identified as a method to remove the fine and rough edges in the finished API product with small crystal size distribution, improved bulk density, flowability, and bulk handling. The introduction of an antisolvent initiates primary nucleation, sometimes with the aid of seeding, and crystal digestion during an aging step. Since mixing within the vessel upon introduction of the second fluid is often not possible prior to crystal nucleation, this prevents optimum crystal quality formation and increases impurity entrainment due to heterogeneous environment [189]. Impinging fluid jet streams are found promising in a continuous crystallizer to achieve homogeneous composition prior to the start of nucleation. This process has been tested on lab scales for the direct crystallization of several APIs such as simvastatin, lovastatin, omeprazole, finasteride, and diltiazem malate to meet the particle size and purity specifications. Controlled crystallization by following a supersaturation profile in the MSZW based on accurate concentration measurements represents that a controlled nucleation barrier to crystallization generally avoids hysterical nucleation [190–192]. It is the concentration–solubility–temperature trajectory bounded by thermodynamic dissolution and kinetic crystallization, which is scale, system, reactor, and operating condition dependent. Consideration of the correct crystallization parameters aids in the production of spheronized particles by avoiding needle-like crystals to improve flow properties for tableting and maintain interparticulate bonds in tablets. It also helps avoiding difficulties in washing, filtering, and drying during manufacturing. Engineering platy particles with better plasticity and incorporating coformers/additives in the crystal structure are successful approaches to increase the crystal free energy, thereby modulating the intrinsic dissolution profiles.

Batch methods are expensive inventories if the batch fails during the final testing but can be overcome by employing continuous manufacturing and allows quicker response for changes required. Simulations of additional processes including recycle loops into continuous processes offer improvements in yield and robustness. Perhaps, continuous operations also run with many hurdles like flow chemistry transformations, difficulties with processing dry solids and solid-laden fluids, lack of equipment at bench and pilot scales, development of control methodologies to guarantee product quality, and breaks in the process, especially between synthesis and formulation. Recently, Novartis-MIT Centre for Continuous Manufacturing demonstrated the first example of an end-to-end, fully integrated continuous manufacturing plant for a pharmaceutical product

[193,194]. It consists of multistep portions of the process. The plant starts from a chemical intermediate and performs all the intermediate reactions, separations, crystallizations, drying, and formulation, which results in a formed final tablet in one tightly controlled process that led to the development of a continuous automated platform for manufacturing the pharmaceutical drug aliskiren hemifumarate. They further developed a reconfigurable, continuous-flow platform featuring custom-designed unit operations for on-demand manufacturing of liquid formulations of APIs. This plug-and-play, first-generation reconfigurable platform succeeded in producing hundreds to thousands of liquid doses per day of diphenhydramine hydrochloride, diazepam, lidocaine hydrochloride, and fluoxetine hydrochloride, each from simple starting materials. Cole and his coworkers at Eli Lilly and Company have recently developed a continuous current good manufacturing practice production process with eight continuous unit operations configured within laboratory for prexasertib monolactate monohydrate to be used in human clinical trials [195].

Less than 1% of drugs are formulated into the market not only due to a lack of efficacy, safety, or unfavorable side effects but majorly due to poor biopharmaceutical properties [196]. The US FDA considers the production of new cocrystals after the successful approval of Entresto [197] by Novartis. Vertex Pharmaceutical production has been using the process of continuous manufacturing since 2015, for the manufacturing of Orkambi (lumacaftor/ivacaftor) [198], i.e., a combination drug for the management of cystic fibrosis. To encourage the process of continuous manufacturing [199], the FDA has started to approve the transfer of production techniques from batch to continuous manufacturing. The first such change was documented by Janssen Pharmaceuticals for the production of Prezista (darunavir), a medication that is used for the treatment of HIV-1 infection [180,199]. Lee et al. [200] reported the cocrystallization of 1:2 ratio of phenazine—vanillin cocrystal using both batch and continuous crystallization approaches with or without agitation in MSMPR method. They reported efficient heat transfer during continuous cocrystallization with advantageous and rapid nucleation rate as well as higher percentage of quantum yield. Similarly, Nishimaru [201] demonstrated the sequence of solution addition into the deposition of undesired crystals during antisolvent crystallization and extended this work to study the effect of thermodynamic and kinetic parameters to develop the multicomponent phase diagram using continuous cocrystallization approaches. In the past decade, the journey of cocrystals from bench- to large-scale extrusion process witnessed a successful movement in the form of marketed drug products. A few examples of cocrystals that are used in clinical trials are discussed in the following section. Many different strategies have been attempted to obtain cocrystal forms, including batch solution cooling crystallization, slow solvent evaporation, slurry conversion, neat grinding, and liquid-assisted grinding,. Svoboda, MacFhionnghaile, and coworkers [181] coupled reactive and antisolvent crystallization to control the supersaturation and manipulated the system for the continuous formation of 2:1 and 1:1 cocrystals of benzoic acid and isonicotinamide. In addition, the antisolvent crystallization process of lactose was studied using design of experiments. The results predicted by the model were in good agreement with the experimental results and provide the basis for the establishment of the optimized α-lactose monohydrate process.

Solvent-free crystallization [202], [31] which includes high-shear granulation [203], twin-screw extrusion (TSE), and hot-melt extrusion (HME) [204], is regarded as alternative effective and environmentally friendly continuous manufacturing strategy in cocrystal synthesis. Kulkarni [205] et al. discussed a novel reproducible control strategy for the stoichiometry in solvent-free continuous cocrystallization. They produced both 1:1 and 2:1 caffeine/maleic acid cocrystals by the appropriate choice of initial process conditions, and the conditions for mutual transformation between these two cocrystals were identified. Besides, it had been set out that parameters of extrusion temperature could tailor the stoichiometry by further investigation. The possibility of controlling the crystallization by an antisolvent method from the perspective of kinetics and thermodynamics has been demonstrated [200]. The purity and production efficiency were improved by certain operation conditions based on multicomponent phase diagrams. Medina et al. [206] presented a manufacturing method for pharmaceutical cocrystals of caffeine/oxalic acid and AMG 517/sorbic acid (1:1) by TSE. They explored TSE parameters such as the screw design, temperature, and residence time in a variety of experiments to investigate high-purity cocrystal production methods. The same research group investigated the scale-up of the AMG 517/sorbic acid cocrystal production by TSE. A high-mixing screw was designed to strengthen the mixing ability in the extruder, which was shown to be the critical factor in forming high-purity cocrystals. Compared with traditional solution crystallization, the cocrystal quality obtained by the TSE method had great advantages in surface area, bulk density, and fluidity. HME was applied by Dhumal et al. [207] to form a cocrystal of ibuprofen and nicotinamide in 1:1 ratio. An appropriate temperature profile and an adaptive screw speed and configuration were considered in the production process of the elasticity to tailor the cocrystal purity. The continuous production of indomethacin/saccharine cocrystals by extrusion processing and antisolvent methods is discussed [208]. The cocrystal product was fully milled and blended, filled into capsules, and amplified at a rate of 50 capsules/min by monitoring of the continuous process using in situ tools.

Polymorphism in cocrystal is also an essential part of drug phase investigation. Zhao et al. [209] carried out an extensive polymorph screen for α-lipoic acid/nicotinamide cocrystal in a small-scale experiment. On the basis of its enhanced thermodynamic stability and solubility differences in contrast to pure α-lipoic acid, they used a continuous oscillatory baffled crystallizer (COBC) to make over 1 kg of cocrystal within 3 h. Besides, the spherical agglomerated product was obtained with good purity and narrow particle size distributions. Powell et al. [210] set up a novel periodic MSMPR crystallizer cascade to select a desired polymorphic form of urea-barbituric acid (UBA) cocrystal. The system consisted of a three-stage MSMPR crystallizer cascade that was characterized by periodic transfer of slurry with high velocities of addition and withdrawal. This new crystallizer enables precise harvesting of the target cocrystal and guaranteed continuous operation for a long time. Continuous process manufacturing delivers a sustainable solution to the rising costs of a new drug development in the pharmaceutical industry.

7. Examples of cocrystals in the market and at advanced clinical stage

Escitalopram oxalate: Approved in 2009 for the treatment of depression and anxiety (Fig. 5.42), it belongs to the selective serotonin reuptake inhibitor [211].

Caffeine citrate: The drug cocrystal is sold under the trade name Cafcit among others and is a medication used to treat a lack of breathing in premature babies. Caffeine citrate (Fig. 5.43) shows high dissolution and low hygroscopicity than caffeine [212].

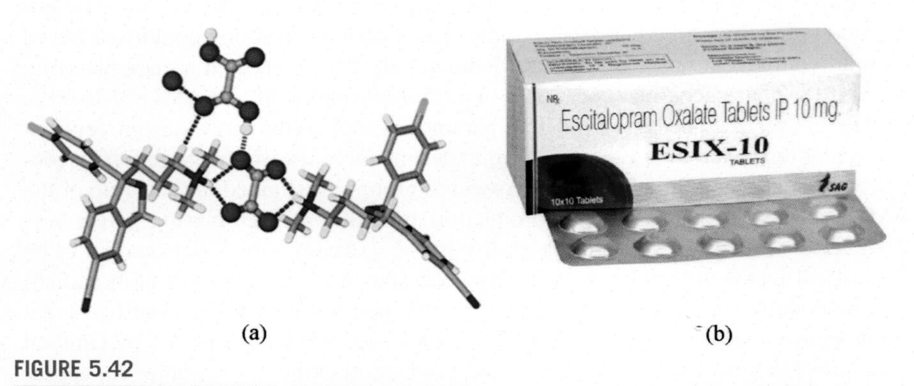

FIGURE 5.42

(A) Escitalopram oxalate cocrystals where one oxalic acid neutral and one is oxalate interact via hydrogen bonds. (B) Escitalopram oxalate cocrystal sold under the trade name ESIX-10.

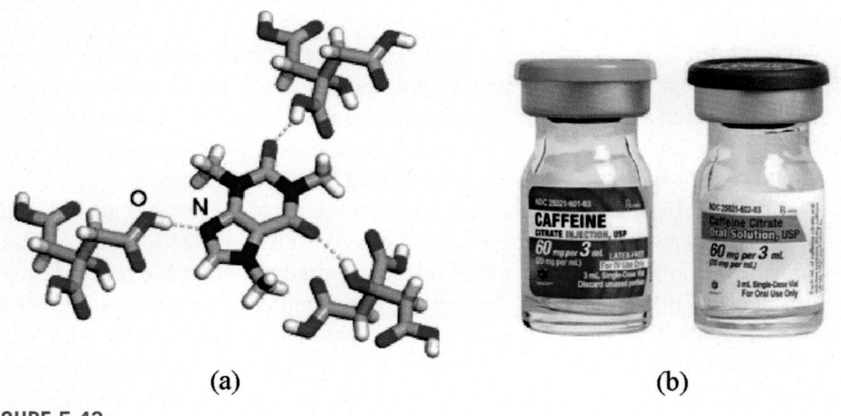

FIGURE 5.43

(A) Crystal structure of caffeine citrate cocrystals. (B) Cafcit is the trade name of caffeine citrate cocrystal.

Depakote: Valproic acid (Depakene) is a liquid at room temperature and is therefore difficult to develop as a solid dosage form and salts are highly hygroscopic. The cocrystal of sodium valproate with valproic acid in a 1:1 stoichiometric ratio has improved stability and comparable pharmacokinetic behavior, marketed as Depakote, is the leading marketed form of valproic acid to treat epilepsy and bipolar disorder and to prevent migraine headaches (Fig. 5.44). Solid substances comprise valproic acid and sodium valproate. Inventor: Bernard Charles Sherman, Canada. Jun. 20, 2000, (US Patent, 6,077,542, 2000.

Entresto: A cocrystal of sacubitril (neprilysin inhibitor) and valsartan (angiotensin receptor blocker) is a prescription medicine (Fig. 5.45) used to reduce the risk of death and hospitalization in people with certain types of symptomatic long-lasting (chronic) heart failure. It is an ICC comprising of six sacubitril and six valsartan moieties in their anionic forms with 18-penta- and hexa-coordinated

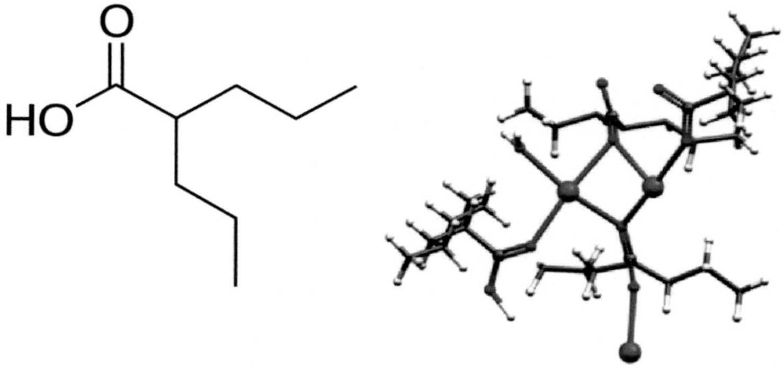

FIGURE 5.44

(A) Molecular structure of valproic acid and (B) crystal structure of Depakote (1:1 sodium valproate with valproic acid).

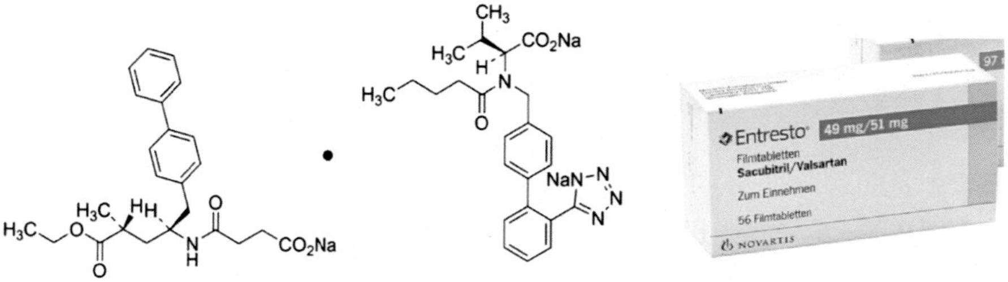

FIGURE 5.45

(A) Molecular structure of monosodium sacubitril and sodium valsartan and (B) cocrystal as drug is marketed under the trade name Entresto.

sodium cations, and 15 water molecules (CSD refcode: NAQLAU). The drug was developed by Novartis and approved under the FDA's priority review process on July 7, 2015 and by EMA in November 2015 [213].

Cocrystal E-58425: Cocrystal E-58425 of tramadol and celecoxib (Fig. 5.46) is considered to treat for moderate-to-severe acute postsurgical pain. Phase III clinical trial is undergoing. The safety of tramadol in cocrystal is improved (low dose), and the bioavailability of celecoxib is increased because of improve dissolution. Celebrex versus tramadol in the treatment of chronic lower back pain is used [214,215].

Ertugliflozin:L-pyroglutamic acid cocrystal: Ertugliflozin is a sodium glucose cotransporter-2 inhibitor that promotes urinary glucose excretion and prevents hyperglycemia and type 2 diabetes mellitus [216,217]. It is highly hygroscopic amorphous solid. Ertugliflozin cocrystal with pyroglutamic acid is under phase III clinical trials led by Pfizer and Merck (Fig. 5.47). Although the FDA executes the continuous cocrystallization process based on science and risk approaches, but the regulatory challenges for quality product development remain similar as in batch processing. Therefore, the plan of sampling for continuous manufacturing needs to be redefined; variability and deviation should be controlled and handled in different ways compared with conventional batch approach.

Trelagliptin succinate: It is an oral dipeptidyl peptidase IV inhibitor originated by Takeda and approved for use in Japan in March 2015 (Fig. 5.48), as once-weekly oral treatment for type 2 diabetes [218].

Suglat: The drug is the first sodium-glucose cotransporter 2 (SGLT2) inhibitor approved for the treatment of type 2 diabetes in Japan (Fig. 5.49). It was jointly

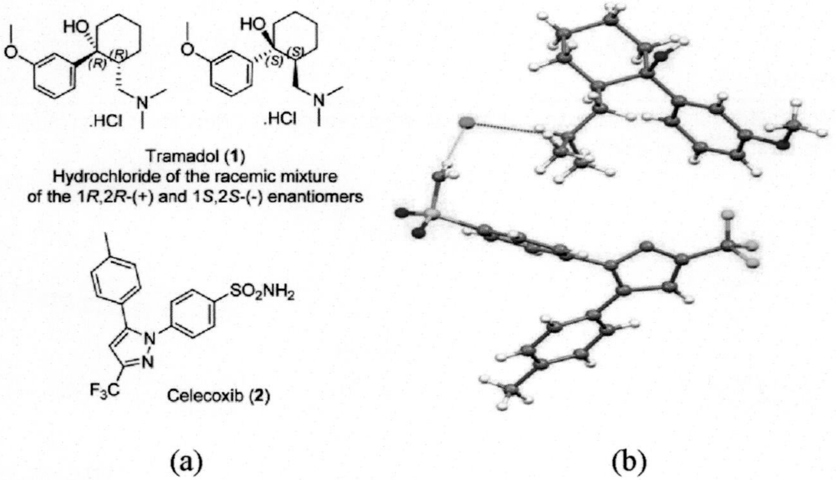

FIGURE 5.46

(A) Tramadol and celecoxib molecular structures. (B) Tramadol, hydrochloride, and celecoxib adduct a combination drug package.

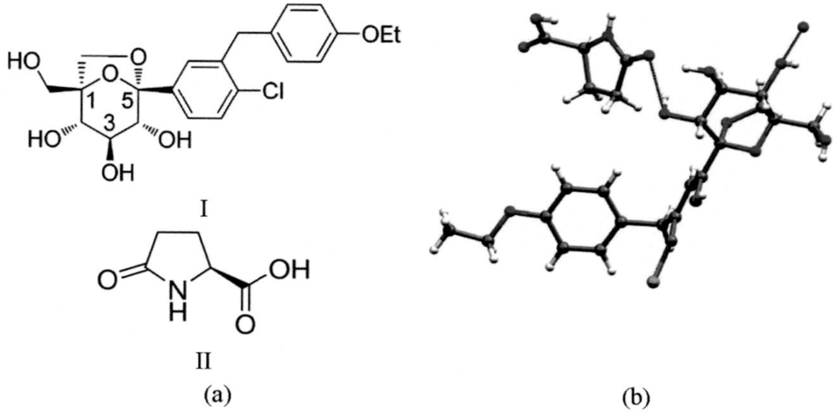

(a) (b)

FIGURE 5.47

Molecular structure of ertugliflozin (I) and L-pyroglutamic acid (II). (B) Crystal structure of ertugliflozin:L-pyroglutamic acid cocrystals.

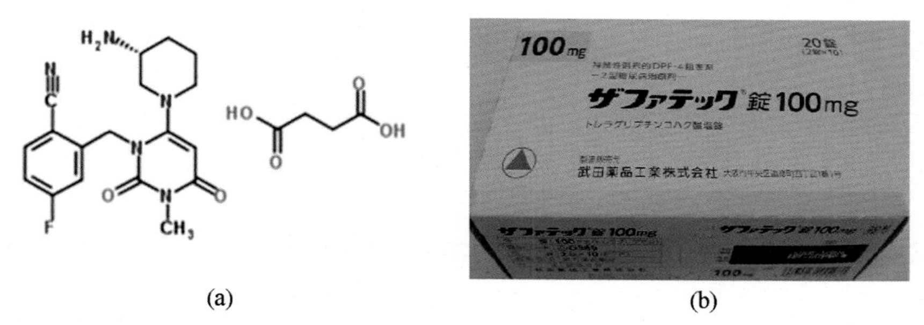

FIGURE 5.48

(A) Structure of the drug trelagliptin (left) and succinic acid conformer (right). (B) Trelagliptin succinate cocrystals as oral dipeptidyl peptidase IV inhibitor.

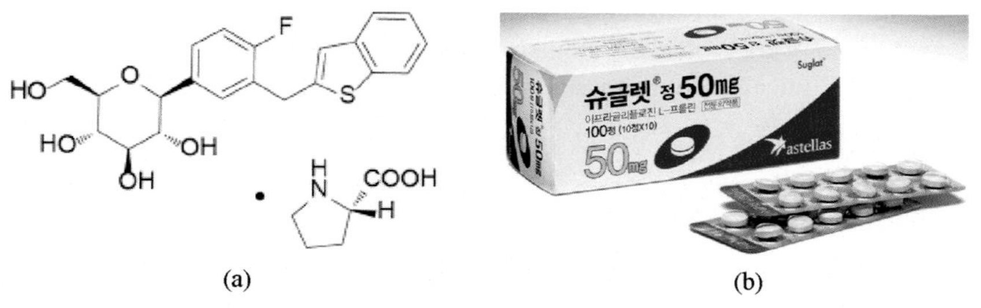

FIGURE 5.49

(A) Structure of the drug ipragliflozin (top) and L-proline. (B) Ipragliflozin and L-proline exist as cocrystal in 1:1 molecular ratio and marketed under the trade name of Suglat.

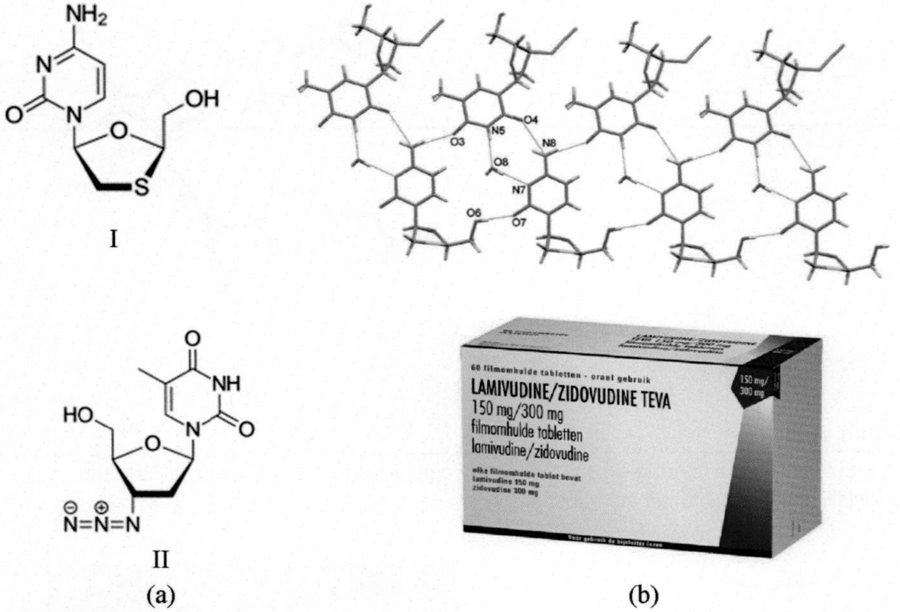

FIGURE 5.50

(A) Molecular structure of lamivudine (I) and zidovudine (II). (B) Crystal structure of lamivudine and zidovudine cocrystal.

developed by Astellas Pharma and Kotobuki The pharmaceutical cocrystal is marketed under the trade name Suglat in doses of 25 and 50 mg. Ipragliflozin and L-proline exists as cocrystal in 1:1 molecular ratio, and the drug product is available as film-coated tablets [219].

Lamivudine and zidovudine cocrystal: Cocrystals of the anti-HIV drugs lamivudine and zidovudine (Fig. 5.50) found to have improved biopharmacokinetic properties [220].

Cocrystal formulation NXP001: A cocrystal formulation NXP001 is a pipeline drug for oncology treatment developed by Nuformix. Nuformix signs agreement with Quotient Clinical to commence a pilot bioequivalence study for NXP001 [221].

8. Outlook and conclusions

With the definition of crystal engineering in the late 1980s together with supramolecular synthon in the 1990s and cocrystal design principles applied via heterosynthons from 2000 onward, the journey to drugs in the market has taken another 15 years. In fact, but for the regulatory approval which took some time, it could

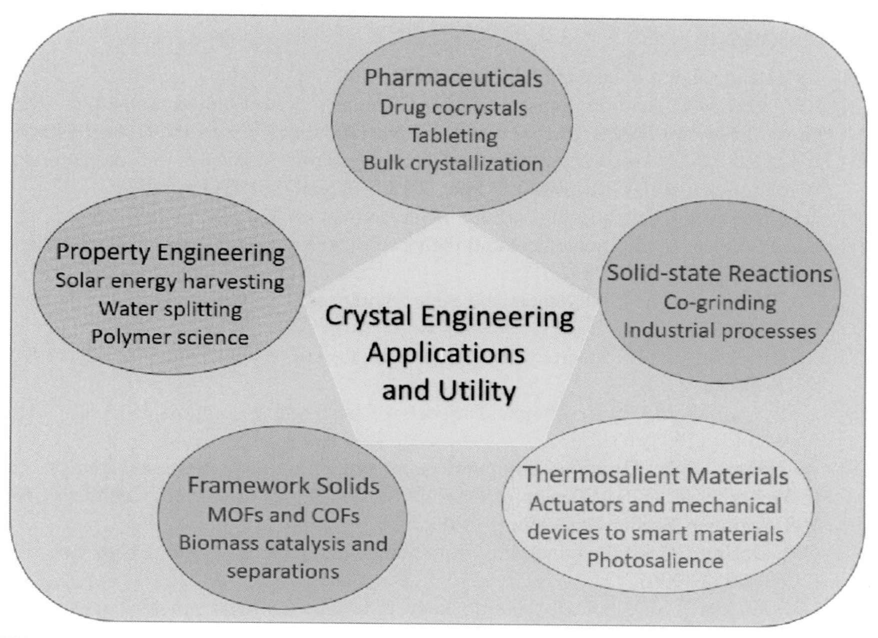

FIGURE 5.51

Five important areas in the future of crystal engineering.

have been possible to launch the early cocrystal drugs even by the 2010s. This chapter has covered the principles of cocrystal and salt design to their applications in pharmaceuticals and manufacturing improvements with continuous-flow processes. The next challenges will be modulation of drug permeability in cocrystals, controlled crystallization of desired polymorph using heteronucleation control, and end-to-end production of pharmaceutical cocrystals and salts from starting materials to drug formulation. The possibilities in crystal engineering are ever growing (Fig. 5.51) [222].

Acknowledgments

BS thanks CSIR-HRDG, India (02(0327)/17/EMR-II), and SERB, India (CRG/2019/004946), for research funding and Tezpur University for infrastructure. AKN thanks CSIR-NCL and DST-SERB (JC Bose fellowship, SR/S2/JCB06/2009 and Pharmaceutical materials and continuous process, CRG/2019/001388) for research funding and School of Chemistry, University of Hyderabad, for providing infrastructure facilities. AKN acknowledges financial and infrastructure support from the University Grants Commission (through UPE and CAS programs) and the Department of Science and Technology (through PURSE and FIST programs).

References

[1] J.M. Lehn, Supramolecular Chemistry, VCH, 1995.
[2] J.W. Steed, J.L. Atwood, Supramolecular Chemistry, John Wiley & Sons, Ltd, 2009, https://doi.org/10.1002/9780470740880. ISBN:9780470512333. ISBN:9780470740880.
[3] J.W. Steed, P.A. Gale, Supramolecular Chemistry: From Molecules to Nanomaterials, 8 Volume Set, Wiley-Blackwell, 9 Feb. 2012. ISBN: 978-0-470-74640-0.
[4] G.R. Desiraju, Chemistry beyond the molecule, Nature 412 (2001) 397–400.
[5] C.J. Pedersen, Cyclic polyethers and their complexes with metal salts, J. Am. Chem. Soc. 89 (1967) 2495–2496.
[6] C.J. Pedersen, Cyclic polyethers and their complexes with metal salts, J. Am. Chem. Soc. 89 (1967) 7017.
[7] C.J. Pedersen, New macrocyclic polyethers, J. Am. Chem. Soc. 92 (2) (1970) 391–394.
[8] L.R. Nassimbeni, Physicochemical aspects of host–guest compounds, Acc. Chem. Res. 36 (8) (2003) 631–637.
[9] A. Nangia, G.R. Desiraju, Supramolecular synthons and pattern recognition, in: E. Weber, et al. (Eds.), Design of Organic Solids, Topics in Current Chemistry, vol. 198, Springer, Berlin, Heidelberg, 1998, pp. 57–95.
[10] G.R. Desiraju, Crystal engineering: from molecule to crystal, J. Am. Chem. Soc. 135 (2013) 9952–9967.
[11] J.F. Stoddart, Mechanically interlocked molecules (MIMs)—molecular shuttles, switches, and machines (Nobel lecture), Angew. Chem. Int. Ed. 56 (2017) 11094–11125.
[12] C.J. Bruns, J.F. Stoddart (Eds.), The Nature of the Mechanical Bond, John Wiley & Sons, 2016.
[13] G.R. Desiraju, Supramolecular synthons in crystal engineering—a new organic synthesis, Angew. Chem. Int. Ed. 34 (1995) 2311–2327.
[14] G.R. Desiraju, J.J. Vittal, A. Ramanan, Crystal Engineering: A Textbook, World Scientific, Singapore, 2011.
[15] R. Pepinsky, Crystal engineering: new concepts in crystallography, Phys. Rev. 100 (1955) 971.
[16] M.D. Cohen, G.M.J. Schmidt, Topochemistry. Part I. A survey, J. Chem. Soc. (1964) 1996–2000.
[17] G.M.J. Schmidt, Photodimerization in the solid state, Pure Appl. Chem. 27 (1971) 647–678.
[18] J.A.R.P. Sarma, G.R. Desiraju, The role of Cl…Cl and C-H…O interactions in the crystal engineering of 4-Å short-axis structures, Acc. Chem. Res. 19 (1986) 222–228.
[19] G.R. Desiraju, Crystal Engineering. The Design of Organic Solids, Elsevier, Amsterdam, 1989.
[20] W.H. Bragg, The structure of organic crystals, Proc. Phys. Soc. London 34 (1921) 33–50.
[21] W.H. Bragg, The crystalline structure of anthracene, Proc. Phys. Soc. 35 (1921) 167–169.
[22] R.G. Dickinson, A.L. Raymond, The crystal structure of hexamethylenetetramine, J. Am. Chem. Soc. 45 (1923) 22–29.
[23] K. Lonsdale, The structure of the benzene ring, Nature 122 (1928) 810.

[24] E.G. Cox, The crystalline structure of benzene, Nature 122 (1928) 401.
[25] K. Lonsdale, The structure of the benzene ring in hexamethylbenzene, Proc. R. Soc. A 123 (1929) 494–515.
[26] K. Lonsdale, X-ray evidence on the structure of the benzene nucleus, Trans. Faraday Soc. 25 (1929) 352–366.
[27] L.O. Brockway, J.M. Robertson, The crystal structure of hexamethylbenzene and the length of the methyl group bond to aromatic carbon atoms, J. Chem. Soc. (1939) 1324–1332.
[28] J.D. Bernal, D. Crowfoot, The structure of some hydrocarbons related to the sterols, J. Chem. Soc. (1935) 93–100.
[29] A.I. Kitaigorodskii, Molecular Crystals and Molecules, Academic Press, New York, 1973.
[30] A.J. Pertsin, A.I. Kitaigorodskii, The Atom-Atom Potential Method, Springer-Verlag, 1987.
[31] J.D. Dunitz, X-Ray Analysis and the Structure of Organic Molecules, Cornell University Press, Ithaca, NY, 1979.
[32] J.D. Dunitz, Phase transitions in molecular crystals from a chemical viewpoint, Pure Appl. Chem. 63 (1991) 177–185.
[33] P. Metrangolo, G. Resnati, Halogen bonding: a paradigm in supramolecular chemistry, Chem. Eur. J. 7 (2001) 2511–2519.
[34] C.B. Aakeroy, S. Panikkattu, P.D. Chopade, J. Desper, Competing hydrogen-bond and halogen-bond donors in crystal engineering, CrystEngComm 15 (2013) 3125–3136.
[35] A. Gavezzotti, Are crystal structures predictable? Acc. Chem. Res. 27 (1994) 309.
[36] M.C. Etter, Encoding and decoding hydrogen-bond patterns of organic compounds, Acc. Chem. Res. 23 (1990) 120–126.
[37] M.C. Etter, Z. Urbanczyk-Lipkowska, M. Zia-Ebrahimi, T.W. Panunto, Hydrogen bond-directed cocrystallization and molecular recognition properties of diarylureas, J. Am. Chem. Soc. 112 (1990) 8415–8426.
[38] H.M. Powell, The structure of molecular compounds. Part IV. Clathrate compounds, J. Chem. Soc. (1948) 61–73.
[39] O. Ermer, Five-fold diamond structure of adamantane-1,3,5,7-tetracarboxylic acid, J. Am. Chem. Soc. 110 (1988) 3147–3754.
[40] T. Steiner, The hydrogen bonds in solid state, Angew. Chem. Int. Ed. 41 (2002) 49–76.
[41] E. Arunan, G.R. Desiraju, R.A. Klein, J. Sadlej, S. Scheiner, I. Alkorta, D.C. Clary, R.H. Crabtree, J.J. Dannenberg, P. Hobza, H.G. Kjaergaard, A.C. Legon, B. Mennucci, D. Nesbitt, Defining the hydrogen bond: an account (IUPAC technical report), J. Pure Appl. Chem. 83 (2011) 1637–1641.
[42] G.R. Desiraju, Angew. Chem. Int. Ed. 50 (2011) 52.
[43] G.R. Desiraju, T. Steiner, The Weak Hydrogen Bond in Structural Chemistry and Biology, 1999.
[44] G. Gilli, P. Gilli, The Nature of the Hydrogen Bond International Union of Crystallography Monographs on Crystallography, 2009.
[45] L. Pauling, The Nature of the Chemical Bond, The Cornell University Press, 1939.
[46] V.R. Thalladi, B.S. Goud, V.J. Hoy, F.H. Allen, J.A.K. Howard, G.R. Desiraju, Supramolecular synthons in crystal engineering. Structure simplification, synthon robustness and supramolecular retrosynthesis, Chem. Commun. (1996) 401–402.
[47] G.R. Desiraju, Hydrogen bridges in crystal engineering: interactions without borders, Acc. Chem. Res. 35 (2002) 565–573.

[48] E.J. Corey, General methods for the construction of complex molecules, Pure Appl. Chem. 14 (1967) 19–38.
[49] E.J. Corey, Robert robinson lecture. Retrosynthetic thinking essentials and examples, Chem. Soc. Rev. 17 (1988) 111–133.
[50] T.R. Shattock, K.K. Arora, P. Vishweshwar, M.J. Zaworotko, Hierarchy of supramolecular synthons: persistent carboxylic Acid···Pyridine hydrogen bonds in cocrystals that also contain a hydroxyl moiety, Cryst. Growth Des. 8 (2008), 4533–4545.
[51] C.B. Aakeröy, M. Baldrighi, J. Desper, P. Metrangolo, G. Resnati, Supramolecular hierarchy among halogen-bond donors, Chem. Eur. J. 19 (2013) 16240–16247.
[52] J.A. Bis, M.J. Zaworotko, The 2-aminopyridinium-carboxylate supramolecular heterosynthon: a robust motif for generation of multiple-component crystals, Cryst. Growth Des. 5 (2005) 1169–1179.
[53] M. Hemamalini, W. Loh, C.K. Quah, H. Fun, Investigation of supramolecular synthons and structural characterisation of aminopyridine-carboxylic acid derivatives, Chem. Cent. J. (2014), https://doi.org/10.1186/1752-153X-8-31.
[54] J.A. McMahon, J.A. Bis, P. Vishweshwar, T.R. Shattock, O.L. McLaughlin, M.J. Zaworotko, Crystal engineering of the composition of pharmaceutical phases. 3. Primary amide supramolecular heterosynthons and their role in the design of pharmaceutical co-crystals, Z. Kristallogr 220 (2005) 340–350.
[55] B.R. Bhogala, A. Nangia, Cocrystals of 1,3,5-cyclohexanetricarboxylic acid with 4,4'-bipyridine homologues: Acid···Pyridine hydrogen bonding in neutral and ionic complexes, Cryst. Growth Des. 3 (2003) 547–554.
[56] T.R. Shattock, K.K. Arora, P. Vishweshwar, M.J. Zaworotko, Hierarchy of supramolecular synthons: persistent carboxylic Acid···Pyridine hydrogen bonds in cocrystals that also contain a hydroxyl moiety, Cryst. Growth Des. 8 (2008) 4533–4545.
[57] N.J. Babu, A. Nangia, Multiple Z' in carboxylic Acid–Pyridine trimer synthon and kagomé lattice in the structure of 5-methylpyrazine-2,3-dicarboxylic acid, Cryst. Growth Des. 6 (2006) 1995–1999.
[58] S. Tothadi, S. Joseph, G.R. Desiraju, Synthon modularity in cocrystals of 4-bromobenzamide with *n*-alkanedicarboxylic acids: type I and type II Halogen···Halogen interactions, Cryst. Growth Des. 13 (2013) 3242–3254.
[59] L. Rajput, K. Biradha, Design of cocrystals via new and robust supramolecular synthon between carboxylic acid and secondary amide: honeycomb network with jailed aromatics, Cryst. Growth Des. 9 (2009) 40–42.
[60] J.A.R.P. Sarma, F.H. Allen, V.J. Hoy, J.A.K. Howard, R. Thaimattam, K. Biradha, G.R. Desiraju, Design of an SHG-active crystal, 4-iodo-4′-nitrobiphenyl: the role of supramolecular synthons, Chem. Commun. (1997) 101–102.
[61] N.J. Babu, L.S. Reddy, A. Nangia, Amide-N-oxide heterosynthon and amide dimer homosynthon in cocrystals of carboxamide drugs and pyridine N-oxides, Mol. Pharmaceutics 4 (2007) 417–434.
[62] L.S. Reddy, N.J. Babu, A. Nangia, Carboxamide–pyridine *N*-oxide heterosynthon for crystal engineering and pharmaceutical cocrystals, Chem. Commun. (2006) 1369–1371.
[63] N.R. Goud, N.J. Babu, A. Nangia, Sulfonamide–Pyridine-N-oxide cocrystals, Cryst. Growth Des. 11 (2011) 1930–1939.
[64] B. Saikia, V. Khatioda, P. Bora, B. Sarma, Pyridine N-oxides as coformers in the development of drug cocrystals, CrystEngComm 18 (2016) 8454–8464.

[65] G. Bolla, S. Mittapalli, A. Nangia, Celecoxib cocrystal polymorphs with cyclic amides: synthons of a sulfonamide drug with carboxamide coformers, CrystEngComm 16 (2014) 24–27.
[66] G. Bolla, S. Mittapalli, A. Nangia, Modularity and three-dimensional isostructurality of novel synthons in sulfonamide–lactam cocrystals, IUCrJ 2 (2015) 389–401.
[67] G. Bolla, A. Nangia, Supramolecular synthon hierarchy in sulfonamide cocrystals with *syn*-amides and *N*-oxides, IUCrJ 6 (2019) 751–760.
[68] G. Bolla, A. Nangia, Multicomponent ternary cocrystals of the sulfonamide group with pyridine-amides and lactams, Chem. Commun. 51 (2015) 15578–15581.
[69] M. Baldrighi, G. Cavallo, M.R. Chierotti, R. Gobetto, P. Metrangolo, T. Pilati, G. Resnati, G. Terraneo, Halogen bonding and pharmaceutical cocrystals: the case of a widely used preservative, Mol. Pharmaceutics 10 (2013) 1760–1772.
[70] C.B. Aakeröy, M. Fasulo, N. Schultheiss, J. Desper, C. Moore, Structural competition between hydrogen bonds and halogen bonds, J. Am. Chem. Soc. 129 (2007) 13772–13773.
[71] D. Cinčić, T. Friščić, W. Jones, Isostructural materials achieved by using structurally equivalent donors and acceptors in halogen-bonded cocrystals, Chem. Eur. J. 14 (2008) 747–753.
[72] R.D.B. Walsh, M.W. Bradner, S. Fleischman, L.A. Morales, B. Moulton, N. Rodríguez-Hornedo, M.J. Zaworotko, Crystal engineering of the composition of pharmaceutical phases, Chem. Commun. (2003) 186–187.
[73] Aitipamula, et al., Polymorphs, salts, and cocrystals: what's in a name? Cryst. Growth Des. 12 (2012) 2147–2152.
[74] A.V. Trask, W.D.S. Motherwell, W. Jones, Solvent-drop grinding: green polymorph control of cocrystallisation, Chem. Commun. (2004) 890–891.
[75] A.V. Trask, D.A. Haynes, W.D.S. Motherwell, W. Jones, Screening for crystalline salts *via* mechanochemistry, Chem. Commun. (2006) 51–53.
[76] W.W. Porter III., S.C. Elie, A.J. Matzger, Polymorphism in carbamazepine cocrystals, Cryst. Growth Des. 8 (2008) 14–16.
[77] K. Seefeldt, J. Miller, F. Alvarez-Nunez, N. Rodríguez-Hornedo, Crystallization pathways and kinetics of carbamazepine–nicotinamide cocrystals from the amorphous state by *in situ* thermomicroscopy, spectroscopy, and calorimetry studies, J. Pharm. Sci. 96 (2007) 1147–1158.
[78] D.J. Berry, C.C. Seaton, W. Clegg, R.W. Harrington, S.J. Coles, P.N. Horton, M.B. Hursthouse, R. Storey, W. Jones, T. Friščić, N. Blagden, Applying hot-stage microscopy to co-crystal screening: a study of nicotinamide with seven active pharmaceutical ingredients, Cryst. Growth Des. 8 (2008) 1697–1712.
[79] F. Wohler, Untersuchungen über das Chinon, Ann. Chem. Pharm. 51 (1844) 145.
[80] A.R. Ling, J.L. Baker, XCVI.—halogen derivatives of quinone. Part III. Derivatives of quinhydrone, J. Chem. Soc. 63 (1893) 1314–1327.
[81] M. Haruo, O. Kenji, N. Isamu, Crystal structure of quinhydrone, C12H10O4, Bull. Chem. Soc. Jpn. 31 (1958) 611–620.
[82] https://www.fda.gov/food/food-ingredients-packaging/generally-recognized-safe-gras. [Accessed 2 May 2020].
[83] S.S. Davis, Scientific principles in design of drug dosage formulations, Br. Med. J. 1 (1972) 102–106.
[84] D.P. Elder, R. Holm, H.L.D. Diego, Use of pharmaceutical salts and cocrystals to address the issue of poor solubility, Int. J. Pharm. 45 (2013) 88–100.

[85] S.L. Morissette, Ö. Almarsson, M.L. Peterson, J.F. Remenar, M.J. Read, A.V. Lemmo, S. Ellis, M.J. Cima, C.R. Gardner, High-throughput crystallization: polymorphs, salts, co-crystals and solvates of pharmaceutical solids, Adv. Drug Delivery Rev. 56 (2004) 275–300.

[86] J.F. Remenar, S.L. Morissette, M.L. Peterson, B. Moulton, J.M. MacPhee, H.R. Guzmán, O. Almarsson, Crystal engineering of novel cocrystals of a triazole drug with 1,4-dicarboxylic acids, J. Am. Chem. Soc. 125 (2003) 8456–8457.

[87] S.A. Myz, T.P. Shakhtshneider, K. Fucke, A.P. Fedotov, E.V. Boldyreva, V.V. Boldyrev, N.I. Kuleshova, Synthesis of co-crystals of meloxicam with carboxylic acids by grinding, Mendeleev Commun. 19 (2009) 272–274.

[88] N.A. Tumanov, S.A. Myz, T.P. Shakhtshneider, E.V. Boldyreva, Are meloxicam dimers really the structure-forming units in the 'meloxicam–carboxylic acid' co-crystals family? Relation between crystal structures and dissolution behavior, CrystEngComm 14 (2012) 305–313.

[89] S.A. Myz, T.P. Shakhtshneider, N.A. Tumanov, E.V. Boldyreva, Preparation and studies of the co-crystals of meloxicam with carboxylic acids, Russ. Chem. Bull. 61 (2012) 1798–1809.

[90] R.W. David, L.C. Miranda, S. Ning, H. Mazen, M.J. Zaworotko, S. Vasyl, S. Shijie, R.S. Juan, Improving solubility and pharmacokinetics of meloxicam via multiple-component crystal formation, Mol. Pharmaceutics 9 (2012) 2094–2102.

[91] A. Fahr, X. Liu, Drug delivery strategies for poorly water-soluble drugs, Exp. Opin. Drug Delivery 4 (2007) 403–416.

[92] Y.A. Abramov, C. Loschen, A. Klamt, Rational coformer or solvent selection for pharmaceutical cocrystallization or desolvation, J. Pharm. Sci. 10 (2012) 3687–3697.

[93] T.P. Shakhtshneider, S.A. Myz, M.A. Dyakonova, V.V. Boldyrev, E.V. Boldyreva, A.I. Nizovskii, A.V. Kalinkin, R. Kumar, Mechanochemical preparation of organic-inorganic hybrid materials of drugs with inorganic oxide, Acta Phys. Pol. A 120 (2011) 272–278.

[94] A. Alhalaweh, L. Roy, N. Rodríguez-Hornedo, S.P. Velaga, pH-dependent solubility of indomethacin-saccharin and carbamazepine-saccharin cocrystals in aqueous media, Mol. Pharmaceutics 9 (2012) 2605–2612.

[95] M.B. Hickey, M.L. Peterson, L.A. Scoppettuolo, S.L. Morrisette, A. Vetter, H. Guzmán, J.F. Remenar, Z. Zhang, M.D. Tawa, S. Haley, M.J. Zaworotko, O. Almarsson, Performance comparison of a co-crystal of carbamazepine with marketed product, Eur. J. Pharm. Biopharm. 67 (2007) 112–119.

[96] E. Lekšić, G. Pavlović, E. Meštrović, Cocrystals of lamotrigine based on coformers involving carbonyl group discovered by hot-stage microscopy and DSC screening, Cryst. Growth Des. 12 (2012) 1847–1858.

[97] R. Chadha, A. Saini, P. Arora, D.S. Jain, A. Dasgupta, T.N.G. Row, Multicomponent solids of lamotrigine with some selected coformers and their characterization by thermoanalytical, spectroscopic and X-ray diffraction methods, CrystEngComm 13 (2011) 6271–6284.

[98] M.L. Cheney, N. Shan, E.R. Healey, M. Hanna, L. Wojtas, M.J. Zaworotko, V. Sava, S. Song, J.R. Sanchez-Ramos, Effects of crystal form on solubility and pharmacokinetics: a crystal engineering case study of lamotrigine, Cryst. Growth Des. 10 (2010) 394–405.

[99] L. Rajput, P. Sanphui, G.R. Desiraju, New solid forms of the anti-HIV drug etravirine: salts, cocrystals, and solubility, Cryst. Growth Des. 13 (2013) 3681–3690.

[100] P. Sanphui, M.K. Mishra, U. Ramamurty, G.R. Desiraju, Tuning mechanical properties of pharmaceutical crystals with multicomponent crystals: voriconazole as a case study, Mol. Pharmaceutics 12 (2015) 889–897.

[101] N.J. Babu, A. Nangia, Solubility advantage of amorphous drugs and pharmaceutical cocrystals, Cryst. Growth Des. 11 (2011) 2662–2679.

[102] G. Bolla, P. Sanphui, A. Nangia, Solubility advantage of tenoxicam phenolic cocrystals compared to salts, Cryst. Growth Des. 13 (2013) 1988–2003.

[103] A. Mukherjee, S. Tothadi, G.R. Desiraju, Halogen bonds in crystal engineering: like hydrogen bonds yet different, Acc. Chem. Res. 47 (2014) 2514–2524.

[104] C.B. Aakeröy, A.M. Beatty, B.A. Helfrich, "Total synthesis" supramolecular style: design and hydrogen-bond-directed assembly of ternary supermolecules, Angew. Chem. Int. Ed. 40 (2001) 3240–3242.

[105] R. Kuroda, Y. Imai, N. Tajima, Generation of a co-crystal phase with novel coloristic properties via solid state grinding procedures, Chem. Commun. (2002) 2848–2849.

[106] M.C. Etter, A new role for hydrogen-bond acceptors in influencing packing patterns of carboxylic acids and amides, J. Am. Chem. Soc. 104 (1982) 1095–1096.

[107] M.C. Etter, Hydrogen bonds as design elements in organic chemistry, J. Phys. Chem. 95 (1991) 4601–4610.

[108] S. Tothadi, A. Mukherjee, G.R. Desiraju, Shape and size mimicry in the design of ternary molecular solids: towards a robust strategy for crystal engineering, Chem. Commun. 47 (2011) 12080–12082.

[109] S. Aitipamula, A.B.H. Wong, P.S. Chow, R.B.H. Tan, Novel solid forms of the anti-tuberculosis drug, Isoniazid: ternary and polymorphic cocrystals, CrystEngComm 15 (2013) 5877–5887.

[110] S. Tothadi, G.R. Desiraju, Designing ternary cocrystals with hydrogen bonds and halogen bonds, Chem. Commun. 49 (2013) 7791–7793.

[111] S. Tothadi, G.R. Desiraju, Synthon modularity in 4-hydroxybenzamide–dicarboxylic acid cocrystals, Cryst. Growth Des. 12 (2012) 6188–6198.

[112] F. Topic, K. Rissanen, Systematic construction of ternary cocrystals by orthogonal and robust hydrogen and halogen bonds, J. Am. Chem. Soc. 138 (2016) 6610–6616.

[113] B. Bolla, A. Nangia, Binary and ternary cocrystals of sulfa drug acetazolamide with pyridine carboxamides and cyclic amides, IUCrJ 3 (2016) 152–160.

[114] S. Aitipamula, P.S. Chow, R.B.H. Tan, Polymorphism in cocrystals: a review and assessment of its significance, CrystEngComm 16 (2014) 3451–3465.

[115] S. Aitipamula, P.S. Chow, R.B.H. Tan, Dimorphs of a 1 : 1 cocrystal of ethenzamide and saccharin: solid-state grinding methods result in metastable polymorph, CrystEngComm 11 (2009) 889–895.

[116] S. Aitipamula, P.S. Chow, R.B.H. Tan, Conformational and enantiotropic polymorphism of a 1 : 1 cocrystal involving ethenzamide and ethylmalonic acid, CrystEngComm 12 (2010) 3691–3697.

[117] S. Aitipamula, P.S. Chow, R.B.H. Tan, Trimorphs of a pharmaceutical cocrystal involving two active pharmaceutical ingredients: potential relevance to combination drugs, CrystEngComm 11 (2009) 1823–1827.

[118] R. Khatioda, P. Bora, B. Sarma, Trimorphic ethenzamide cocrystal: in vitro solubility and membrane efflux studies, Cryst. Growth Des. 18 (2018) 4637–4645.

[119] N. Takata, K. Shiraki, R. Takano, Y. Hayashi, K. Terada, Cocrystal screening of stanolone and mestanolone using slurry crystallization, Cryst. Growth Des. 8 (2008) 3032–3037.

[120] P. Sanphui, J.N. Babu, A. Nangia, Temozolomide cocrystals with carboxamide coformer, Cryst. Growth Des. 13 (2013) 2208–2219.
[121] S. Aitipamula, A.B.H. Wong, P.S. Chow, R.B.H. Tan, Polymorphism and phase transformations of a cocrystal of nicotinamide and pimelic acid, CrystEngComm 11 (2012) 8193–8198.
[122] S.L. Childs, K.I. Hardcastle, Cocrystals of chlorzoxazone with carboxylic acids, CrystEngComm 9 (2007) 364–367.
[123] T. Ueto, N. Takata, N. Muroyama, A. Nedu, A. Sasaki, S. Tanida, K. Terada, Polymorphs and a hydrate of furosemide–nicotinamide 1:1 cocrystal, Cryst. Growth Des. 12 (2012) 485–494.
[124] A.J. Cruz-Cabeza, G.M. Day, W. Jones, Towards prediction of stoichiometry in crystalline multicomponent complexes, Chem. Eur. J. 14 (2008) 8830–8836.
[125] L. Orola, M.V. Veidis, Nicotinamide fumaric acid supramolecular cocrystals: diversity of stoichiometry, CrystEngComm 11 (2009) 415–417.
[126] B. Saikia, P. Bora, R. Khatioda, B. Sarma, Hydrogen bond synthons in the interplay of solubility and membrane permeability/diffusion in variable stoichiometry drug cocrystals, Cryst. Growth Des. 15 (2015) 5593–5603.
[127] R. Thakuria, B. Sarma, Drug-drug and drug-nutraceutical cocrystal/salt as alternative medicine for combination therapy: a crystal engineering approach, Crystals 8 (2018) 101.
[128] S. Nakao, S. Fujii, T. Sakaki, K.I. Tomita, The crystal and molecular structure of the 2:1 molecular complex of theophylline with phenobarbital, Acta Crystallogr. B 33 (1977) 1373–1378.
[129] M.L. Cheney, D.R. Weyna, N. Shan, M. Hanna, L. Wojtas, M.J. Zaworotko, Coformer selection in pharmaceutical cocrystal development: a case study of a meloxicam aspirin cocrystal that exhibits enhanced solubility and pharmacokinetics, J. Pharm. Sci. 100 (2011) 2172–2181.
[130] S. Kakkar, B. Bhattacharya, C.M. Reddy, S. Ghosh, Tuning mechanical behaviour by controlling the structure of a series of theophylline co-crystals, CrystEngComm 20 (2018) 1101–1109.
[131] P.M. Bhatt, Y. Azim, T.S. Thakur, G.R. Desiraju, Cocrystals of the anti-HIV drugs lamivudine and zidovudine, Cryst. Growth Des. 9 (2009) 951–957.
[132] D.O. Corrigan, O.I. Corrigan, A.M. Healy, Physicochemical and in vitro deposition properties of salbutamol sulphate/ipratropium bromide and salbutamol sulphate/excipient spray dried mixtures for use in dry powder inhalers, Int. J. Pharm. 322 (2006) 22–30.
[133] P. Grobelny, A. Mukherjee, G.R. Desiraju, Drug-drug co-crystals: temperature-dependent proton mobility in the molecular complex of isoniazid with 4-aminosalicylic acid, CrystEngComm 13 (2011) 4358–4364.
[134] S.C.R. Plata, C.S. Videla, N. Tesson, C.M. Trilla, Co-Crystals of Venlafaxine and Celecoxib, Patent EP 2,515,892, October 31, 2012.
[135] I. Nugrahani, S. Asyarie, S.N. Soewandhi, S. Ibrahim, The antibiotic potency of amoxicillin-clavulanate co-crystal, Int. J. Pharmacol. 3 (2007) 475–481.
[136] J. Lu, A.J. Cruz-Cabeza, S. Rohani, M.C. Jennings, A 2:1 sulfamethazine-theophylline cocrystal exhibiting two tautomers of sulfamethazine, Acta Crystallogr. C. 67 (2011) o306–o309.
[137] A.O.L. Évora, R.A.E. Castro, T.M.R. Maria, M.T.S. Rosado, R.M. Silva, A. Matos Beja, J. Canotilho, M.E.S. Eusébio, Pyrazinamide-diflunisal: a new dual-drug co-crystal, Cryst. Growth Des. 11 (2011) 4780–4788.

[138] R. Roy, J. Deb, S.S. Jana, P. Dastidar, Exploiting supramolecular synthons in designing gelators derived from multiple drugs, Chem. Eur. J. 20 (2014) 15320–15324.

[139] O. Lebel, M. Perron, T. Maris, S.F. Zalzal, A. Nanci, J.D. Wuest, A new class of selective low-molecular-weight gelators based on salts of diaminotriazinecarboxylic acids, Chem. Mater. 18 (2006) 3616–3626.

[140] S. Mittapalli, M.K.C. Mannava, R. Sahoo, A. Nangia, Cocrystals, salts, and supramolecular gels of nonsteroidal anti-inflammatory drug niflumic acid, Cryst. Growth Des. 19 (2019) 219–230.

[141] S. Casalini, C.A. Bortolotti, F. Leonardi, F. Biscarini, Chem. Soc. Rev. 46 (2017) 40–71.

[142] J. Aizenberg, A.J. Black, G.M. Whitesides, Self-assembled monolayers in organic electronics. Control of nucleation by patterned self-assembled monolayers, Nature 398 (1999) 495–498.

[143] A. Ulman, Formation and structure of self-assembled monolayers, Chem. Rev. 96 (1996) 1533–1554.

[144] J. Aizenberg, A.J. Black, G.M. Whitesides, Oriented growth of calcite controlled by self-assembled monolayers of functionalized alkanethiols supported on gold and silver, J. Am. Chem. Soc. 121 (1999) 4500–4509.

[145] A.Y. Lee, I.S. Lee, S.S. Dette, J. Boerner, A.S. Myerson, Crystallization on confined engineered surfaces: a method to control crystal size and generate different polymorphs, J. Am. Chem. Soc. 127 (2005) 14982–14983.

[146] K. Kim, I.S. Lee, A. Centrone, T.A. Hatton, A.S. Myerson, Formation of nanosized organic molecular crystals on engineered surfaces, J. Am. Chem. Soc. 131 (2009) 18212–18213.

[147] P. Bora, B. Saikia, B. Sarma, Oriented crystallization on organic monolayers to control concomitant polymorphism, Chem. Eur. J. 26 (2020) 699–710.

[148] J. Liu, C. Wöll, Surface-supported metal–organic framework thin films: fabrication methods, applications, and challenges, Chem. Soc. Rev. 46 (2017) 5730–5770.

[149] G. Bolla, A.S. Myerson, SURMOF induced morphological crystal engineering of substituted benzamides, Cryst. Growth Des. 18 (2018) 7048–7058.

[150] G. Bolla, A.S. Myerson, SURMOF induced polymorphism and crystal morphological engineering of acetaminophen polymorphs: advantage of heterogeneous nucleation, CrystEngComm 20 (2018) 2084–2088.

[151] D. Braga, F. Grepioni, O. Shemchuk, Organic–inorganic ionic co-crystals: a new class of multipurpose compounds, CrystEngComm 20 (2018) 2212–2220.

[152] T. Wang, J.S. Stevenes, T. Vetter, G.F.S. Whitehead, I.J. Vitorica-Yrezabal, H. Hao, A.J. Cruz-Cabeza, Salts, cocrystals, and ionic cocrystals of a "simple" tautomeric compound, Cryst. Growth Des. 18 (2018) 6973–6983.

[153] D. Braga, F. Grepioni, L. Maini, S. Prosperi, R. Gobetto, M.R. Chierotti, From unexpected reactions to a new family of ionic co-crystals: the case of barbituric acid with alkali bromides and caesium iodide, Chem. Commun. 46 (2010) 7715–7717.

[154] D. Braga, F. Grepioni, G.I. Lampronti, L. Maini, A. Turrina, Ionic co-crystals of organic molecules with metal halides: a new prospect in the solid formulation of active pharmaceutical ingredients, Cryst. Growth Des. 11 (2011) 5621–5627.

[155] L. Song, O. Shemchuk, K. Robeyns, D. Braga, F. Grepioni, T. Leyssens, Ionic cocrystals of etiracetam and levetiracetam: the importance of chirality for ionic cocrystals, Cryst. Growth Des. 19 (2019) 2446–2454.

[156] L. Song, K. Robeyns, T. Leyssens, Crystallizing ionic cocrystals: structural characteristics, thermal behavior, and crystallization development of a piracetam-$CaCl_2$ cocrystallization process, Cryst. Growth Des. 18 (2018) 3215–3221.

[157] S. Cherukuvada, A. Nangia, Eutectics as improved pharmaceutical materials: design, properties and characterization, Chem. Commun. 50 (2014) 906–923.

[158] T. Friščić, A.V. Trask, W. Jones, W.D.S. Motherwell, Screening for inclusion compounds and systematic construction of three-component solids by liquid-assisted grinding, Angew. Chem. Int. Ed. 45 (2006) 7546–7550.

[159] N. Yamamoto, T. Taga, K. Machida, Drug solid solutions – a method for tuning phase transformations, Acta Crystallogr. Sect. B Struct. Sci. 45 (1989) 162.

[160] L.S. Reddy, P.M. Bhatt, R. Banerjee, A. Nangia, G.J. Kruger, Variable-temperature powder X-ray diffraction of aromatic carboxylic acid and carboxamide cocrystals, Chem. Asian J. 2 (2007) 505–513.

[161] N.B. Singh, S.S. Das, N.P. Singh, T. Agrawal, Computer simulation, thermodynamic and microstructural studies of benzamide–benzoic acid eutectic system, J. Cryst. Growth 310 (2008) 2878–2884.

[162] M. Dabros, P.R. Emery, V.R. Thalladi, A supramolecular approach to organic alloys: cocrystals and three and four component solid solutions of 1,4-diazabicyclo[2.2.2]octane and 4X-phenols (X=Cl, CH_3, Br), Angew. Chem. Int. Ed. 46 (2007) 4132–4135.

[163] I. Sathisaran, S.V. Dalvi, Crystal engineering of curcumin with salicylic acid and hydroxyquinol as coformers, Cryst. Growth Des. 17 (2017) 3974–3988.

[164] P. Sanphui, N.R. Goud, U.B.R. Khandavilli, S. Bhanoth, A. Nangia, New polymorphs of curcumin, Chem. Commun. 47 (2011) 5013–5015.

[165] P. Sanphui, N.R. Goud, U.B.R. Khandavilli, A. Nangia, Fast dissolving curcumin cocrystals, Cryst. Growth Des. 11 (2011) 4135–4145.

[166] S.N. Wong, S. Hu, W.W. Ng, X. Xu, K.L. Lai, W.Y.T. Lee, A.H.L. Chow, C.C. Sun, S.F. Chow, Cocrystallization of curcumin with benzenediols and benzenetriols via rapid solvent removal, Cryst. Growth Des. 18 (2018) 5534–5546.

[167] K. Suresh, C.M.K. Mannava, A. Nangia, Cocrystals and alloys of nitazoxanide: enhanced pharmacokinetics, Chem. Commun. 52 (2016) 4223–4226.

[168] W. Zhaomeng, S. Mengchi, L. Tian, G. Zisen, Y. Qing, T. Xiao, H.Y. Xian, S. Jin, W. Dun, H. Zhonggui, Co-amorphous solid dispersion systems of lacidipine-spironolactone with improved dissolution rate and enhanced physical stability, Asian J. Pharm. Sci. 14 (2019) 95–103.

[169] Q. Shi, S.M. Moinuddin, T. Cai, Advances in coamorphous drug delivery systems, Acta Pharm. Sin. B 9 (2019) 19–35.

[170] P. Katrine, T. Jensen, K. Löbmann, T. Rades, H. Grohganz, Improving co-amorphous drug formulations by the addition of the highly water soluble amino acid, Pharmaceutics 6 (2014) 416–435.

[171] M. Su, Y. Xia, Y. Shen, W. Heng, Y. Wei, L. Zhang, Y. Gao, J. Zhang, S. Qian, A novel drug–drug coamorphous system without molecular interactions: improve the physicochemical properties of tadalafil and repaglinide, RSC Adv. 10 (2020) 565–583.

[172] W. Wu, Y. Wang, K. Löbmann, H. Grohganz, T. Rades, Transformations between co-amorphous and co-crystal systems and their influence on the formation and physical stability of co-amorphous systems, Mol. Pharm. 16 (2019) 1294–1304.

[173] K. Suresh, M.K.C. Mannava, A. Nangia, A novel curcumin–artemisinin coamorphous solid: physical properties and pharmacokinetic profile, RSC Adv. 4 (2014) 58357–58361.

[174] S. Allu, K. Suresh, G. Bolla, M.K.C. Mannava, A. Nangia, Role of hydrogen bonding in cocrystals and coamorphous solids: indapamide as a case study, CrystEngComm 21 (2019) 2043–2048.
[175] N.J. Babu, P. Sanphui, A. Nangia, Crystal engineering of stable temozolomide cocrystals, Chem. Asian J. 7 (2012) 2274–2285.
[176] Kusuma, K. Kanakaraju, V. Lavanya, et al., Temozolomide cocrystals exhibit drug sensitivity in glioblastoma cells, Proc. Natl. Acad. Sci. India Sect. A Phys. Sci. 84 (2014) 321–330.
[177] E. Sravani, M.K.C. Mannava, D. Kaur, B.R. Annapurna, R.A. Khan, K. Suresh, S. Mittapalli, A. Nangia, B.D. Kumar, Preclinical bioavailability–bioequivalence and toxico-kinetic profile of stable succinic acid cocrystal of temozolomide, Curr. Sci. 108 (2015) 1097–1106.
[178] A. Charalabidis, M. Sfouni, C. Bergström, P. Macheras, The Biopharmaceutics classification system (BCS) and the Biopharmaceutics drug disposition classification system (BDDCS): beyond guidelines, Int. J. Pharm. 566 (2019) 264–281.
[179] P. Sanphui, G. Bolla, A. Nangia, V. Chernyshev, Acemetacin cocrystals and salts: structure solution from powder X-ray data and form selection of the piperazine salt, IUCrJ 1 (2014) 136–150.
[180] L. Yu, Continuous Manufacturing Has a Strong Impact on Drug Quality, FDA Voice, 2016, p. 12. http://blogs.fda.gov/fdavoice/index.php/2016/04/continuous-manufacturing-has-a-strong-impact-on-drug-quality/.
[181] V. Svoboda, P. MacFhionnghaile, J. McGinty, L.E. Connor, I.D. Oswald, J. Sefcik, Continuous cocrystallization of benzoic acid and isonicotinamide by mixing-induced supersaturation: exploring opportunities between reactive and antisolvent crystallization concepts, Cryst. Growth Des. 17 (2017) 1902–1909.
[182] Y. Ma, S. Wu, E.G.J. Macaringue, T. Zhang, J. Gong, J. Wang, Recent progress in continuous crystallization of pharmaceutical products: precise preparation and control, Org. Process Res. Dev. XXX (2020) XXX, https://doi.org/10.1021/acs.oprd.9b00362.
[183] D.L. Hughes, Applications of flow chemistry in the pharmaceutical industry—highlights of the recent patent literature, Org. Process Res. Dev. XXXX (2020) XXX, https://doi.org/10.1021/acs.oprd.0c00156.
[184] H. Zhang, J. Quon, A.J. Alvarez, J. Evans, A.S. Myerson, B. Trout, Development of continuous anti-solvent/cooling crystallization process using cascaded mixed suspension, mixed product removal crystallizers, Org. Process Res. Dev. 16 (2012) 915–924.
[185] S. Lawton, G. Steele, P. Shering, L. Zhao, I. Laird, X. Ni, Continuous crystallization of pharmaceuticals using a continuous oscillatory baffled crystallizer, Org. Process Res. Dev. 13 (2009) 1357–1363.
[186] Zhang, et al., Advanced continuous flow platform for on-demand pharmaceutical manufacturing, Chem. Eur. J. 24 (2018) 2776–2784.
[187] B. Wood, K.P. Girard, C.S. Polster, D.M. Croker, Progress to date in the design and operation of continuous crystallization processes for pharmaceutical applications, Org. Process Res. Dev. 23 (2019) 122–144.
[188] S. Kim, C. Wei, S. Kiang, Crystallization process development of an active pharmaceutical ingredient and particle engineering via the use of ultrasonics and temperature cycling, Org. Process Res. Dev. 7 (2003) 997–1001.
[189] C. Darmali, S. Mansouri, N. Yazdanpanah, M.W. Woo, Mechanisms and control of impurities in continuous crystallization: a review, Ind. Eng. Chem. Res. 58 (2019) 1463–1479.

[190] A.J. Alvarez, A.S. Myerson, Continuous plug flow crystallization of pharmaceutical compounds, Cryst. Growth Des. 10 (2010) 2219–2228.
[191] J.L. Quon, H. Zhang, A. Alvarez, J. Evans, A.S. Myerson, B.L. Trout, Continuous crystallization of aliskiren hemifumarate, Cryst. Growth Des. 12 (2012) 3036–3044.
[192] H. Zhang, R. Lakerveld, P.L. Heider, M. Tao, M. Su, C.J. Testa, A.N. D'Antonio, P.I. Barton, R.D. Braatz, B.L. Trout, A.S. Myerson, K.F. Jensen, J.M.B. Evans, Application of continuous crystallization in an integrated continuous pharmaceutical pilot plant, Cryst. Growth Des. 14 (2014) 2148–2215.
[193] A. Adamo, R.L. Beingessner, M. Behnam, J. Chen, T.F. Jamison, K.F. Jensen, J.M. Monbaliu, A.S. Myerson, E.M. Revalor, D.R. Snead, T. Stelzer, N. Weeranoppanan, S.Y. Wong, P. Zhang, On-demand continuous-flow production of pharmaceuticals in a compact, reconfigurable system, Science 352 (2016) 61–67.
[194] K. Tacsi, H. Pataki, A. Domokos, B. Nagy, I. Csontos, I. Markovits, F. Farkas, Z.K. Nagy, G. Marosi, Direct processing of a flow reaction mixture using continuous mixed suspension mixed product removal crystallizer, Cryst. Growth Des. 20 (2020) 4433–4442.
[195] Cole, et al., Kilogram-scale prexasertib monolactate monohydrate synthesis under continuous-flow CGMP conditions, Science 356 (2017) 1144–1150.
[196] T. Loftsson, M.E. Brewster, Pharmaceutical applications of cyclodextrins: basic science and product development, J. Pharm. Pharmacol. 62 (2010) 1607–1621, https://doi.org/10.1111/j.2042-7158.2010.01030.x.
[197] https://www.accessdata.fda.gov/drugsatfda_docs/label/2015/207620Orig1s000lbl.pdf.
[198] F-D-C Reports Inc, The Gold Sheet, Pharmaceutical & Biotechnology Quality Control, Attention Turns to the Business Case of Quality by Design, January 2009.
[199] M. Ende, T. Bernhard, V. Lubezyk, U. Dressler, et al., Risk Management and Lifestyle Efficiency within Process Development, Leadership Forum 2008, Washington DC, May 29, 2008.
[200] T. Lee, H.R. Chen, H.Y. Lin, H.L. Lee, Continuous co-crystallization as a separation technology: the study of 1: 2 co-crystals of phenazine-Vanillin, Cryst. Growth Des. 12 (2012) 5897–5907.
[201] M. Nishimaru, S. Kudo, H. Takiyama, Cocrystal production method reducing deposition risk of undesired single component crystals in anti-solvent cocrystallization, J. Ind. Eng. Chem. 36 (2016) 40–43.
[202] S.A. Ross, D.A. Lamprou, D. Douroumis, Engineering and manufacturing of pharmaceutical co-crystals: a review of solvent-free manufacturing technologies, Chem. Commun. 52 (2016) 8772–8786.
[203] S. Korde, S. Pagire, H. Pan, C. Seaton, A. Kelly, Y. Chen, Q. Wang, P. Coates, A. Paradkar, Continuous manufacturing of cocrystals using solid state shear milling technology, Cryst. Growth Des. 18 (2018) 2297–2304.
[204] M.M. Crowley, F. Zhang, M.A. Repka, S. Thumma, S.B. Upadhye, K. Battu, J.W. McGinity, C. Martin, Pharmaceutical applications of hot-melt extrusion: part I, Drug Dev. Ind. Pharm. 33 (2007) 909–926.
[205] C. Kulkarni, C. Wood, A.L. Kelly, T. Gough, N. Blagden, A. Paradkar, Stoichiometric control of co-crystal formation by solvent free continuous co-crystallization (SFCC), Cryst. Growth Des. 15 (2015) 5648–5651.
[206] C. Medina, D. Daurio, K. Nagapudi, F. Alvarez-nunez, Manufacture of pharmaceutical co-crystals using twin screw extrusion: a solvent-less and scalable process, J. Pharm. Sci. 99 (2010) 1693–1696.

[207] R.S. Dhumal, A.L. Kelly, P. York, P.D. Coates, A. Paradkar, Cocrystalization and simultaneous agglomeration using hot melt extrusion, Pharmaceut. Res. 27 (2010) 2725–2733.
[208] M. Lee, N. Chun, I. Wang, J.J. Liu, M. Jeong, G.J. Choi, Understanding the formation of indomethacin–saccharin cocrystals by anti-solvent crystallization, Cryst. Growth Des. 13 (2013) 2067–2074.
[209] L. Zhao, V. Raval, N.E.B. Briggs, R.M. Bhardwaj, T. McGlone, I.D.H. Oswald, A.J. Florence, From discovery to scale-up: α-lipoic acid : nicotinamide co-crystals in a continuous oscillatory baffled crystallizer, CrystEngComm 16 (2014) 5769–5780.
[210] K.A. Powell, G. Bartolini, K.E. Wittering, A.N. Saleemi, C.C. Wilson, C.D. Rielly, Z.K. Nagy, Toward continuous crystallization of urea-barbituric acid: a polymorphic co-crystal system, Cryst. Growth Des. 15 (2015) 4821–4836.
[211] W.T.A. Harrision, H.S. Yathirajan, S. Bindya, H.G. Anilkumar, Devaraju, Escitalopram oxalate: co-existence of oxalate dianions and oxalic acid molecules in the same crystal, Acta Crystallogr. C 63 (2007) o129–o131.
[212] S. Karki, T. Friščić, W. Jones, W.D.S. Motherwell, Screening for pharmaceutical cocrystal hydrates via neat and liquid-assisted grinding, Mol. Pharmaceutics 4 (2007) 347–354.
[213] D.P. Kale, S.S. Zode, A.K. Bansal, Challenges in translational development of pharmaceutical cocrystals, J Pharm. Sci 106 (2017) 457–470.
[214] N. Gascon, C. Almansa, M. Merlos, J.M. Vela, G. Encina, A. Morte, K. Smith, C. Plata-Salamán, Co-crystal of tramadol-celecoxib: preclinical and clinical evaluation of a novel analgesic, Expert Opin. Invest. Drugs 28 (2019) 399–409.
[215] C. Almansa, R. Merce, N. Tesson, J. Farran, J. Tomas, C.R. Plata-Salaman, Co-crystal of tramadol Hydrochloride–Celecoxib (ctc): a novel API– API co-crystal for the treatment of pain, Cryst. Growth Des. 17 (2017) 1884–1892.
[216] D. Bernhardson, T.A. Brandt, C.A. Hulford, R.S. Lehner, B.R. Preston, K. Price, et al., Development of an early-phase bulk enabling route to sodium-dependent glucose cotransporter 2 inhibitor ertugliflozin, Chem. Commun. 52 (2016) 640–655.
[217] C.K. Chung, P.G. Bulger, B. Kosjek, K.M. Belyk, N. Rivera, M.E. Scott, G.R. Humphrey, J. Limanto, D.C. Bachert, K.M. Emerson, Process development of C–N cross-coupling and enantioselective biocatalytic reactions for the asymmetric synthesis of niraparib, *Org. Process Res. Dev.* 18 (2014) 215–227.
[218] K. McKeage, Trelagliptin: first global approval, Drugs 75 (2015) 1161–1164.
[219] R.M. Poole, R.T. Dungo, Ipragliflozin: first global approval, Drugs 74 (2014) 611–617.
[220] P.M. Bhatt, T.S. Thakur, G.R. Desiraju, Co-crystals of the anti-HIV drugs lamivudine and zidovudine, Cryst. Growth Des. 9 (2009) 951–957.
[221] https://www.prnewswire.com/news-releases/vistagen-and-nuformix-announce-agreement-to-develop-novel-patentable-cocrystalline-forms-of-av-101-for-treatment-of-multiple-cns-conditions-301065803.html.
[222] A.K. Nangia, G.R. Desiraju, Crystal engineering. An outlook for the future, Angew. Chem. Int. Ed. 58 (2019) 4100–4107.

CHAPTER

From molecular electrostatic potential surfaces to practical avenues for directed assembly of organic and metal-containing crystalline materials

Christer B. Aakeröy[1], Marijana Đaković[2]

[1]*Department of Chemistry, Kansas State University, Manhattan, KS, United States;* [2]*Department of Chemistry, Faculty of Science, University of Zagreb, Zagreb, Croatia*

1. Introduction

From discrete and dispersed, to condensed and organized, intermolecular interactions provide the links between molecular structure and bulk properties of crystalline materials [1]. They also hold the key to effective bottom-up approaches toward design and synthesis of both single-component and multicomponent organic solids, as well as of coordination compounds with desirable structural features. The deceptively simple act of molecular recognition creates the means whereby different discrete chemical components recognize, bind, and aggregate, and such events are controlled by a delicate balance and competition between intermolecular forces that inhabit a broad range of energy scales. The primary challenge hampering reliable synthesis of materials with precise structural metrics and desired chemical composition is the shortage of transferable guidelines for preplanned crystal engineering based on noncovalent interactions.

The term *crystal engineering* [2] was employed by G.M. Schmidt in the late 1960s [3] in the context of the design and control of photochemical dimerization reactions in the solid state. Some 20 years later, the seminal book on crystal engineering by G.R. Desiraju served as a focal point and inspiration for a large number of studies, comprising theory and experiment, fundamental, and applied sciences, which helped to transform this field into a well-established and highly interdisciplinary scientific discipline. The introduction of the concept (and terminology) of supramolecular synthons in 1995 [4] further illustrates how structural chemistry and the availability of reliable and curated experimental data in the Cambridge

Hot Topics in Crystal Engineering. https://doi.org/10.1016/B978-0-12-818192-8.00003-2

Structural Database [5] have provided a foundation for the development of supramolecular synthetic tools that in some cases begin to approach organic named reactions in terms of reliability [6–9].

To transition crystal engineering from the assembly of specific structural features [10–12] to an ability to deliver new practical applications by dialling-in or fine-tuning physical properties [13–17], the completion of two crucial steps will be required [18]. These involve the delivery of (1) robust and reliable supramolecular synthetic guidelines for producing a crystalline solid with a targeted topology and (2) precise and well-defined correlations between a topology/crystal structure and property or performance of a solid.

Although the supramolecular synthon concept arose primarily from observed trends of structural behavior of functional groups capable of forming hydrogen bonds, the same analysis can readily be applied to noncovalent interactions that are typically brought together under the umbrella of σ-hole interactions. Consequently, the effective and versatile crystal engineering of the future will also rely on the considered and deliberate use of halogen bonds, chalcogen bonds, and pnictogen bonds.

The area of crystal engineering, broadly defined, has historically been dominated by the structural chemistry and reactivity of purely organic solid-state systems, whereas their inorganic counterparts have received substantially less attention. As a result, molecular-recognition preferences and structural trends as displayed by functional groups such as oximes, carboxylic acids, and amides have been classified in some detail, especially in neutral single-component and multicomponent ("cocrystal") systems [19–25]. Such systematic structural studies have allowed the identification of robust synthons that form the center piece of many successful strategies of supramolecular synthesis. These synthetic vectors have now also begun to find their way into inorganic crystal engineering, especially in the assembly of various 0-D as well as infinite 1-D and 2-D structural constructs in metal-containing systems [26–35].

There is an added incentive for applying the full arsenal of synthon-driven supramolecular synthesis to inorganic and metal–organic systems because metal ions can facilitate access to important advances in, e.g., catalysis, magnetism, photophysics, electronics [36], and mechanically responsive materials [37]. From a supramolecular synthetic perspective, metal–organic systems are often more difficult to control when compared with their purely organic correspondents. They frequently present building units with a considerable degree of structural complexity as well as multiple acceptor sites for noncovalent bonding. At the same time, many counterions are structurally isotropic, which hampers their use for imparting directional control over the assembly process. Furthermore, the anions that are needed for balancing the charge on metal cations may compete with and subsequently disrupt the intended (self-) complementary hydrogen/halogen bond–based synthons that comprise a particular design strategy.

At this point, it should be clear that there are numerous challenges associated with forging reversible (and often competing) noncovalent interactions into effective strategies for predictable and reliable supramolecular synthesis. With this in mind, it is important to identify tools that can offer practical guidelines for robust and versatile crystal engineering using a combination of hydrogen bonds and a variety of σ-hole interactions. Against this background we will focus, in this chapter, on the use of calculated molecular electrostatic potentials (MEPs) for developing synthetic strategies and for rationalizing observed structural features in both organic and metal-containing solids. A simple electrostatic picture of intermolecular chemical bonds is available by focusing on classical $\delta^{+}\cdots\delta^{-}$-driven interactions since the partial charges are visible by examining the unperturbed MEP mapped out on the isodensity surfaces of the isolated monomers. In this chapter, we hope to demonstrate that calculated MEPs can provide simple and generally understandable qualitative descriptions of the structural consequences of supramolecular synthons that can be of practical benefits to crystal engineering.

2. Using molecular electrostatic potentials for evaluating the balance between intermolecular forces

Electron density and the electrostatic potential are both experimentally observable quantities, and it makes intuitive sense to use such molecular features as a basis for describing and rationalizing the structural consequences of molecular recognition events.

The correlations between MEPs mapped onto molecular surfaces and experimentally determined thermochemical properties have been systematically evaluated and established by Politzer and Murray [38, 39]. The use of MEPs has also played an integral part in the development of concepts such as σ-hole and π-hole, which have been very useful for identifying and classifying intermolecular interactions between closed-shell systems [40–42]. Despite the relative simplicity of this electrostatic view of noncovalent bonding, calculated MEPs have provided considerable insight into the structure and properties of molecules and molecular aggregates [43–47] in a diversity of systems, from small molecules [48–50] to large molecular complexes [51]. Finally, MEPs have been used as a computational tool in rational drug design starting in the early 1980s [52], but despite this progress, it was not until quite recently that MEPs found more widespread use in rationalization supramolecular interactions and for developing versatile and practical crystal engineering strategies.

The electrostatic potential, which can be determined either experimentally (by diffraction techniques) or computationally, is essentially an application of Coulomb's law, where the electrostatic force ($\boldsymbol{F}$) is given by

$$\boldsymbol{F}=\frac{1}{4\pi\varepsilon_o}\frac{q_1\,q_2}{r^2}\boldsymbol{i} \tag{6.1}$$

When this is applied to molecules, a collection of point charges (nuclei) and electrons (not stationary, which average number in each volume element $d\boldsymbol{r}$ is given by the electron density function $\rho(\boldsymbol{r})$), the electrostatic potential at a point $\boldsymbol{r}$ is given as

$$V(r) = \sum_A \frac{Z_A}{|\boldsymbol{R}_A - \boldsymbol{r}|} - \int \frac{\rho(\boldsymbol{r}')d\boldsymbol{r}'}{|\boldsymbol{r}' - \boldsymbol{r}|} \tag{6.2}$$

with Z_A presenting the charge on nucleus A located at $\boldsymbol{R}_A$, with $\boldsymbol{R}_A - \boldsymbol{r}$ and $\boldsymbol{r}' - \boldsymbol{r}$ being the distances between $\boldsymbol{r}$ and the nucleus or each electronic charge increment $\rho(\boldsymbol{r}')d\boldsymbol{r}'$, respectively.

The relative magnitudes of the electrostatic potential around a discrete molecule can be displayed in a variety of ways (at a specific point in space ($\boldsymbol{r}$), along some axis through the molecule, in a two-dimensional space, etc.), but it is often most convenient to visualize it by mapping its calculated values, $V(\boldsymbol{r})$, on a molecular surface defined by an outer contour of electron density [53]. An isoelectronic density surface has advantages over other molecular surfaces (e.g. a van der Waals surface) as it can clarify and highlight specific molecular features such as lone pairs and π electrons, which are of key importance when evaluating intermolecular interactions. The calculated electrostatic potentials thus provide the Coulombic interaction energies (expressed in kJ/mol) between a positive point probe and the surface of the molecule at that particular point. Positive and negative values for the interaction energy indicate a positive and negative surface potential, respectively; locally the most positive surface values correspond to electron-deficient zones and can be associated with, e.g., hydrogen or halogen bond donor sites, whereas negative values reflect regions with pronounced electron density that are subsequently associated with the corresponding bond accepting sites.

Hunter and coworkers have demonstrated that MEP surface values can be effectively employed for ranking the relative hydrogen bond donor and acceptor strengths of different molecular entities [54]. It was established that experimental thermochemical data on hydrogen bond interactions of simple molecules in carbon tetrachloride [55,56] could be extrapolated to a broad range of chemical functionalities in any solvent as well as to the solid state. This model relies on the electrostatic contributions to intermolecular interactions at various single points on the molecular surface, and despite its relative simplicity, it has proved to be a very practical and useful instrument for dissecting a broad range of challenges related to understanding the balance between competing noncovalent interactions both in solution and in the solid state. Moreover, the fact that calculations can be performed at a relatively low level of theory makes the MEP a versatile and readily accessible tool for practical crystal engineering. Solid-state structures comprising both hydrogen and halogen bonds as key synthetic vectors have been treated using Hunter's approach, and although it was originally applied to purely organic systems, it has recently been extended to metal-containing solids (see Chapter 6 and references therein).

3. Using molecular electrostatic potentials for analyzing key structural features in the solid state

3.1 Molecular electrostatic potential in hydrogen-bonded systems

An early example of some informative use of MEPs for crystal engineering purposes was a survey of the Cambridge Crystallographic Database (CSD) focusing on the geometry of hydrogen bonds involving metal-bound halogens as hydrogen bond acceptors [57]. The survey clearly demonstrated that metal-bound fluorine (M–F) exhibits substantially different behavior from that of its heavier congeners. Notably, M–F moieties generally participate in markedly shorter hydrogen bonds with substantially different directionality, i.e., $\angle(\text{M–F}\cdots\text{H}) = 120–160$ degrees, compared with the other halogen atoms $\angle(\text{M–X}\cdots\text{H}) = 90–130$ degrees ($X = \text{Cl, Br, I}$). This behavior was rationalized by analyzing the calculated MEPs mapped out on the surface around the metal-bound halogen atom of the model molecules ($PdX(Me)(PH_3)_2$, X = F, Cl, Br, I). The results clearly supported an explanation based upon electrostatic factors, as opposed to steric influences, for the experimentally observed geometric features. The overall order of the hydrogen bond strength ($\text{M–F}\cdots\text{H} >> \text{M–Cl}\cdots\text{H} > \text{M–Br}\cdots\text{H} > \text{M–I}\cdots\text{H}$), as indirectly implied by the $\text{H}\cdots\text{X}$ distances (retrieved from the CSD), was consistent with the relative magnitudes of the potential minima calculated on the halogen atoms in the model complexes. In addition, the observed angles were in excellent agreement with the anisotropy observed in the MEPs around the halide atoms bound to metal ions (Fig. 6.1).

Another example that nicely demonstrates the benefits of calculated MEPs in crystal engineering leads to the identification of new hydrogen-bonding synthons in inorganic systems [58]. It was found that hydrogen bond donors displayed a pronounced preference for approaching the metal-bound chlorides almost orthogonally

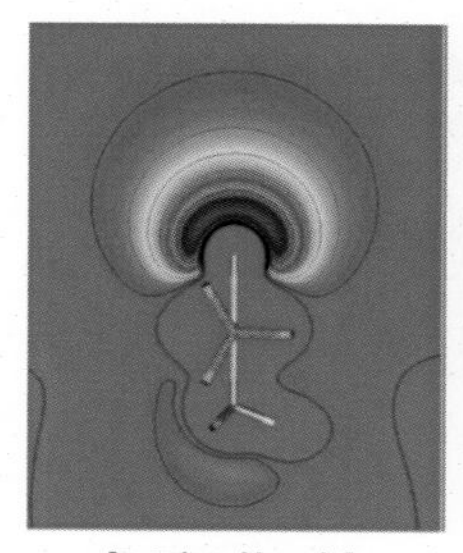

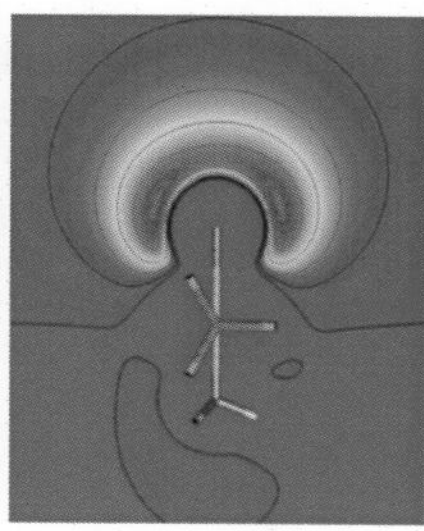

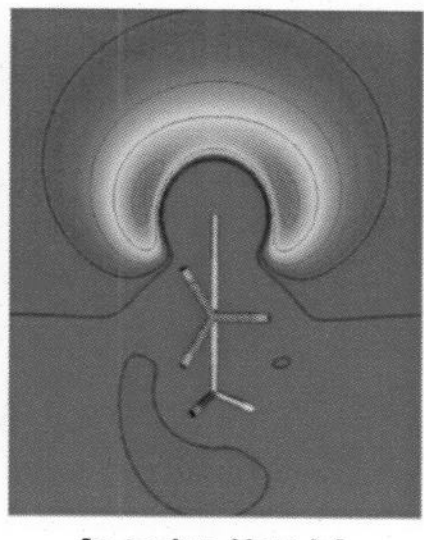

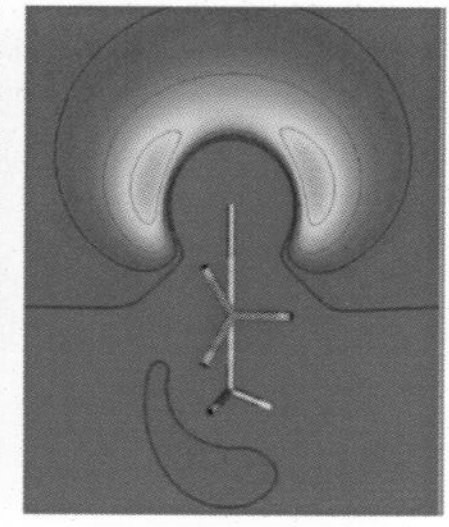

FIGURE 6.1

Calculated electrostatic potentials in the metal coordination plane of the model compounds *trans*-$[PdX(Me)(PH_3)_2]$, X = F, Cl, Br, I.

With permission from L. Brammer, E.A. Bruton, P. Sherwood, Fluoride ligands exhibit marked departures from the hydrogen bond acceptor behaviour of their heavier halogen congeners, New J. Chem. 23 (1999) 965–968.

to the M—Cl bond. By way of an explanation for this behavior, it was proposed that when metal-bound chlorides are located close to each other (*cis*-positioned in square-planar, tetrahedral, and octahedral coordination geometry, or *fac* in octahedral complexes), significant synergy and cooperativity might arise whereby the MEP minima located on neighboring metal-bound chlorides could reinforce each other. Here again, calculated MEPs mapped onto the isodensity surfaces of model complexes were used to locate the electrostatic potential minima. For *cis*-square-planar complex, a distinct electrostatic potential minimum located between neighboring chloride ligands can be viewed as an effective "binding pocket." A similar structural feature was noted for each pair-wise combination of the three chloride ligands in the *fac*-octahedral complex. In contrast to the binding pockets found in the *cis*-square-planar complexes, three global minima (located between each pair of neighboring chlorides) in octahedral complexes resulted in a larger region of minimum potential energy located between them.

The observation formed the basis for the subsequent successful implementation of the synthesis of targeted supramolecular systems constructed from cationic hydrogen bond donors and anionic perhallometalate anions as hydrogen bond acceptors in a comprehensive study combining theory and experiments. The CSD survey was carried out to identify any close contacts between N—H groups and three different chlorometallate anions, planar $[MCl_4]^{n-}$, tetrahedral $[MCl_4]^{n-}$, and octahedral $[MCl_6]^{n-}$, and summarized in the 3-D contour maps (Fig. 6.2:Ia-b, IIa-b, IIIa-b, respectively), while MEPs were calculated for representative anions ($[PdCl_4]^{2-}$, $[PdCl_6]^{2-}$, and $[CdCl_4]^{2-}$, Fig. 6.2:Ic-d and IIc-d, respectively). Both structural and computational data led to consistent and complementary conclusions (Fig. 6.2) and reliably predicted structural outcomes.

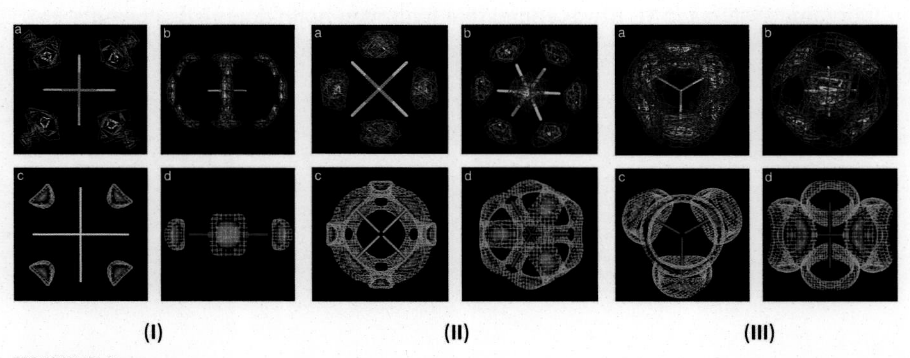

FIGURE 6.2

Population density of hydrogen atoms from N—H groups in the vicinity of square-planar $[MCl_4]^{n-}$ (Ia-b), octahedral $[MCl_6]^{n-}$ (IIa-b), and tetrahedral $[MCl_4]^{n-}$ (IIIa-b) anions obtained from the CSD. Calculated negative electrostatic potential in the vicinity of the $[PdCl_4]^{2-}$ (Ic-d), $[PdCl_6]^{2-}$ (IIc-d), $[CdCl_4]^{2-}$ (IIIc-d).

With permission from L. Brammmer, J.K. Swearingen, E.A. Bruton, P. Sherwood, Hydrogen bonding and perhalometallate ions: a supramolecular synthetic strategy for new inorganic materials, Proc. Natl. Acad. Sci. U. S. A. 99 (2002) 4956–4961.

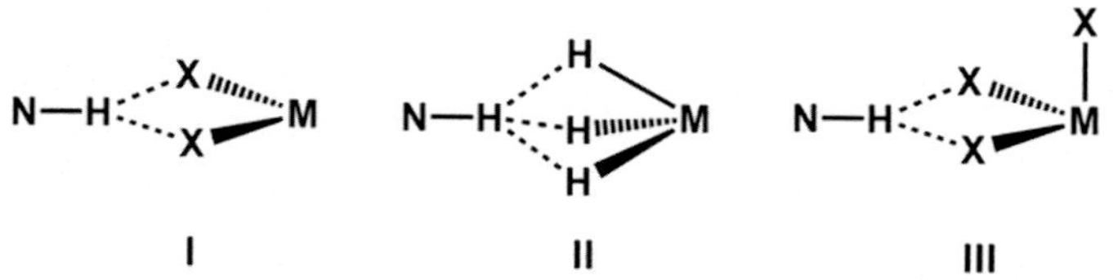

FIGURE 6.3

Three different synthons (I—III) comprising N—H hydrogen bonds to metal—halide (M—X) acceptor groups.

Modified from L. Brammmer, J.K. Swearingen, E.A. Bruton, P. Sherwood, Hydrogen bonding and perhalometallate ions: a supramolecular synthetic strategy for new inorganic materials, Proc. Natl. Acad. Sci. U. S. A. 99 (2002) 4956–4961.

The presence of these electrostatically generated hydrogen bond acceptor sites in *cis*-square-planar and *fac*-octahedral complexes gives rise to three plausible hydrogen-bonded motifs (**I—III**), Fig. 6.3, as supported by both theory and structural data.

For a tetrahedral geometry, where the angle between chloride ligands (109.47 degrees) being markedly larger than that in square-planar and octahedral geometry (90 degrees), the cooperativity can be expected to be less pronounced, and both experimental and theoretical data clearly show that a trifurcated hydrogen bond interaction is not favored. Instead, in MEP maps, two local minima were found on each of the six edges of a tetrahedron (Fig. 6.2:IIIa-b). A close inspection of relevant crystal structures shows that many of the N—H···X hydrogen bonds are actually bifurcated with considerable geometric asymmetry involving one short and one substantially longer contact from the hydrogen bond donor to neighboring halide acceptors, thus favoring the synthon **III** as the preferred choice.

Similar structurally important "binding pockets" created by adjacent regions of negative electrostatic potentials have also been identified in a number of octahedral $[M(\beta\text{-dik})_2L_2]$ [59], and square-pyramidal $[M(\beta\text{-dik})_2L]$ [60], β-diketonato complexes (β-dik = pentane-2,4-dion (*acac*), 1,1′,1″-trifluoropentane-2,4-dion (*tfac*), 1,1′,1″,6,6′,6″-hexafluoropentane-2,4-dion (*hfac*)). Typically, two oxygen atoms (from two coordinated β-diketonato ligands) positioned close to each other combine to offer the most negative region on the MEP surface located in the space between them. This cooperative feature leads to an electrostatically based acceptor site, which promotes the formation of bifurcated hydrogen bonds (Fig. 6.4).

3.2 Molecular electrostatic potential in halogen-bonded systems: bifurcated halogen bonds

The presence and nature of bifurcated halogen bonds in both purely organic and metal—organic systems was explored and rationalized with the help of calculated MEP values mapped onto isodensity surfaces of unperturbed individual molecules and coordination complexes.

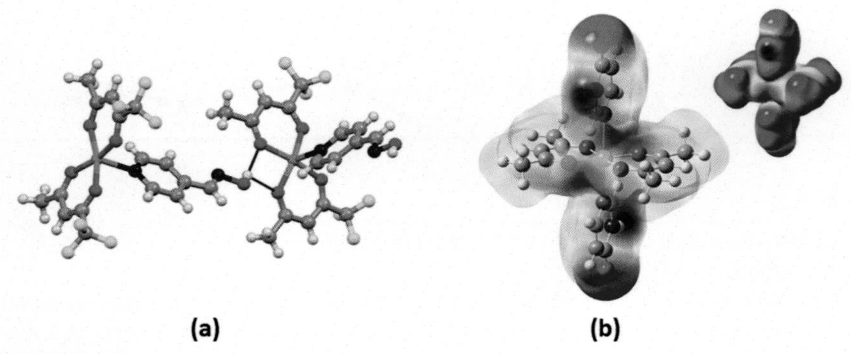

FIGURE 6.4

(A) Bifurcated O—H···O hydrogen bonds to the pocket between two *tfac* oxygen atoms in the crystal structure of *acac*-based complex of copper(II). (B) Calculated MEP surface highlighting the electrostatic nature of the acceptor site. *MEP*, molecular electrostatic potential.

(A) Modified from I. Kodrin, M. Borovina, L. Šmital, J. Valdés-Martínez, C.B. Aakeröy, M. Đaković, Exploring and predicting intermolecular binding preferences in crystalline Cu(II) coordination complexes, Dalton Trans. 48 (2019) 16222–16232. (B) Modified from M. Borovina, I. Kodrin, M. Đaković, Predicting supramolecular connectivity of metal-containing solid-state assemblies using calculated molecular electrostatic potential surfaces, Cryst. Growth Des. 19 (2019) 1985–1995.

The iodo···nitro (I···O_2N) interaction was one of the first examples of a synthon involving a bifurcated halogen bond, and interestingly, it displays close structural analogies with the bifurcated ethynyl···nitro (C≡C—H···O_2N) hydrogen bond. Although the significance of the two motifs in crystal engineering had been highlighted earlier [61–64], a description of the halogen···nitro synthon in the context of its electronic effects was provided in 2005 by Nangia and Jaskólski [65]. They showed that the driving force for the X···O_2N interaction is electrostatic in nature, and this was evident because of the orientation of the electropositive region of the halogen atom (positive lobe, depicted in blue) with respect to the electronegative pocket created by the "lone pairs" of the two oxygen atoms in the nitro group. It was also shown that the differences in bond strengths between the I···O_2N and Cl···O_2N interactions result from significant differences in the relative magnitudes of the electropositive regions of the respective halogen atoms (Fig. 6.5) – a moderately positive σ-hole on chlorine results in a much weaker bonding compared with that of iodine.

Following the iodo···nitro example, a bifurcated halogen bond was subsequently detected in a variety of other systems. An interesting illustration was the bifurcated halogen bond involving heteroatoms of extended aromatic systems (*o*-phenanthroline) as halogen bond acceptors that was reported in the solid state (Fig. 6.6A) [66] and fully examined through different theoretical methods a decade thereafter (Fig. 6.6B) [67,68]. It is noteworthy that in all instances, regardless of the level of theory [67] or experiment [66], a bifurcated halogen bond was likely to display pronounced asymmetry (with one contact shorter and more linear and the other one longer and more inclined).

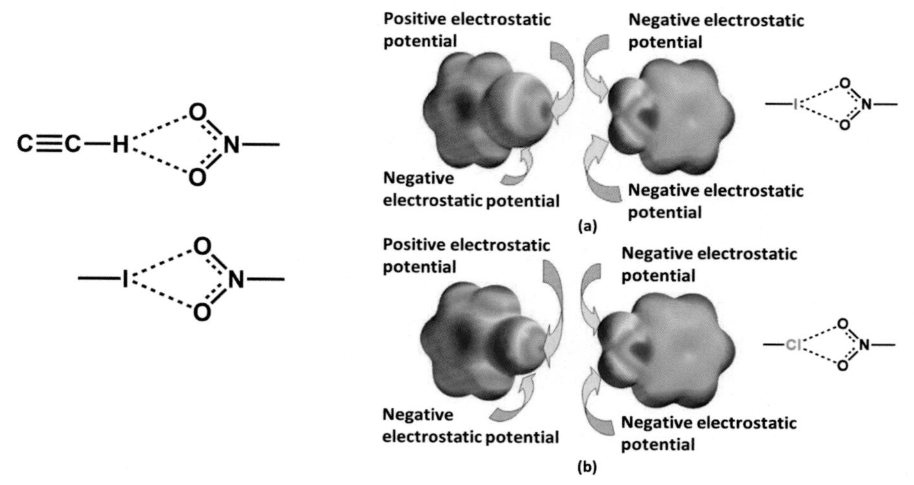

FIGURE 6.5

(A) Analogy between hydrogen- and halogen-bonded synthons involving the nitro group, ethynyl···nitro and iodo···nitro; (B) calculated MEPs showing the electropositive region of the halogen atom (blue tips at the extension of the C—X bond; iodine (upper) and chlorine (lower)) oriented toward the electronegative "pocket" formed by two oxygen atoms of the nitro group.

A part of the figure with permission from B.K. Saha, A. Nangia, M. Jaskólski, Crystal Engineering with hydrogen bonds and halogen bonds, CrystEngComm 7 (2005) 355–358.

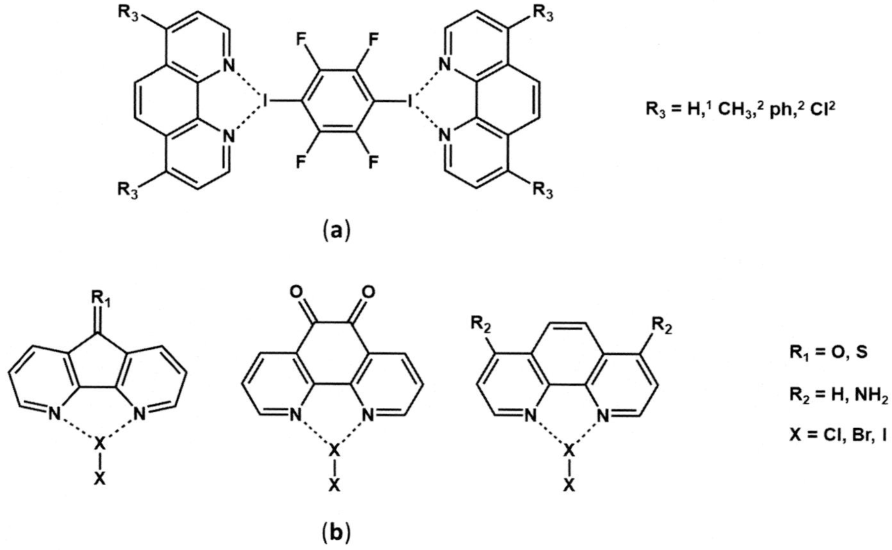

R_3 = H,[1] CH_3,[2] ph,[2] Cl[2]

(a)

R_1 = O, S

R_2 = H, NH_2

X = Cl, Br, I

(b)

FIGURE 6.6

Bifurcated halogen bonds: (A) observed experimentally, and (B) explored via a variety of theoretical methods.

(A) From R. Liu, Y.J. Gao, W.J. Jin, Colour-tunable phosphorescence of 1,10-phenanthrolines by 4,7-methyl/-diphenyl/-dichloro substituents in cocrystals assembled via bifurcated C–I···N halogen bonds using 1,4-diiodotetrafluorobenzene as a bonding donor, Acta Crystallogr. B73 (2017) 247–254. (B) From E. Bartashevich, E. Troitskaya, Á.M. Pendás, V. Tsirelson, Understanding the bifurcated halogen bonding N···Hal···N in bidentate diazaheterocyclic compounds, Comput. Theor. Chem. 1053 (2015) 229–237.

To engineer symmetrically bifurcated halogen bonds, Ji and coworkers built their synthetic strategy on the electrostatic arguments only [69] that (1) it is the electrostatic character that favors bifurcation (especially its symmetrical mode) and (2) it is the anisotropic charge distribution on halogen atom (a *relatively small* positive region in the extension of the σ-bond) that drives the bond toward linearity. This suggests that for producing a symmetrically bifurcated halogen bond, two acceptor sites residing on the acceptor molecule must be located as close as possible to each other.

Therefore, they employed calculated MEP surfaces as guidance for selecting among four potential acceptor molecules (Fig. 6.7). The MEP clearly showed that

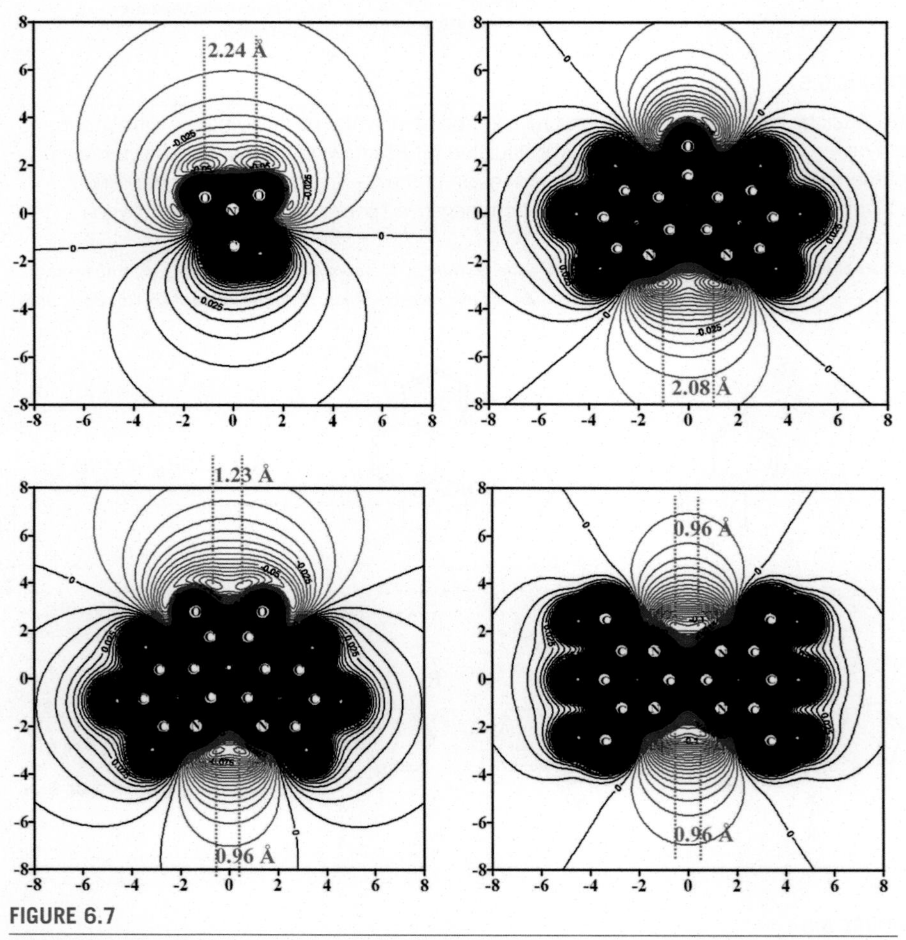

FIGURE 6.7

MEP surface maps of nitromethane, 3,5-diazafluoren-9-one (DAFONE), 1,10-phenanthroline-5,6-dione (PDONE), and 4,4′,6,6′-tetramethyl-2,2′-bipyrimidine (TMBPM) presented in their respective molecular planes (positive regions presented in black, negative in blue; the contour interval is 0.005 a.u.). Numbers in red represent the distance between two local most negative electrostatic potentials. *MEP*, molecular electrostatic potential.

With permission from B. Ji, W. Wang, D. Deng, Y. Zhang, Symmetrical bifurcated halogen bond: design and synthesis, Cryst. Growth Des. 11(2011) 3622–3628.

1,10-phenanthroline-5,6-dione (PDONE) and 4,4′,6,6′-tetramethyl-2,2′-bipyrimidine (TMBPM) display substantially shorter distances between two sites displaying the most negative electrostatic potentials (less than 1.5 Å) than is the case in nitromethane and 3,5-diazafluoren-9-one (DAFONE) (more than 2.0 Å). Subsequent cocrystallizations and structure elucidations did, in fact, reveal symmetrically bifurcated halogen bonds in the PDONE:1,4-DITFB (1,4-diiodotetrafluorobenzene) and TMBPM:1,4-DITFB cocrystals, while conventional single halogen bonds linked the constituent molecules in the DAFONE:1,4-DITFB cocrystal (Fig. 6.8).

Recently, the same synthon was employed for synthesizing halogen-bonded cocrystals with color-tunable phosphorescence (Fig. 6.6A) [70]. In tailoring the experiment, the authors again used MEP surfaces to identify two different types of electrostatically positive regions on the 1,4-DITFB molecule, described as a σ-hole and π-hole, respectively (Fig. 6.9). The most negative region on the MEP surfaces of potential acceptor molecules was successfully used as a basis for predicting the most likely types of intermolecular interactions in the solid state.

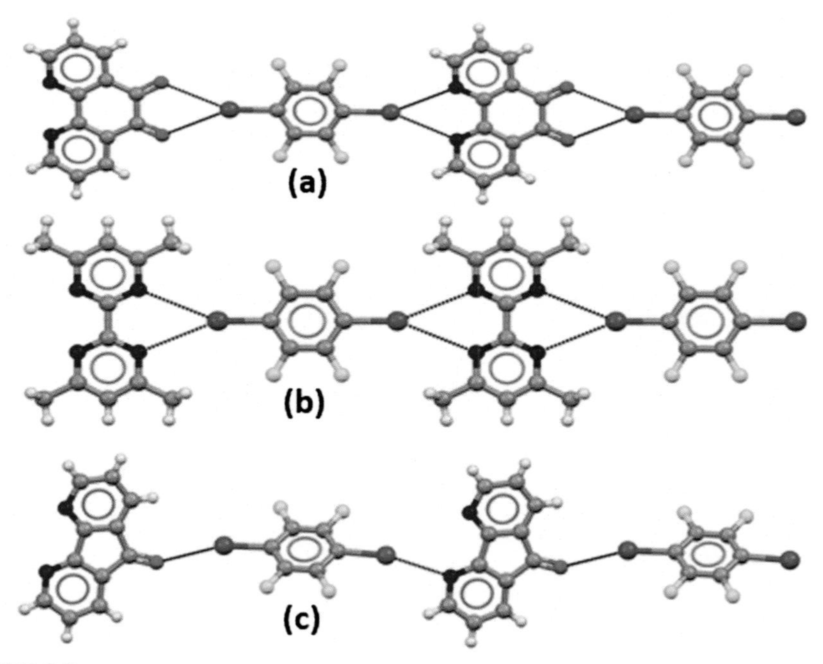

FIGURE 6.8

Symmetrically bifurcated halogen bonds in the (A) PDONE:1,4-DITFB and (B) TMBPM:1,4-DITFB cocrystals tailored on the distances between two potential acceptor sites as revealed by calculated MEP surfaces (Fig. 6.7), and a conventional single halogen bond in the DAFONE:1,4-DITFB cocrystal (C). *1,4-DITFB,* 1,4-diiodotetrafluorobenzene; *DAFONE,* 3,5-diazafluoren-9-one; *MEP,* molecular electrostatic potential; *PDONE,* 1,10-phenanthroline-5,6-dione.

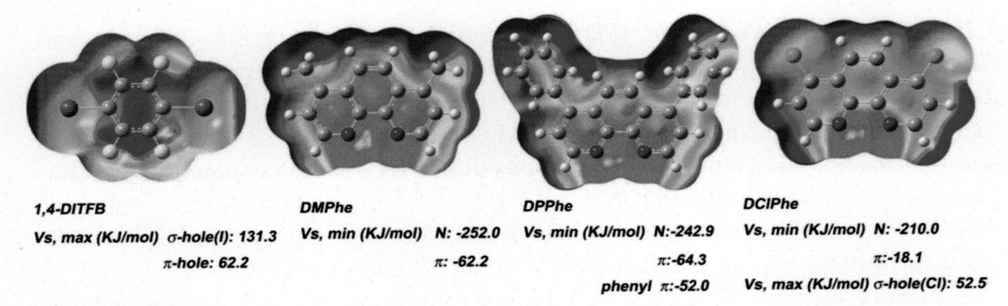

FIGURE 6.9

Calculated MEP surfaces for 1,4-DITFB and potential acceptor molecules mapped on the isodensity surface. *1,4-DITFB*, 1,4-diiodotetrafluorobenzene; *MEP*, molecular electrostatic potential.

With permission from R. Liu, Y.J. Gao, W.J. Jin, Colour-tunable phosphorescence of 1,10-phenanthrolines by 4,7-methyl/-diphenyl/-dichloro substituents in cocrystals assembled via bifurcated C–I···N halogen bonds using 1,4-diiodotetrafluorobenzene as a bonding donor, Acta Crystallogr. B73 (2017) 247–254.

Bifurcated halogen bonds have also been observed in metal–organic systems, mostly in β-diketonato complexes involving divalent metal centers (M = Co, Ni, Cu, Pd, Pt) and a variety of pentane-2,4-one derivatives [71–74].

Even though the two different acceptor atoms are closer together than the relevant acceptor atoms in organic systems (which typically result in symmetrically bifurcated halogen bonds), the majority of bifurcated halogen bonds in metal–organic systems are asymmetric (Fig. 6.10). The only symmetric structures were observed in cocrystals of square-planar complexes of [Cu(*acac*)$_2$] with 1,4-DITFB and 1,4-DBTFB [74] and [Pd(*acac*)$_2$] with INPN (INPN = 4-iodo-5-nitrophthalonitrile) [75]. All other cases of cocrystals or single-component discrete metal complexes presented asymmetrically bifurcated halogen bonds. The difference between symmetrical and asymmetrical bifurcation was nicely depicted by calculated MEP surfaces [74].

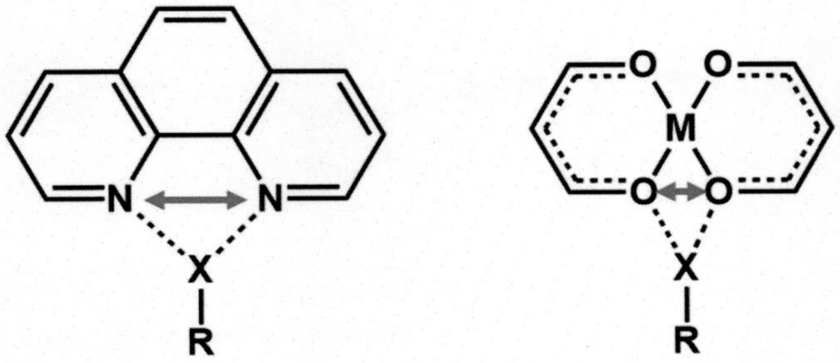

FIGURE 6.10

Comparison of distances between two respective halogen bond acceptor sites in organic and metal–organic settings that produce bifurcated halogen bonds.

4. From molecular electrostatic potentials to interaction energies

The use of electrostatic potentials for ranking which molecular recognition events are most likely to take place in competitive system is helpful for devising supramolecular synthetic strategies. However, being able to convert those potentials into interaction energies for a wide range of synthons would offer additional value to the use of MEPs in crystal engineering. Hunter and coworkers [76] have proposed protocols for achieving this transformation based on a hierarchical view of hydrogen bond interactions (i.e., the best hydrogen bond donor binds to the best acceptor and forms the strongest interaction, while the second-best donor interacts with the second-best acceptor, and so on until all the interaction sites are exhausted). By combining all local minima and maxima on the MEP surface of a single unperturbed component, a set of hydrogen bond interaction sites can be identified that completely describes the interactions that a molecule can make with its nearest neighbors in the solid state. The method works on the assumption that all hydrogen bond donors/acceptors are independent of each other and capable of finding their ideal partner in the crystal, without considering steric effects, "packing forces," or possible cooperativity between interactions sites that are close to each other.

Two parameters α_i and β_i (describing hydrogen bond donors and hydrogen bond acceptors, respectively) were used to calculate interaction site pairing energies (where the highest α_i interacts with the highest β_i, the second-highest α_i with the second-highest β_i, etc.) of the solid:

$$E = -\sum_i \alpha_i \beta_i$$

The calculated MEPs are needed to obtain hydrogen bond parameters using the following equations:

$$\alpha = 0.0000162\, MEP_{\max}^2 + 0.00962\, MEP_{\max}$$

$$\beta = 0.000146\, MEP_{\min}^2 - 0.00930\, MEP_{\min}$$

where $MEP_{\max}$ and $MEP_{\min}$ are local maxima and minima on the MEP (0.002 electron Å^{-3}) electron isodensity surface (in kJ/mol).

This approach is versatile and can be applied to multicomponent as well as to single-component solids. In the case of a cocrystal, the hydrogen bond donor and acceptor parameters (α_i and β_i, respectively) of both components are combined into a single parameter list, and the highest α_i is then paired with the highest β_i, the second-highest α_i with the second-highest β_i, and so on until no more pairwise interactions are possible. Cocrystal stoichiometries different from 1:1 can be also handled by adding the appropriate number of α_i and β_i values to the combined single parameter list. The probability of cocrystal formation is then estimated by examining the difference in interaction site pairing energies (ΔE) of the potential cocrystal (E_{cc}) and the two homomeric forms (E_1 and E_2):

$$\Delta E = E_{cc} - nE_1 - mE_2$$

where coefficients m and n define the cocrystal stoichiometry. The larger the ΔE value, the stronger the interaction energy between the two cocrystal components, and the more likely the formation of a cocrystal (Fig. 6.11).

This protocol is fast and easy to implement, and it is chemically intuitive and understandable, which means that it can offer important guidance for practical crystal engineering. The validity of the concept was tested by comparing the results from an in silico cocrystal screening of several active pharmaceutical ingredients (APIs) against a large number of potential coformers, with literature data on experimental cocrystal screens of APIs [77]. The results showed that the use of ΔE, albeit not perfect, provides a very useful guide for narrowing and refining the experimental space when designing a cocrystal screen.

A pairwise energy-based approach was also used for calculating hydrogen bond energies for predicting the likelihood of cocrystal formation of six multifunctional custom-designed thiazoles [78]. Cocrystal predictions based on calculated energies were compared with those obtained from structural informatics tools such as hydrogen propensity calculations [79–81], and the two methods were found to produce comparable results. A similar approach was utilized for rationalizing supramolecular connectivity and association energies in a non–hydrogen-bonded system, comprising aryl-functionalized 1,2,3,5,-dithiadiazoyl radicals [82]. The outcome provided valuable insight into the balance between several structural features, dimerization mode, torsion angles, and favorable electrostatic interactions. Furthermore, the results from the pairing-energy calculations were consistent with calculations on relevant experimentally obtained structures.

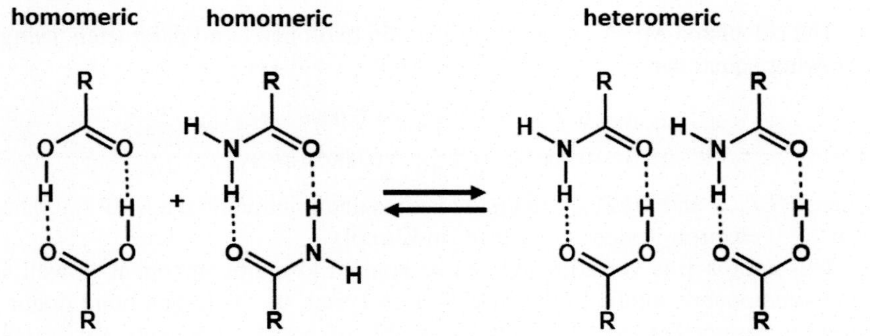

homomeric homomeric heteromeric

$2\alpha\beta = -2(3.5 \times 5.0) = -35$ kJ/mol

$2\alpha\beta = -2(2.9 \times 8.0) = -46$ kJ/mol

$\Sigma\alpha\beta = -2(3.5 \times 8.0 + 2.9 \times 5.0) = -85$ kJ/mol

FIGURE 6.11

The interaction site pairing energy for a carboxylic acid – amide cocrystal formation starting from the respective single components, carboxylic acid, and amide.

Modified from R.B. Walsh, C.W. Padgett, P. Metrangolo, G. Resnati, T.W. Hangs, W.T. Pennington, Crystal engineering through halogen bonding: complexes of nitrogen heterocycles with organic iodides, Cryst. Growth Des. 1 (2001) 167–175.

5. Using molecular electrostatic potential for driving supramolecular synthetic strategies in the organic solid state

5.1 An incentive for effective and predictable crystal engineering

To successfully combine discrete molecules into larger intermolecular aggregates of desirable structural complexity and dimensionalities, the chosen synthetic strategy should be strong enough to resist potentially interfering interactions. On the other hand, the assembly process should not be too narrow in range, as this would limit any versatility and potential responsiveness to external changes and stimuli. In short, structural singularity is undesirable if we want to promote functional diversity and adaptability in the resulting materials. To overcome these apparently contradictory requirements, it is important to have access to reliable strategies for encoding and manipulating structural subunits into higher-level molecular aggregates through appropriate interactional methodologies. The ability to (1) position chemical constructs exactly where we want them to be; (2) build multicomponent architectures with desirable metrics; and (3) translate intermolecular communication into blueprints for materials design and for constructing viable biological mimics represent highly significant long-term goals of interest to a wide range of scientists. To successfully realize the implicit advantages of directed assembly, we have to rely on the inherent information encoded onto individual components, as these features ultimately determine the interactions between them. In addition, the deliberate assembly of distinct chemical components into specific constructs of well-defined topology and spatial orientation spans all dimensions and length scales. The dramatic changes in length scale that are inevitable in bottom-up approaches to materials synthesis can only be harnessed with the help of a hierarchy of synthetic tools that operate in a synergistic manner with a minimum amount of mutual interference. Therefore, understanding, developing, and utilizing these tools in directed-assembly processes are critically important for the synthesis and engineering of multifunctional and tunable materials and devices.

5.2 Hydrogen bond–based strategies for organic crystal engineering

A particular challenge with noncovalent synthesis is that chemically different molecular entities (the reactants in the synthesis) are held together by reversible intermolecular interactions, and therefore, synthetic procedures have to take place via a one-pot process. A supramolecular "intermediate" can rarely be prepared and isolated, and then added to another reactant in a sequential protocol, and supramolecular synthesis of heteromeric aggregates is therefore essentially limited to one-pot reactions. To overcome this challenge, a hierarchical approach to noncovalent assembly is necessary, and this means that we need to simultaneously utilize and combine several different synthons side by side without structural interference and overlap [83]. Aakeröy and coworkers demonstrated that a ranking of hydrogen bond—based synthons using the pK_a/pK_b values of the functional groups residing on the participating molecules could be used as a basis for synthesis ternary cocrystals with considerable reliability and reproducibility (Fig. 6.12) [84].

However, this method of ranking is only appropriate when the synthetic drivers are chemically similar, i.e., a series of different carboxylic acids. A more broadly applicable protocol requires the use of MEPs for ranking the relative importance/strength of hydrogen bonds involving different functional groups. The contradictory results obtained by using pK_a/pK_b and MEPs, respectively, for ranking synthons is illustrated in the cocrystal of 4-hydroxybenzoic acid:isonicotinamide (Fig. 6.13) [85].

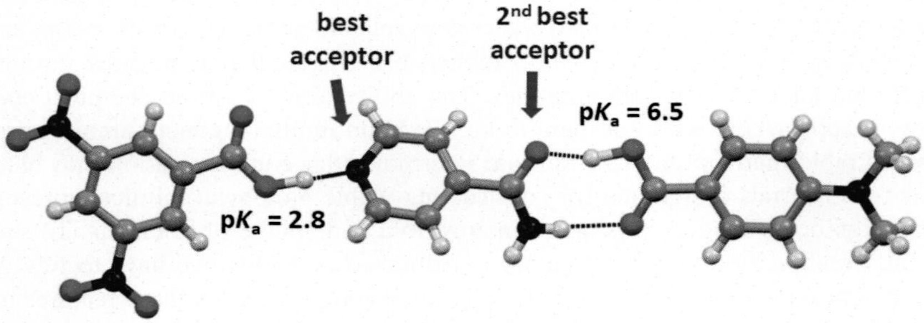

FIGURE 6.12

Ternary supramolecule constructed from two carboxylic acids and isonicotinamide. The stronger acid···pyridine interaction being the "best" hydrogen bond and the weaker acid···amide interaction being the "second-best" hydrogen bond (*refcode*: BUFBIP).

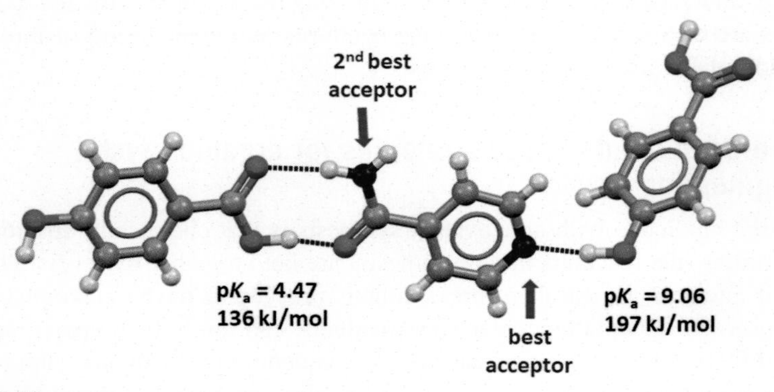

FIGURE 6.13

Comparison of MEPs and pK_a values for two most influential hydrogen bond donors (*refcode*: VAKTOR). *MEPs*, molecular electrostatic potentials.

Based on pK_a values for the two hydrogen bond donors (pK_a(−COOH) = 4.47; pK_a(−OH) = 9.06; Fig. 6.13), the expectation would be that the carboxylic acid (as the better donor) would bind with the pyridine nitrogen atom (the best acceptor), but this was not borne out by experiment. However, if the two moieties are ranked by calculated MEP values (197 kJ/mol and 136 kJ/mol for −OH and −COOH functionalities, respectively) [83], the best donor/best acceptor guideline is adhered to.

Numerous systematic structural studies have supported the usefulness and reliability of this synthetic design principle, and one illustrative example can be found in a series of cocrystallizations of biimidazole-based symmetric ditopic molecules with different aliphatic carboxylic diacids (Fig. 6.14) [86].

The calculated MEPs for the two different acceptor sites, imidazole nitrogen (N_{im}) and pyridine nitrogen (N_{py}), indicate that in all cases, N_{py} is the best acceptor (MEP range −153 kJ/mol to −187 kJ/mol), leaving the N_{im} as the second-best acceptor (MEP range: −125 kJ/mol to −132 kJ/mol). In all 12 crystal structures that were obtained in this study, the O−H···N hydrogen bond, which drives the cocrystal assembly, utilizes the N_{py} acceptor site only, which underscores that a simplified electrostatic view of hydrogen bonding can provide a practical tool for predicting selectivity of intermolecular interactions in a structurally competitive environment (Fig. 6.15).

A substantial body of new structural data [87] has lent support to the suggestion that, for systems with multiple potential structural outcomes, the dominant molecular recognition events can often be predicted correctly using Etter's "best donor–best acceptor" guidelines [88], where the ranking is based on calculated MEP values. It is important to note that these are guidelines formulated as a result of

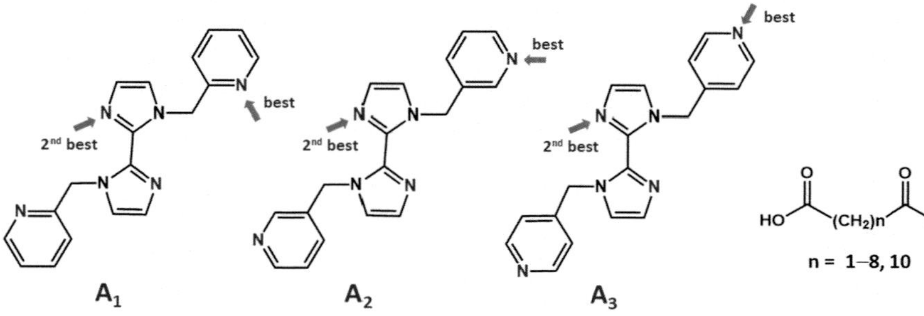

MEP /kJ mol^{-1}	A_1	A_2	A_3
N_{py}	−153	−182	−187
N_{im}	−125	−128	−132

FIGURE 6.14

Hydrogen bond acceptors and hydrogen bond donors in cocrystallizations intended to test an MEP-based ranking for developing into reliable supramolecular synthetic strategies. *MEP*, molecular electrostatic potential.

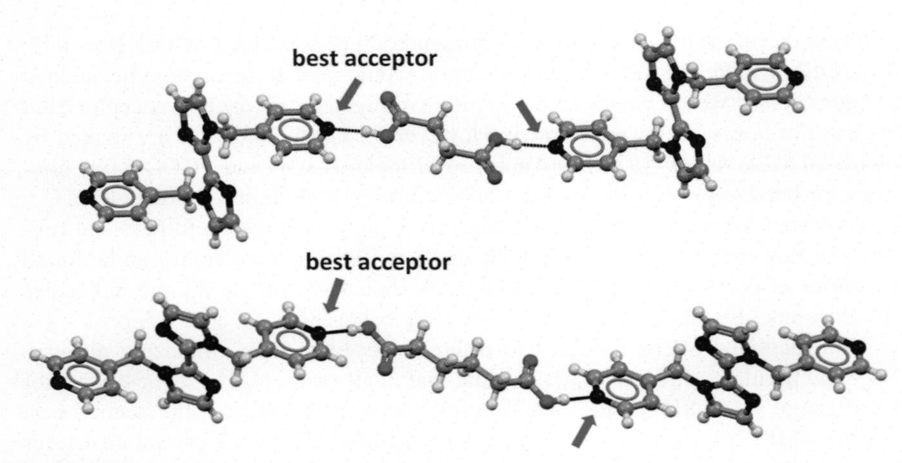

FIGURE 6.15

Crystal structures of cocrystals between biimidazole-based symmetric ditopic molecules and carboxylic acids. In each structure, the hydrogen bond engages with the best acceptor, the N_{py} atom, as ranked by MEPs (*refcodes*: DOSZAO, top; DOSZES, bottom). *MEPs*, molecular electrostatic potentials.

observed patterns of preferences. Exceptions can be expected since all these assemblies involve relatively weak and reversible interactions. Nevertheless, thanks to the simplicity of the approach and the ease with which MEPs can be obtained, this protocol can offer synthetic strategies of considerable practical value. In some cases, when single-point hydrogen bonds are in competition with two-point aggregation, it is also important to consider synergistic effects. This is particularly important in those cases where the geometry and relative orientation of the reactant species can enhance their binding through "supramolecular chelation" [89]. For example, in a series of six multifunctional custom-designed thiazole-amides [78,90] (**T_1–T_6**, Fig. 6.16) comprising three hydrogen bond acceptors (O, S, and N) and only one

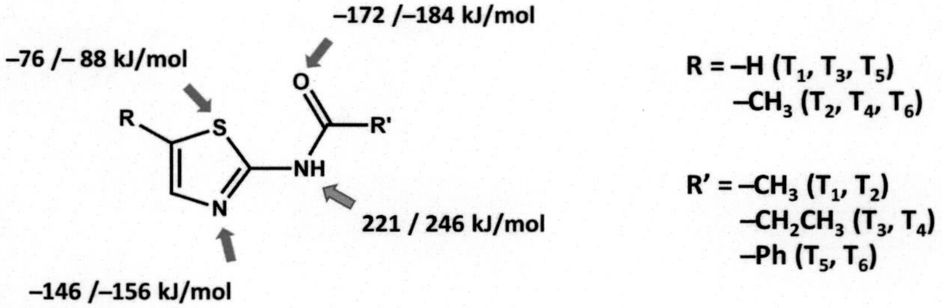

–172 /–184 kJ/mol
–76 /– 88 kJ/mol
221 / 246 kJ/mol
–146 /–156 kJ/mol
R = –H (T_1, T_3, T_5)
–CH_3 (T_2, T_4, T_6)
R′ = –CH_3 (T_1, T_2)
–CH_2CH_3 (T_3, T_4)
–Ph (T_5, T_6)

FIGURE 6.16

Hydrogen bond donor/acceptor sites indicated by blue and red arrows, respectively, and calculated MEP ranges for multifunctional custom-designed thiazole-amides (T_1–T_6). *MEP*, molecular electrostatic potential.

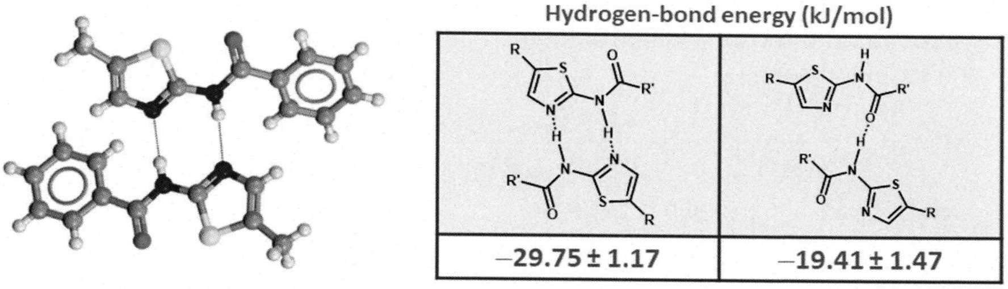

Hydrogen-bond energy (kJ/mol)	
−29.75 ± 1.17	−19.41 ± 1.47

FIGURE 6.17

The self-complementary $R^2_2(8)$ synthon composed of two N—H···N hydrogen bonds between adjacent thiazole-amides in T_6 (left). Calculated hydrogen bond energies of two possible hydrogen bond—based synthons (right).

hydrogen bond donor (amide N—H), the electrostatically based ranking of three acceptor sites revealed their importance as follows: C=O > $N_{aromatic}$ >> $S_{aromatic}$.

If neighboring molecules were to assemble via a single hydrogen bond, the expected synthon would, according to Etter's "best donor—best acceptor" guidelines, be an N—H···O interaction. However, in four of the five reported structures (**T_1**, **T_2**, **T_4**—**T_6**), the self-complementary two-point synthon involving two N—H···N interactions was preferred (Fig. 6.17).

The outcome was further rationalized against a pairwise energy-based approach presented by Hunter and coworkers [76], which uses an electrostatic description of the key hydrogen bond interactions. Although the energy of the single N—H···O hydrogen bond (−19.41 ± 1.47 kJ/mol) was calculated to be greater than each of the individual N—H···N interactions in the synthon formed (Fig. 6.17), the self-complementary synthon was more likely to occur due to synergistic involvement of two hydrogen bonds, giving rise to "supramolecular chelating" [90].

5.3 Exploring the use of halogen bonding in crystal engineering via molecular electrostatic potentials

The halogen bond [91—94] (XB) is a σ-hole interaction of the type R—X···A, where X is typically Cl, Br, or I. The XB is highly directional due to the anisotropic electron density distribution in covalently bonded halogen atoms. The halogen atom presents a positive electrostatic potential along the extension of the covalent bond, which imparts the ability to this atom to act as a Lewis acid, a halogen bond donor. The fact that the halogen atom also has an equatorial region of negative potential means that a single halogen atom can act as a self-complementary halogen-bonded synthon. This interaction is gaining considerable attention and momentum as a versatile, highly directional, and effective tool in practical crystal engineering. The practical and directional importance of the electrostatic component [95] of every halogen-bond donor means that MEPs mapped onto surfaces represent a natural choice for describing and analyzing many aspects of the halogen bond

Table 6.1 Calculated molecular electrostatic potential surface values (mapped on the 0.001 electron bohr^{-3} contour of electron density) at the iodine σ-hole.

MEP($I_{\sigma\text{-hole}}$)/kJ/mol					
C(sp)–I		C(sp^2)–I		C(sp^3)–I	
N≡C–C≡C–I	204.8	C_6F_5–I	139.9	CF_3–I	137.5
		F_2C=CF–I	141.0		
H–C≡C–I	148.9	C_6H_5–I	71.7	CH_3–I	58.3
		H_2C=CH–I	74.2		

Modified from J.-Y. Le Questel, C. Laurence, J. Graton, Halogen-bond interactions: a crystallographic basicity scale towards iodoorganic compounds, CrystEngComm. 15 (2013) 3212–3221.

[41,65,96]. It is well established that the effectiveness of a halogen bond donor is related to the degree of polarization of the halogen atom (the magnitude of its σ-hole). Politzer and coworkers [40] have shown that the electrostatic potential calculated at the halogen atom's surface is directly related to the strength of halogen bonds of a series of RX halides and, in essence, the greater the positive MEP at the σ-hole, the more powerful the halogen bond will be [97,98]. This observation leads to the idea of deliberately strengthening the halogen bond through various means of "activation," whereby the electrophilic region on the halogen atom is enhanced. Activation of a halogen bond donor can be achieved by the addition of electron-withdrawing groups to the molecular scaffold to which the halogen atom is attached, and this has been done mostly commonly via fluorination [99,100] or by the use of nitro substituents [101]. In addition, halogen bond activation can also be accomplished by changing the hybridization of the carbon atom to which the halogen atom is bonded, and this has been particularly effective with substituted iodoacetylenes (Table 6.1) [102].

The impact of hybridization on the magnitude of the σ-hole is as follows: C(sp)–I >> C(sp^2)–I > C(sp^3)–I, and the cumulative electron-withdrawing effect of three (F_2C=CF–I and CF_3–I) and five (C_6F_5–I) fluorine atoms are not sufficient to overcome the sole effect of the triple bond in iodoacetylene [102].

5.3.1 Ranking halogen bond donors

MEP calculations were used extensively in a systematic structural and computational study that offered valuable insight into the ability of haloethynylbenzenes to compete with fluorinated aromatic halogen bond donor functionalities (Fig. 6.18) [103]. Although the ability of the halo-ethynyl moiety to form halogen bonds was known, it was not clear how this activation would affect its relative ranking within the existing library of well-established XB donors.

The MEP values showed that the hybridization of the carbon atom adjacent to the halogen atom can overcome the electron-withdrawing effect of fluorine atoms

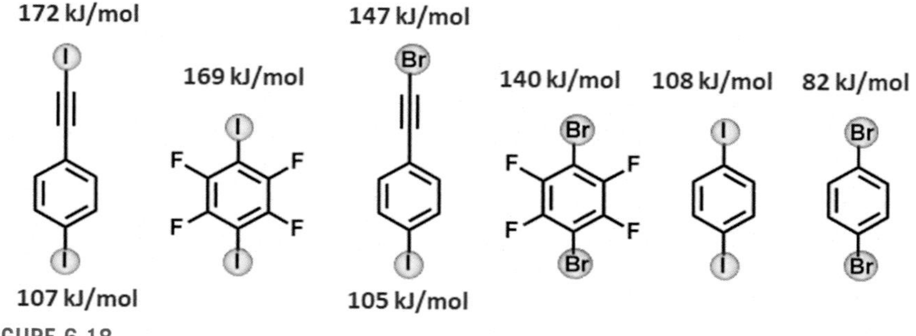

FIGURE 6.18

Relative ranking of haloethynyl moieties, fluorinated, and nonactivated halogen bond donors based on calculated molecular electrostatic potential values.

attached the aromatic core, and the polarizing impact of the sp-hybridized carbon atom affords the ethynyl-based iodine and bromine atoms with σ-holes of comparable magnitudes to those on the analogous diiodo/dibromo-tetrafluorobenzenes.

Aakeröy and coworkers also explored the benefits of a "double activation" on the magnitude of the σ-hole, by combining the electron-withdrawing capabilities of nitromoieties with the polarizing effect of sp-hybridized carbon atoms with the aim of producing highly electrophilic halogen bond donors (Fig. 6.19) [104]. To

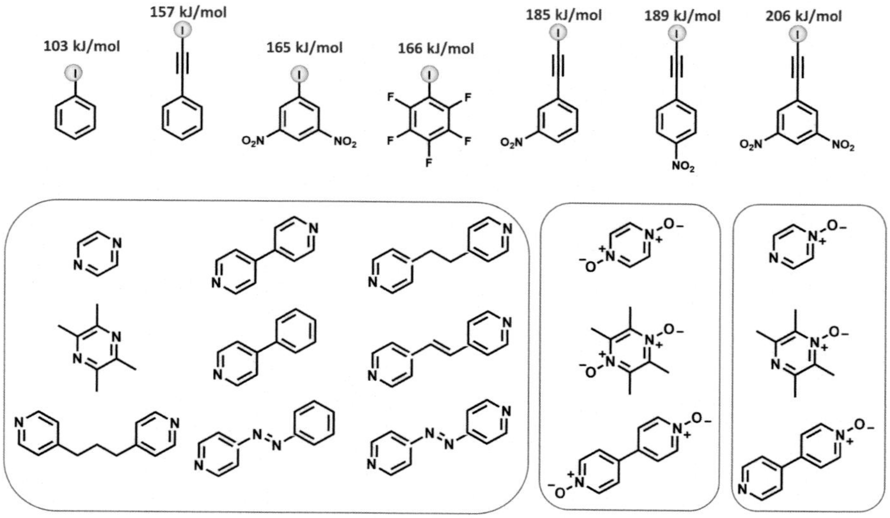

FIGURE 6.19

Relative ranking, based on calculated MEPs, of "doubly activated," "singly activated," and nonactivated halogen bond donors (top); three families of halogen bond acceptors: N-based, O-based, and N-/O-based acceptors.

examine the ability of these halogen bond donors to drive the formation of new cocrystals, an extensive cocrystallization screen was carried out. A library of 15 acceptors was employed, and this included N-based, O-based, and N- and O-based acceptors, which could be further classified as monotopic, symmetric ditopic, and dissymmetric ditopic acceptors.

According to calculated MEP values (Fig. 6.19), the double activation afforded a significant increase in positive potential at σ-hole of the iodine atom; iodoethynylnitrobenzenes displayed even more positive (20–40 kJ/mol) MEP values than other well-known halogen bond donors such as iodopentafluorobenzene.

The cocrystallization experiments, involving three iodoethynylnitrobenzenes (Fig. 6.19) and six bromo- and chloro- analogs with 15 selected acceptors, revealed that the increased depth of the σ-hole reflected the ability of the donor to drive cocrystal synthesis (as demonstrated by the IR spectra of the cocrystallization products compared with the spectra from the individual reactants). The success rates for iodoethynylnitrobenzene were (45/45; 100%) and somewhat smaller for bromoethynylnitrobenzene (9/15; 60%), whereas none of the 15 attempts with chloroethynylnitrobenzene produced a cocrystal. The double activation produced a substantially more effective donor, and even a bromine atom became electrophilic enough to become a practically useful synthetic vector as evidenced by the first reported crystal structure where the bromoethynyl moiety drives a cocrystal synthesis (Fig. 6.20).

The idea of double activation was also explored by Matzger and coworkers [105], and they used MEP to establish how different substitutions on the backbone of a 1,3,5-triiodobenzene core influenced the σ-hole on the iodine atoms (Fig. 6.21). It was found that three nitro groups gave rise to an exceptional halogen bond donor with one of the largest MEP values known.

5.3.2 Ranking halogen bond acceptors

The early examples of utilizing MEP values for ranking the relative efficiency of halogen bond donors created the opportunities for deriving crystal engineering strategies built on a hierarchy of competing halogen bonds. However, a transition from calculated MEPs on individual molecules to reliable and practical synthetic routes to

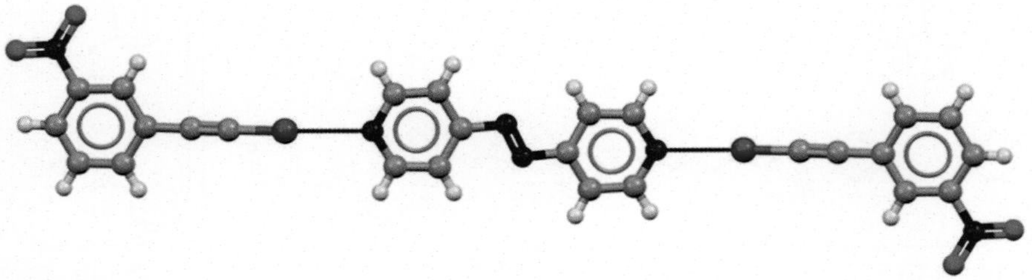

FIGURE 6.20

The product of the first cocrystal synthesis driven by the bromoethynyl moiety (*refcode*: KUKWUN).

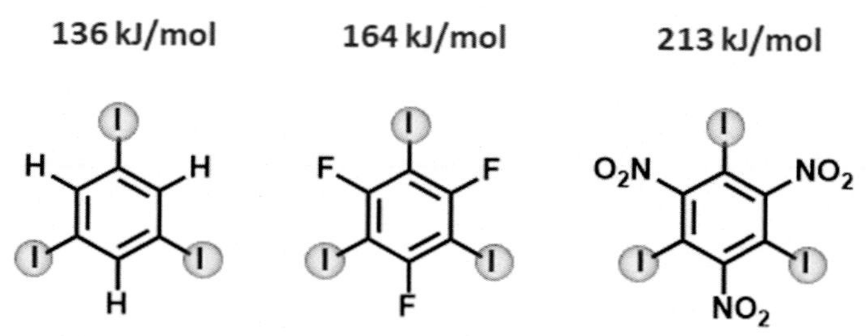

FIGURE 6.21

MEPs of "double activated" iodine atoms. *MEPs*, molecular electrostatic potentials.

cocrystal synthesis does require a very clear understanding of if, or how well, an MEP-based ranking of halogen-bond acceptors is reflected in experimental structural consequences in the resulting solid.

To address this challenge, Aakeröy and coworkers performed a series of cocrystallization experiment of biimidazole-based symmetric ditopic acceptor molecules (the same set as used for predicting selectivity of hydrogen bond in competitive recognition events, i.e., N_{im} vs. N_{py}; Fig. 6.14) with iodo-substituted halogen bond donors, all of which were activated by a fluorinated aromatic scaffold (Fig. 6.22) [106].

A total of 10 crystal structures of new halogen-bonded cocrystals were obtained in this study, and in each and every case, the halogen bond donor, the activated iodine atom, engaged with the nitrogen atom (N_{py}), with the more negative MEP value (Fig. 6.23).

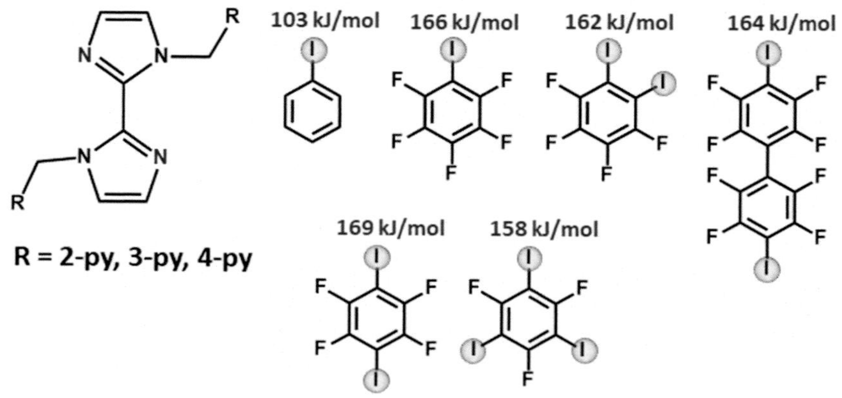

FIGURE 6.22

Halogen bond donors and acceptors in cocrystallization experiments testing whether an MEP-based hierarchy of XB bond acceptors is reflected in the solid state. *MEP*, molecular electrostatic potential.

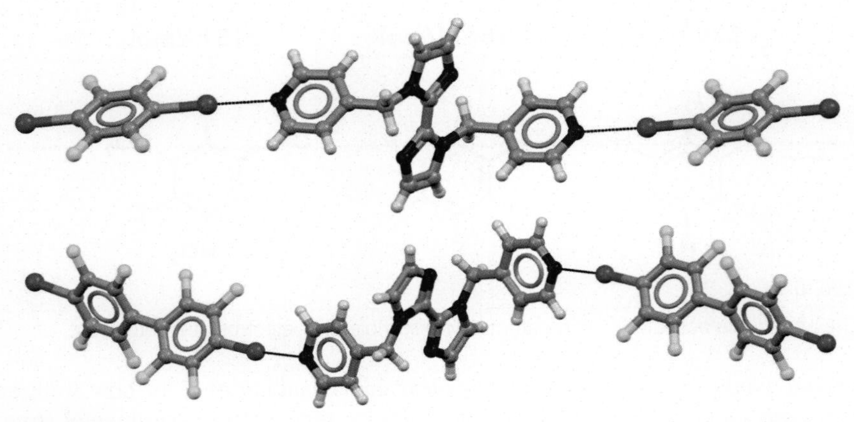

FIGURE 6.23

Two crystal structures showing that the key I···N halogen bond involves the nitrogen atom with the largest negative MEP value (*refcodes*: EFUDUG, top, EFUFAO, bottom). *MEP*, molecular electrostatic potential.

In three cases, the second iodine atom was simultaneously involved in an additional $I\cdots N_{im}$ bifurcated interaction, although these XB distances were substantially longer. To ensure that any steric hindrance was not responsible for limiting the primary $I\cdots N_{im}$ halogen bonds, control experiments were also carried out on a target molecule equipped with a phenyl group in place of pyridine, and in the resulting crystal structures, a halogen bond to the imidazole nitrogen (the only viable acceptor site) was observed. These results add weight to the argument that a ranking based on MEP values can be used as a corner stone of a synthetic strategy for the deliberate assembly of cocrystals with desired synthons and intermolecular connectivities. Furthermore, this also illustrates that Etter's best donor/best acceptor guidelines, which arose from hydrogen-bonded systems, are applicable to halogen-bonded organic crystals.

5.3.3 The electrostatic potential difference as a tool for rationalizing halogen bond selectivity

A large number of natural products are known to contain halogen atoms, and the fact that they are mostly located at the molecular periphery means that they are very likely to play a key role in intermolecular recognition and binding. Furthermore, approximately one-third of all drugs in therapeutic use are halogenated [107], indicating that the importance of halogen atom–based intermolecular interactions has been vastly underestimated up to this point. However, drug development is only one of the many areas where the presence and consequences of halogen bonding are likely to be of increased importance [108]. In fact, halogen bonds have been studied in the context of cf. biological systems [109], catalysis [110], liquid crystals [111], anion sensing [112], separation [113], and semiconductors [114]. Against this background, in systems with multiple halogen bond acceptors, these can be ranked according to their calculated MEP values [115]. However, a subsequent systematic cocrystal screen showed that even though the observed halogen bond obeyed the "best donor/best acceptor rule," the detailed halogen bond connectivity did not always materialize as intended [116]. As a result, Aakeröy and coworkers explored the structural influence of competing halogen bond interactions to define more clearly to what extent electrostatically based halogen bond preferences can be relied upon to produce a specific synthons [117]. Thus, 12 ditopic acceptors presenting two binding sites with different MEP values (within the range 22−175 kJ/mol) were cocrystallized with nine fluorinated aliphatic and aromatic halogen bond donors of comparable halogen bond capability (all donors in the series displaying only a 10 kJ/mol MEP difference). Structurally characterized cocrystals were organized into subcategories according to the specific nature of the halogen bonding connectivity observed in each case. Three structurally distinct groups emerged: (1) the XB donor selectively binds only to the better/best acceptor site, (2) the XB donor binds to both acceptor sites ("no selectivity"), and (3) the XB donor has no preference for any one of the acceptor sites (a *diffuse* area where all supramolecular outcomes are observed) (Scheme 6.1). The halogen bond acceptors in each group could be clearly differentiated based on the MEP differences of the two acceptor sites they bear.

This is an important example, which clearly demonstrates that a difference in MEP values (ΔE) between halogen bond donor sites can be used for rationalizing and, potentially predicting, dominant supramolecular outcomes.

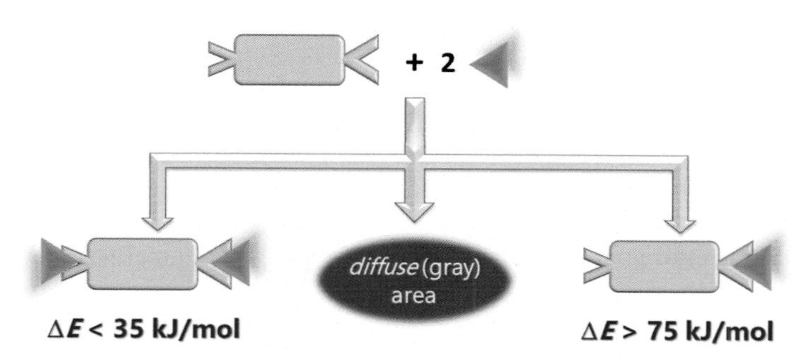

SCHEME 6.1

Three groups of the supramolecular outcomes of cocrystallization between asymmetric ditopic acceptors and halogen bond donors.

Modified from C.B. Aakeröy, T.K. Wijethunga, J. Desper, M. Đaković, Electrostatic potential differences and halogen-bond selectivity, Cryst. Growth Des. 16 (2016) 2662–2670.

5.4 Hydrogen and halogen bonding in competitive supramolecular systems

The deliberate combination of halogen bonds and hydrogen bonds in supramolecular synthesis has so far only been examined in relatively few papers. The explanation for this dearth of information is partly due to the relatively recent recognition of halogen bonding as being much more than an esoteric intermolecular interaction, but also due to the fact that relatively few molecules capable of acting as strong halogen bond/hydrogen bond donors simultaneously have been examined. Quite often, hydrogen and halogen bonds were found to readily interfere with each other in many different systems in both solutions [118–120] and the solid state [121,122]. The main question that arises at the outset of this section is, how do conventional hydrogen bond donors "stack up" against halogen bond donors as far as their structural influence go? Furthermore, is it possible to use MEPs mapped onto distinctly different elements such as hydrogen and iodine, as a way of ranking or estimating the relative competitiveness of these supramolecular drivers? From a practical crystal engineering perspective, it is essential to understand how these two interactions mutually operate in a complex setting, and to establish robust and reliable means of ranking so they can be employed simultaneously in a hierarchical assembly process for delivery of architectures with desired structural features and chemical composition [123–125].

Evaluation and rationalization of the supramolecular outcomes, with the aim of establishing a relative ranking of two most important interactions, were attempted by (1) ranking the multiple competing acceptor sites upon the calculated MEP values and comparing their MEP differences, and (2) by applying the same approach to competing hydrogen and halogen bond donor sites. While the former was explored on a substantially larger number of cases [99,126,127], the latter includes a limited number of reports [128].

To refine the understanding of competing intermolecular forces against the backdrop of an electrostatic canvas, 10 molecules, equipped with hydrogen and halogen bond donor sites on the same molecular backbone, were cocrystallized with 20 symmetric ditopic and asymmetric ditopic acceptors (Fig. 6.24) [129]. The structural outcome was interpreted with the help of MEPs on the different donors and acceptor sites with the overall goal of identifying reliable synthetic guidelines for practical crystal engineering.

A total of 200 cocrystallizations were attempted, half of these produced cocrystals, and 24 new crystal structures were ultimately obtained. The crystal structures were organized into three subgroups, according to acceptor type (monotopic, symmetric ditopic, and dissymmetric ditopic; Fig. 6.25).

The first group comprised five crystal structures, four of which displayed only hydrogen bonds, while in one case, both hydrogen and halogen bonds participated in the assembly process. The second group, which intended to probe if two donors were of comparable strength, yielded 13 structures. In eight of these, both hydrogen and halogen bonds were formed, whereas in the five remaining structures, only hydrogen bonds were observed. Finally, in the third group, which was designed to explore how two donors would compete for acceptors with different MEP values,

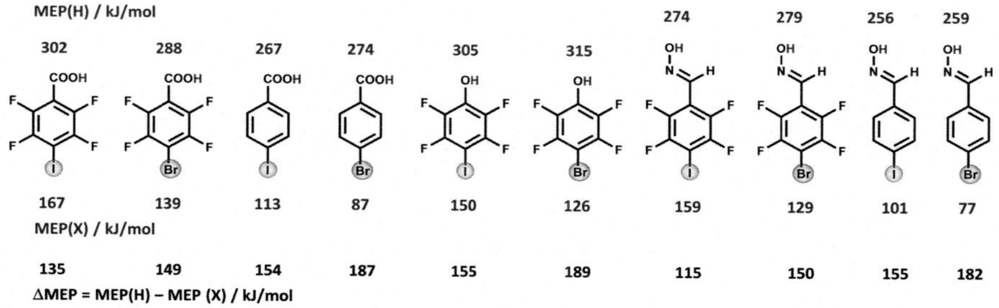

FIGURE 6.24

Donor molecules each equipped with one hydrogen and one halogen bond donor site, accompanied with the calculated MEP surface values. The MEP differences between the two donor sites are listed in black. *MEP*, molecular electrostatic potential.

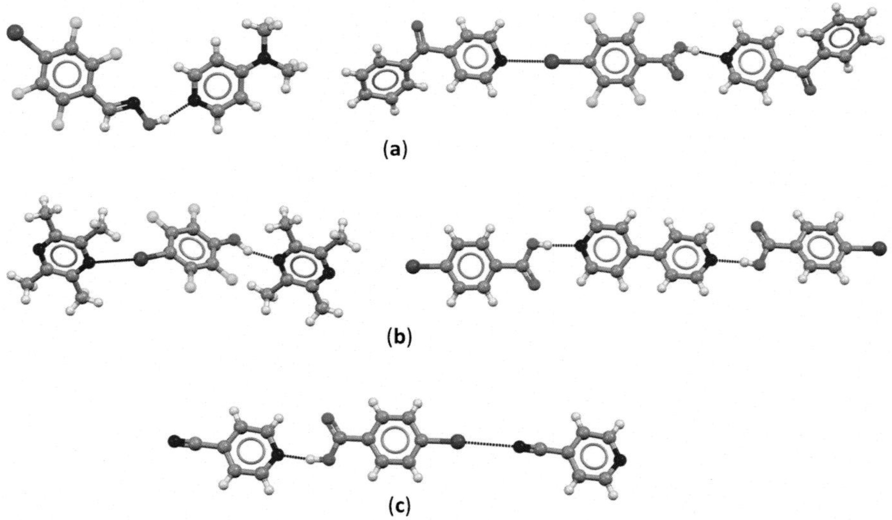

FIGURE 6.25

Cocrystallization products of hydrogen/halogen bond donors and (A) monotopic acceptors (*refcodes*: BUNDIZ, left, BUNFEX, right), (B) ditopic symmetric acceptors (*refcodes*: BUNFUN, left, BUNHEZ, right), and (C) ditopic asymmetric acceptors (*refcode*: BULTUZ).

six crystal structures were obtained, and in four of these, it was possible to identify a distinct binding preference. In one case, the HB donor linked with the best acceptor and the XB donor linked with the second-best acceptor (ranking based on calculated MEPs), while in the remaining three structures, the same hierarchy of interactions was observed (ranking based upon extensive crystallographic data). It is worth noting that in none of the cases (24 outcomes), a halogen bond formed solely.

A close analysis of the electrostatic potential values showed that although there is not a clear boundary between, (1) structures that contained only hydrogen bonding and (2) structures that contained both hydrogen and halogen bonding, the data implied the following conclusions: (1) the greater the MEP difference of the two donor sites, the more probable formation of only hydrogen bond, and (2) the smaller the MEP difference of the two donor sites, the more likely that both interactions will be present.

Based on extensive systematic structural studies of molecules covering a range of molecular weights and conformational flexibility, it is clear that calculated MEPs can be utilized as an informative and easy-to-use platform for organic crystal engineering practitioners. The ability to rank chemical entities has a fundamental role to play in supramolecular chemistry, where a detailed knowledge of intermolecular interactions—and therefore the interacting preference of one chemical moiety when confronted with several possible partners—is crucial for the effective recognition and sequestration of specific targets, and for the preparation of selective sensors or, in fact, for any chemical device that relies on an intermolecular binding event for "function" and "performance."

6. Molecular electrostatic potentials as a tool for assembling metal–organic systems

Despite the fact that metal–organic systems that are dominated by strucutral features of lower dimensionalities (i.e., not including 3-D metal–organic frameworks, MOFs), are not yet fully explored and mapped out, they offer a a very valuable and informative testing ground for verifying and improving guidelines and principles derived for assembling their metal-free analogous.

To make the inherently very challenging inorganic systems more amenable to directed assembly using noncovalent interactions, it is undoubtedly helpful to limit the number of ways in which individual building blocks can (or are likely) to aggregate. At the very least, if the metal–ligand coordination is predictable, and the main structural features display specific discrete 0-D building units or infinite 1-D architectures, the challenge is less daunting. In this context, the use of halide anions and β-diketonates has emerged as highly useful avenues for controlling the coordination geometry as well as the building block dimensionality. Here, we present illustrative examples of coordination complexes where the magnitude of the calculated MEPs of key acceptor sites is highly sensitive to a variety of structural features. In addition, a number of chemical handles for *dialing-in* a desirable supramolecular link have been recognized.

6.1 Molecular electrostatic potential in evaluation of electronic effects of metal-based building units

An interesting MEP sensitivity–related feature was noticed in hydrogen-bonded octahedral bis-β-diketonato complexes [60]. In contrast to analogous β-diketonato complexes of Ni(II) and Co(II) where the β-diketonato ligands occupy the equatorial plane, in Cu(II) complexes, due to the Jahn–Teller effect, the β-diketonato ligands chelate differently and occupy one equatorial and one axial site. Moreover, the *hfac* (1,1,1,6,6,6-hexafluoropentane-2,4-dione) complexes continuously presented an elongated coordination octahedron, while for *tfac* (1,1,1-trifluoropentane-2,4-dione), a compressed octahedron was found (for each member of the whole family; Fig. 6.26). The feature was subsequently reflected on the MEP values of two potential acceptor sites, making them hierarchically different.

The more distant oxygen atoms, axial (*hfac*) and equatorial (*tfac*), displayed more negative values of the calculated MEP ($\Delta MEP(O_{ax}-O_{eq})_{hfac} = 7-15$ kJ/mol; $\Delta MEP(O_{eq}-O_{ax})_{tfac} = 3-7$ kJ/mol), thus being categorized as the better acceptor sites. A selectivity for the better hydrogen bond acceptor was demonstrated in all the complexes with the $O_{oxime}-H\cdots O_{\beta\text{-dik}}$ hydrogen links, i.e., $O_{oxime}-H\cdots O_{ax}$ (*hfac*) and $O_{oxime}-H\cdots O_{eq}$ (*tfac*) links were formed. The data suggests that, when targeting a particular supramolecular outcome, we may also take advantage of specific electronic features of metal centers (displayed in slight distortions of coordination geometry) for directing the assembly process toward the desired supramolecular product.

Another handle for fine-tuning the calculated MEP of prospective acceptor sites was observed as a result of metal ion exchange in a series of halogen-bonded isostructural acetylacetonato (*acac*) complexes. The idea was introduced by Brammer and coworkers [130] for rationalization of the observed decrease of halogen bond lengths ($C-X\cdots N{\equiv}C-M$; X = Br, I; M = Cr, Fe, Co) in a familiy of isostructural halopyridinium hexacyanometallate salts. As the trend (in decreasing halogen bond length, and implicitly the increase in halogen bond strength) correlated well with the increase of the metal d-electron count, it was suggested that the increased metal d-electron count in turn increases the π-backdonation to the cyanide ligands (through the increase of population of the t_{2g} energy level). And, that the increased

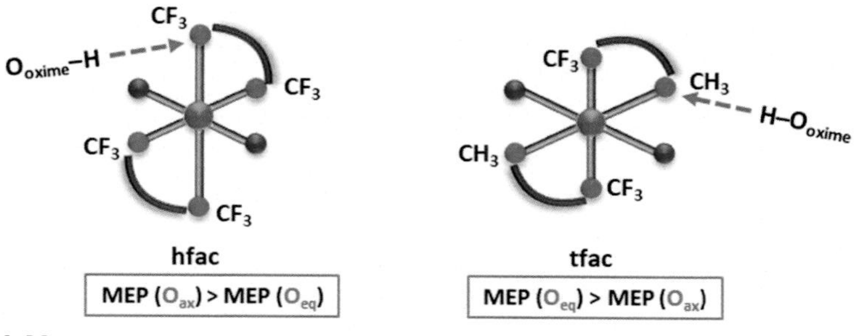

FIGURE 6.26

Jahn–Teller elongation (left) and compression (right) of the coordination octahedron in 1,1,1,6,6,6-hexafluoropentane-2,4-dione (*hfac*) and 1,1,1-trifluoropentane-2,4-dione (*tfac*) complexes of copper(II), respectively. MEPs displayed more negative value for the more distant β-diketonato oxygen atoms, regardless of the geometry as well as the substituent (CH_3/CF_3) in the vicinity of the respective coordination site. *MEPs*, molecular electrostatic potentials.

π-backdonation consequently increases the partial negative charge and therefore negative electrostatic potential associated with the cyanide ligand. This was consistent for isostructural *acac*-based complexes of Co(II) and Ni(II) where for each Co(II)/Ni(II) isostructural pair, the *acac* oxygen atoms displayed more negative electrostatic potential (13–18 kJ/mol) upon moving from Co(II) → Ni(II) (Fig. 6.27).

Since these systems carry two rather weak donor sites (carbon-bound halogen and aromatic hydrogen atoms) with relatively small MEP differences, they were highly sensitive to even slight structural/electronic changes. Even subtle electronic changes introduced via a simple change of metal cations (Co(II) → Ni(II)) resulted in sufficient electron perturbation, observed in the reverse relevance of the potential donor sites, for the targeted link (C–H···O vs. C–X···O) to occur. The metal cation exchange thus revealed an additional handle for switching between different supramolecular interactions and architectures through fine-tuning of electronic effects around prominent donor sites.

Inspired by work on organic solid-state systems, tunability of the donor and acceptor power through the covalent modification in the vicinity of prospective interacting sites was also examined. The influence of the additional heterocyclic nitrogen atom and its position in the aromatic ring on the magnitude of halogen atom σ-hole was explored (Fig. 6.28). It was presented that the introduction of nitrogen atom *para* to pyridine coordination site (i.e., pyridine ↔ pyrazine exchange) and *ortho* to halogen atom does not enhance its halogen bond donating power. Instead, a barely noticeable reduction of the σ-hole was observed (3-Xpy → 2-Xpz: 1–5 kJ/mol). In contrast, when nitrogen atom is positioned *meta* to both coordination site and halogen atom (i.e., pyridine ↔ pyrimidine exchange), substantial "activation" of halogen atom was achieved (by approximately 20 kJ/mol).

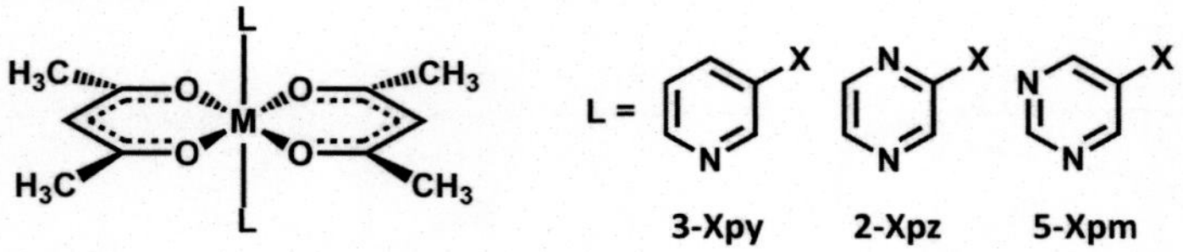

	$MEP(O)_{Co}$ / kJ/mol	$MEP(O)_{Ni}$ / kJ/mol
[**Co/Ni**(acac)$_2$(3-Clpy)$_2$]	–170	–183
[**Co/Ni**(acac)$_2$(3-Ipy)$_2$]	–167	–181
[**Co/Ni**(acac)$_2$(2-Clpz)$_2$]	–139	–157
[**Co/Ni**(acac)$_2$(2-Brpz)$_2$]	–138	–155
[**Co/Ni**(acac)$_2$(2-Ipz)$_2$]	–140	–155

FIGURE 6.27

MEPs of the acetylacetonato oxygen atoms in isostructural acetylacetonato complexes of Co(II) and Ni(II). Higher metal d-electron count is reflected in more negative values of the *acac* oxygen atom in Ni(II) than in Co(II) complexes (X = Cl, Br, I). MEPs, molecular electrostatic potentials.

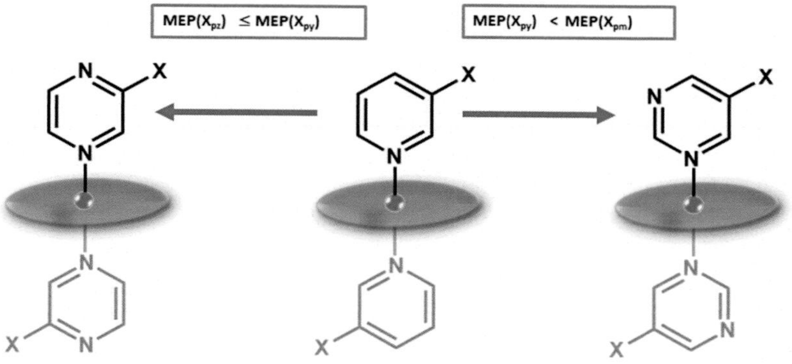

MEP(X_{pz}) ≤ MEP(X_{py})   MEP(X_{py}) < MEP(X_{pm})

FIGURE 6.28

Comparison of the donor power (envisaged in the MEP values) of three differently surrounded halogen atoms (i.e., attached to the pyridine, pyrazine, and pyrimidine ligands) in acetylacetonato complexes of Co(II) and Ni(II). *MEP*, molecular electrostatic potential.

Perturbation of the electron density throughout the whole system (building unit) was observed on a similar class of compounds, *acac*-based complexes with pyrimidinone and quinazolinone ligands. *Only one* change at the time was introduced to the building unit and the calculated MEP values at prospective hydrogen bond donors and acceptors were mapped out. Each of the modifications, the $-CH_3 \leftrightarrow -CF_3$ exchange at the *acac*-based scaffold, exchange of the metal cation, or "extending" the aromatic system to additional fused aromatic ring (i.e., pyrimidinone ↔ quinazolinone ligand exchange), resulted not only in a substantial alteration of the donating/accepting power of prospective hydrogen bond donor or acceptor site in close vicinity but also had notable effects on the whole system (Fig. 6.29). The most dramatic

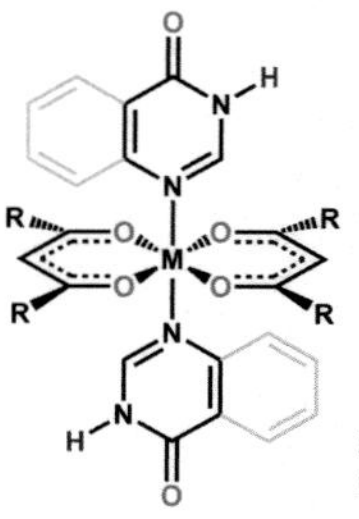

ΔMEP	$-CH_3 \leftrightarrow -CF_3$	Co ↔ Ni	Co ↔ Cu	Ni ↔ Cu	pym ↔ quz
O_{lac}	20–60	≤ 2	4–9	5–9	0–8
H_{lac}	20–50	≤ 2	5–13	7–14	10–24
O_{acac}	70–90	≤ 4	3–23	7–25	14–45

pym = 4(3H)-pyrimidinone; quz = 4(3H)-quinazolinone; lac = lactame functionality
ΔMEP = difference in calculated molecular electrostatic potential of respective hydrogen-bond donor/acceptor sites as a result of indicated change only (in kJ/mol)

M = Co, Ni, Cu
R = $-CH_3$, $-CF_3$

FIGURE 6.29

Differences in calculated MEP values at indicated hydrogen bond donor (lactam hydrogen atom, H_{lac}) and acceptor sites (lactam, O_{lac}, and β-diketonato, O_{acac}, oxygen atoms) caused by *only one* change introduced to the metal-based building unit ($-CH_3$/-CF_3, Co/Ni/Cu/, pym/quz), while all the other structural parameters were kept constant [59]. *MEP*, molecular electrostatic potential.

effect was observed for the $-CH_3 \leftrightarrow -CF_3$ exchange, where the electron withdrawing effect was noticed at all potential hydrogen bond interacting sites making them all substantially more electropositive (O_{acac} by 70–90 kJ/mol, O_{lac} by 20–60 kJ/mol, and H_{lac} by 20–50 kJ/mol, depending on the metal cation and pyrimidinone/quinazolinone ligand involved; here, we list only intervals in which the differences are manifested with intention to depict only trends in changes, not exact values, which are in turn dependent on the method used. For more details on the calculation and results, we direct the reader to the original paper) [59].

The application of the fact that the strength of potential hydrogen bond donor and acceptor could be fine-tuned via a variety of *handles* was nicely demonstrated on the family of Co(II) and Ni(II) β-diketonato complexes with the pyridine–oxime ligands where the oxime functionality was intended for promoting assembly process to targeted 1-D supramolecular chains. It has been shown that the reactants (metal-based building units) could be adjusted through the covalent modification so that the difference in the calculated MEP values of the competing acceptor sites (oxime and β-diketonato oxygen atoms) allows the formation of the targeted supramolecular outcome (Fig. 6.30).

Each time an unintended supramolecular product, typically a disruptive *acac*-oxime hydrogen bond, was delivered, covalent adjustments were made in the vicinity of interacting sites (**1**: H/CH_3 exchange; **2**: *meta*/*para* positioning of the oxime moiety; **3**: phenyl/CF_3 exchange) to alter the balance between hydrogen bond donors and acceptors until the intended supramolecular target was achieved with a satisfactory frequency of occurrence (Fig. 6.31). The final successful synthetic protocol was developed in much the same way as conventional organic syntheses is refined and optimized in response to synthetic outcomes and product yields.

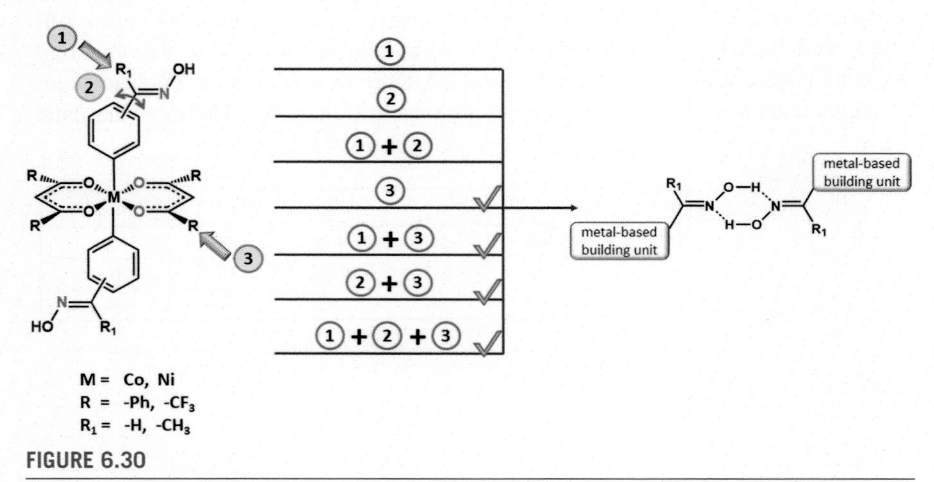

FIGURE 6.30

Covalent modification of the supramolecular reactant, metal-based building unit, performed to adjust the difference in calculated MEP values between the competing hydrogen bond acceptor sites (oxime nitrogen and β-diketonato oxygen atoms) so to deliver targeted supramolecular product, 1-D supramolecular chains realized via self-complementary oxime motif. *MEP*, molecular electrostatic potential.

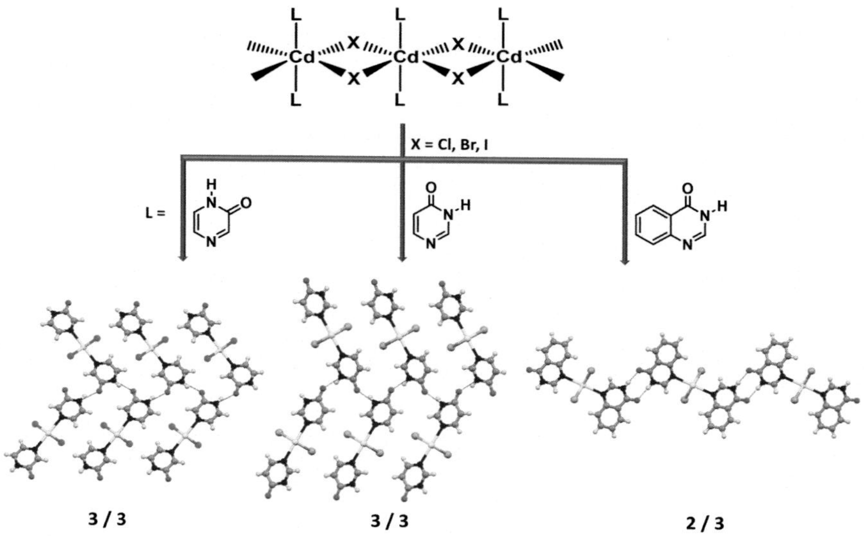

FIGURE 6.31

Supramolecular products together with frequency of occurrence. 1-D metal-containing structural scaffold was constructed from cadmium(II) cations and halide anions (X = Cl, Br, I), while the supramolecular recognition points were ensured through employment of pyrazine, pyrimidone, and quinazolinone ligands bearing the lactam moiety as the intended supramolecular vectors.

6.2 Paving the way toward predictive assembly of metal—organic systems

6.2.1 Hydrogen-bonded metal—organic systems

Deriving supramolecular synthetic strategies for assembling metallo-supramolecular systems in a more predictive way is still far behind that what is derived for their organic counterparts. Several metallo-supramolecular systems of different dimensionalities were recently explored to that end [59,131,132]. The approach used was to examine if supramolecular protocols already derived for organic crystalline systems could be directly translated to metal—organic systems, or additional restrictions and modifications need to be introduced. For that purpose, all the explored systems were designed to ensure (1) control of the details of coordination chemistry around each metal cation and (2) change of only one parameter at the time when mapping out their solid-state binding preferences.

6.2.2 1-D metal-based systems

The need for keeping metal—organic systems substantially simple and preorganized, while trying to understand how those systems organize themselves in the solid state, is effectively met through utilization of 1-D building units. Those building blocks, while ensuring control of the coordination geometry details around the metal centers, provide a balance between a preorganization of the systems in hand and building units with significant freedom to communicate and assemble into a 3-D architecture.

It is well known that halide ions, when combined with appropriately selected metal cations, readily produce 1-D polymeric units [133]. That was recently employed for delivery of custom-designed probe molecules, one-dimensional Cd(II)-based polymers [131]. The Cd(II)-halide structural scaffold was further decorated with small and rigid heterocyclic ligands equipped with the lactam functionality as the supramolecular vector for further propagation of the assembly (Fig. 6.31).

The intended control of the details of the coordination chemistry and building block dimensionality was very satisfying; the 1-D building units were delivered in eight of nine structures (8/9; 89%), while the supramolecular connectivity (the intended lactam–lactam link) displayed a 100% success rate. In all nine structures, the N–H···O hydrogen bond was found as the primary interaction between the metal-based building units, while in eight instances (all having 1-D building blocks and metal cations displaying the octahedral coordination geometry), the $R^2_2(8)$ and $C(4)$ [134,135] amide/lactam supramolecular motifs were found as the primary supramolecular connection.

The supramolecular connectivity was straightforwardly interpreted based upon the calculated MEP surface values (Table 6.2). In all cases, the carbonyl oxygen atom and the hydrogen atom bound to the endocyclic nitrogen atom presented the sites with the most negative and the most positive electrostatic potential values, respectively.

The results demonstrate that transferability of the supramolecular links from organic to metal–organic solid state is achievable as long as metal cations and their charge-balancing anions do not perturb the relative importance and ranking of

Table 6.2 Calculated molecular electrostatic potential (MEP) (kJ/mol) values at prospective hydrogen bond donor and acceptor sites, the lactam oxygen and hydrogen atom, and metal-bound halide anions.

Compound	O_{lac}	X	$\lvert MEP(O_{lac}) - MEP(X) \rvert$	$H(N)_{lac}$
$[CdCl_2(2\text{-pyz})_2]_n$	−215	−181	34	298
$[CdBr_2(2\text{-pyz})_2]_n$	−212	−150	62	301
$[CdI_2(2\text{-pyz})_2]_n$	−200	−120	80	296
$[CdCl_2(4\text{-pym})_2]_n$	−188	−116	72	305
$[CdBr_2(4\text{-pym})_2]_n$	−188	−90	98	310
$[CdI_2(4\text{-pym})_2]_n$	−187	−61	126	311
$[CdCl_2(4\text{-quz})_2]_n$	−205	−120	85	256
$[CdBr_2(4\text{-quz})_2]_n$	−200	−97	103	256
$[CdI_2(4\text{-quz})_2]$	−193	−78	115	252

2-pyz, *2(1H)-pyrazinone;* **4-pym**, *4(3H)-pyrimidinone;* **4-quz**, *4(3H)-quinazolinone.*

respective hydrogen bond donors and acceptors. The study provided an interesting example that clearly illustrated that supramolecular synthetic strategies built upon the "best donor–best acceptor" guidelines derived for organic crystalline solids can be successfully implemented for directing the supramolecular assembly of metal–organic solid-sate systems. Also, the study showed that ranking of potential hydrogen bonding acceptors and donors on the basis of calculated MEP values could have considerable value in metal–organic crystal engineering.

In addition, the study elucidated the importance of spatial requirements as well as secondary interactions for transferability of particular supramolecular motifs, i.e., $R^2_2(8)$ versus $C(4)$, from organic to metal–organic solid-state systems.

6.2.3 0-D building units

The success in understanding the assembly process of 1-D systems provided the encouragement for testing the same approach to metal-containing systems with lower preorganization (i.e., 0-D building units). The aim was to understand if the findings on 1-D systems are just inherent to a particular dimensionality (i.e., 1-D), or if they are more universally applicable, i.e., if they have a potential for being translated into practical supramolecular synthetic strategies for assembling the metal complexes in predictive ways [59]. For that purpose, the octahedral coordination geometry around the metal centers and the supramolecular synthetic vectors were retained, while the lower preorganization was achieved via exchanging halide anions for β-diketonato ligands.

Monomeric (0-D) building units were created by employing *acac*-based ligands (*acac* and *hfac*) in combination with several metal cations (Co(II), Ni(II), and Cu(II)) and pyrazinone/quinazolinone ligands (Fig. 6.32). The idea was to explore if the supramolecular synthetic products of 0-D metal-containing solid-state systems comprising multiple acceptor sites (A_1 and A_2), and only one good hydrogen bond donor, residing on the same building block (Fig. 6.34), can be reliably predicted based *solely* on the relative strength of the two competing acceptor sites.

The strategy was again similar to the one derived for the organic setting [117]. The difference in the calculated MEP values between two competing acceptor sites (ΔE) was taken as a criterion for predicting the supramolecular product (Table 6.3). It was surmised that (1) if the difference (ΔE) is substantial, the hydrogen bond

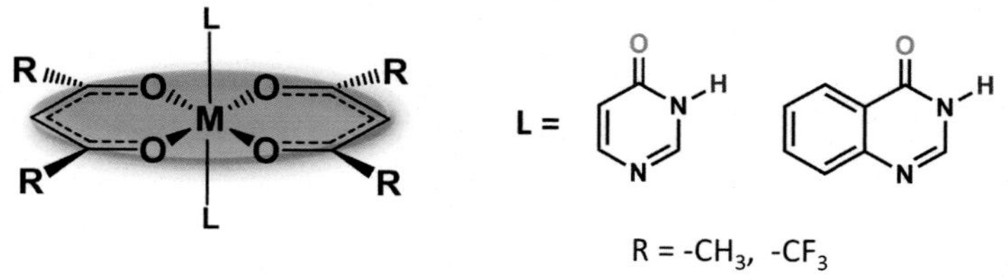

FIGURE 6.32

Discrete (0-D) metal-based building units constructed from metal cations (Co(II), Ni(II), Cu(II)), β-diketonato ligands (pentan-2,4-dione, *acac*; 1,1,1,6,6,6,-pentan-2,4-dione, *hfac*) and 4(3*H*)-pyrazinone (*pyz*)/4(3*H*)-quinazolinone (*quz*) ligands.

Table 6.3 Molecular electrostatic potential (MEP) differences at two competing hydrogen bond acceptor sites (lactam, O_{lac} (A_1), and *acac*-based oxygen atoms, O_{acac} (A_2)), followed by the expected supramolecular outcome and the observed supramolecular motifs.

Compound	MEP(O_{lac}) − MEP(O_{acac})/ kJ/mol	Expected supramolecular outcome	Supramolecular motif observed
[Ni(*acac*)$_2$(4-pym)$_2$]	11	***No selectivity***	N−H···O_{acac}
[Cu(*acac*)$_2$(4-pym)$_2$]	13	Any motif possible	−
[Co(*acac*)$_2$(4-pym)$_2$] - I	15		$R^2{}_2(8)$
[Co(*acac*)$_2$(4-pym)$_2$] - II	15		$C(4)$
[Cu(*acac*)$_2$(4-quz)$_2$]	32	***Transferability of the***	−
[Co(*acac*)$_2$(4-quz)$_2$]	46	***synthons***	$R^2{}_2(8)$
[Ni(*acac*)$_2$(4-quz)$_2$]	49	(exclusively $R^2{}_2(8)$ or	$R^2{}_2(8)$
[Cu(*hfac*)$_2$(4-pym)$_2$]	64	$C(4)$ motifs formed)	$R^2{}_2(8)$
[Co(*hfac*)$_2$(4-pym)$_2$]	71		$C(4)$
[Ni(*hfac*)$_2$(4-pym)$_2$]	77		$C(4)$
[Cu(*hfac*)$_2$(4-quz)$_2$]	77		$R^2{}_2(8)$
[Ni(*hfac*)$_2$(4-quz)$_2$]	97		$R^2{}_2(8)$
[Co(*hfac*)$_2$(4-quz)$_2$]	98		$R^2{}_2(8)$

hfac, *1,1,1,6,6,6-hexafluoroacetylacetone; −, no coordination product obtained.*

Modified from M. Borovina, I. Kodrin, M. Đaković, Predicting supramolecular connectivity of metal-containing solid-state assemblies using calculated molecular electrostatic potential surfaces, Cryst. Growth Des. 19 (2019) 1985−1995.

donor will be selective for only one acceptor (the stronger one), while (2) if the MEP difference is relatively small, both acceptors will be equally competitive and *no selectivity* will be observed. In case of the latter (2), both acceptor sites will be engaged in linking neighboring building units, and a variety of supramolecular motifs is to be expected. If the MEP difference (ΔE) is substantial (1) and in favor of the acceptor residing on the supramolecular functionality (A_1), the metal-containing building units will be assembled via the same supramolecular link and will display the same motifs as observed in metal-free systems, i.e., $R^2{}_2(8)$ or $C(4)$.

For differentiating between two potential outcomes, (1) and (2), the MEP difference (ΔE) was compared with a working cutoff value (ΔE_{cutoff}) that was "borrowed" from the related 1-D systems ($\approx$30 kJ/mol) as the smallest value observed for which the transferability of the lactam (amide) synthon from organic to metal−organic setting was observed.

The premise was successfully proven since in 10 of 12 intended complexes (10/12, 83%) the correct supramolecular link was properly assigned, while for other two cases (Table 6.3), the coordination chemistry failed and conseqeuently prevented testing of all the 12 (three metal cations × two *acac*-based ligands × two lactam molecules) predicted outcomes (Fig. 6.33). Despite the fact that the working ΔE_{cutoff}

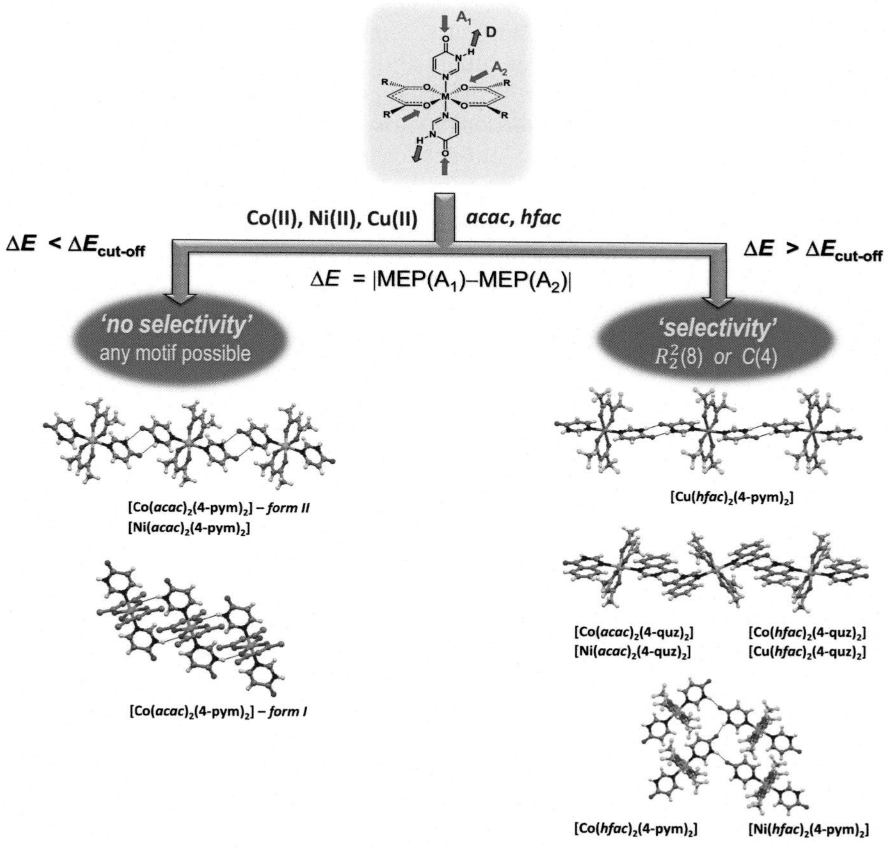

FIGURE 6.33

A discrete 0-D metal-based building unit involving just one good hydrogen bond donor (D) and two competing acceptor sites (A_1 and A_2; top). Assembling of metal complexes via supramolecular motifs typically observed in metal-free systems (amide/lactam self-complementary $R^2{}_2(8)$ ring and C(4) chain motifs). The prediction of the supramolecular products is made on the difference in calculated MEP values between two competing hydrogen bond acceptor sites (lactam and *acac*-based oxygen atoms). *MEP*, molecular electrostatic potential.

value worked well for the examined system, it is expected that it might undergo certain tuning with the availability of more relevant data. The research provided another verification of the "gray zone" concept.

The results showed that supramolecular products involving low-dimensional (0-D) metal-based building units could also be *predicted* employing a simple electrostatic view of the hydrogen bond. Moreover, the study suggested two additional criteria for *fine-tuning* of MEP-based protocol when planning the supramolecular synthesis that relies on the transferability of supramolecular synthons from organic to metal–organic systems is intended. It was demonstrated that (1) the introduction of metal cations and charge-compensating entities to the metal-free systems must not disrupt the relative importance of the hydrogen bond donor and acceptor sites (i.e., the acceptor residing at the functionality that is intended to drive the supramolecular assembly, must even after introduction of the metal center and charge-balancing entity remain the best acceptor on the building unit) and (2) the difference in MEP values (ΔE) between the competing acceptor sites (residing on the same building unit) is greater than a certain cutoff value (ΔE_{cutoff}).

The research also brought to surface an interesting notion—the possibility that the cutoff value (ΔE_{cutoff}), the value for deciding between different supramolecular outcomes could be associated with the supramolecular functionality itself rather than with other structural factors (such as building unit dimensionality, employment of metal cations or ligands for compensation of charges, or maintaining the coordination geometry, steric requirements, and so on). The working cutoff value (ΔE_{cutoff}) that was used to differentiate between the possible supramolecular events was "borrowed" from a related system having solely the functionality as a mutual characteristic, but the system of different dimensionality (1-D). It proved to be a valid decisive criterion despite new structural features (metal cations, additional ligands) was introduced to the system. This may also suggest that probably other organic functionalities might behave similarly; the idea unquestionably requires more attention and systems to be tested.

An additional study based on 0-D metal-containing building units was performed to examine the versatility of the MEP-based guidelines and their translation to metal–organic systems [132].

Here again, the system was kept as simple as possible, allowing for changing just one parameter at the time. Discrete 0-D building units with octahedral coordination geometry were employed, and those were built up from Co(II) and Ni(II) cations, *acac*-based ligands (1,1,1,6,6,6-hexafluoroacetylacetone, *hfac*, and dibenzoilmethane, *dbm*) and small aromatic ligands equipped with the oxime functionality to drive the assembly process (Fig. 6.34).

The results showed that when the MEP differences between two competing hydrogen bond acceptor sites were substantial (greater than the certain cutoff value;

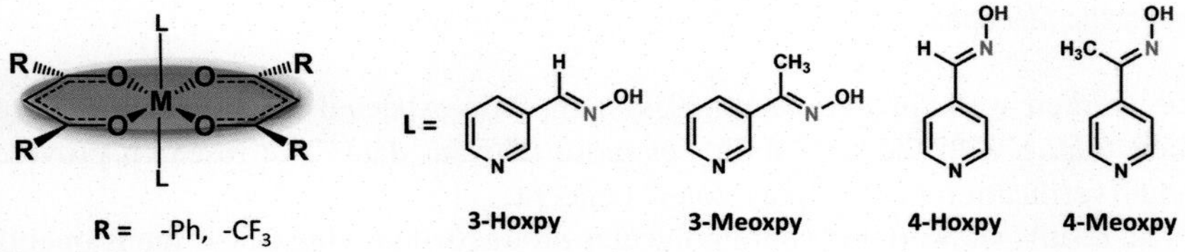

FIGURE 6.34

Discrete 0-D octahedral building units constructed from the *acac*-based and oxime ligands, coordinated to the Co(II) and Ni(II) cations.

Table 6.4 Calculated molecular electrostatic potential (MEP) surface values at two competing hydrogen bond acceptor sites, the oxime nitrogen and β-diketonato oxygen atoms, for a family of 0-D metal-containing building units, along with the MEP difference and the supramolecular link materialized.

Compound[a]	N_{oxime}	O_{acac}	$MEP(O_{acac})-MEP(N_{oxime})$/kJ/mol	Supramolecular link
$[Co(dbm)_2(4\text{-Hoxpy})_2]$	−83	−158	80	$O{-}H\cdots O_{\beta\text{-}dik}$
$[Ni(dbm)_2(4\text{-Hoxpy})_2]$	−84	−164	76	
$[Co(dbm)_2(4\text{-Meoxpy})_2]$	−90	−163	75	
$[Ni(dbm)_2(4\text{-Meoxpy})_2]$	−94	−170	73	
$[Co(dbm)_2(3\text{-Hoxpy})_2]$	−88	−149	67	
$[Ni(dbm)_2(3\text{-Hoxpy})_2]$	−89	−156	61	
$[Co(dbm)_2(3\text{-Meoxpy})_2]$	−95	−143	58	$O{-}H\cdots N_{oxime}$ $R^2_2(8)$
$[Ni(dbm)_2(3\text{-Meoxpy})_2]$	−96	−154	58	
$[Co(hfac)_2(4\text{-Hoxpy})_2]$	−54	−112	57	
$[Co(hfac)_2(4\text{-Meoxpy})_2]$	−62	−119	48	
$[Co(hfac)_2(3\text{-Meoxpy})_2]$	−65	−104	39	

[a] dbm, *dibenzoylmethane;* hfac, *1,1,1,6,6,6-hexafluoroacetylacetone.*

$\Delta E > 60$ kJ/mol) and in favor of the acceptor residing at the charge-balancing ligand (i.e., β-diketonato oxygen atom), the link involving the better acceptor, e.g., $N{-}H\cdots O_{\beta\text{-}dik}$ hydrogen bond, was formed. Consequently, translation of the supramolecular synthon from organic to metal–organic system was not achieved (Table 6.4). On the other hand, when the MEP difference was smaller than the cutoff value ($\Delta MEP < 60$ kJ/mol), the direct switch between the supramolecular links occurred, and the link involving N_{oxime} as hydrogen bond acceptor was the sole supramolecular product. No *diffuse* region (where both outcomes are possible, i.e., $N{-}H\cdots O_{\beta\text{-}dik}$ and $N{-}H\cdots O_{oxime}$) within the examined family was noticed.

6.3 Halogen bonding

Systematic structural studies on halogen bonding in metal–organic systems are, despite their recently growing number, still quite limited. Especially scarce are those reports that mimic MEP-based protocols in metal-free systems. The fact is that halogen-equipped metal-based building units "suffer" from some restrictions that the typical organic systems are free from.

For organic systems, it has been shown that the ability of a halogen atom to participate in the halogen link can be enhanced by "activation" through covalent modification(s) in the molecular scaffold, typically via introduction of the electron-withdrawing substituents [103–105]. On the other hand, in metal–organic system, depletion of the electron density from a potential halogen-bearing ligand can, in turn, hamper its ability to effectively coordinate to a metal cation. Moreover, in contrast to hydrogen-bonded synthons that often display multipoint interactions, halogen-bonded analogues typically involve single-point interactions, which can make them kinetically and thermodynamically less effective. This may introduce additional hurdles for producing single-component metal-based solid-state assemblies primarily relying on halogen bonds, as other interactions tend to prevail. To weaken the impact of halogen "activation" via electron-withdrawing substituents on the coordination site [136], Aakeröy and coworkers tried a different approach. They tailored ligands to contain two "separated" aromatic rings each bearing one functionality, a halogen bond donor and the coordination site. But despite carefully designed structural characteristics of the ligands, π-stacking interactions dominated the overall architecture. To limit the impact of π-stacking on the resulting crystal structure, Đaković and coworkers then employed small ligands having a single aromatic ring and fine-tuned the electron density at the halogen atom σ-hole by introduction of the additional nitrogen on the aromatic ring and by changing its position in the ring (Fig. 6.35) [72].

A moderate "activation" of the halogen atom produced the desired halogen bond interaction only in a limited number of supramolecular products; instead, two other supramolecular links, a C—X···O halogen and a C—H···O hydrogen bond, were found to have a larger influence on the overall supramolecular assembly (Fig. 6.36).

The roles of the two prominent links were explained on the basis of calculated MEP values. Irrespective of the supramolecular connectivity (C—X···O/C—H···O), the *acac* oxygen atom was always found to be the best acceptor site, displaying the most negative MEP value, while the two competing donor sites did not display any clear trend in their MEP values (Table 6.5). In contrast, the MEP differences between the two donors presented a clearer explanation for the observed supramolecular outcomes. For the two cases at each end, substantially larger MEP difference in favor of halogen bond donor and substantially larger MEP difference in favor of hydrogen bond acceptor, the halogen link and hydrogen link formed, respectively. For all the other cases with smaller MEP differences, a *diffuse* region with both structural outcomes was present.

Although the system examined presented two rather weak competing donor sites (halogen and hydrogen atoms) with relatively small MEP differences, the MEP-based protocol derived for organic solid-state systems readily provided a rationale for the supramolecular outcomes.

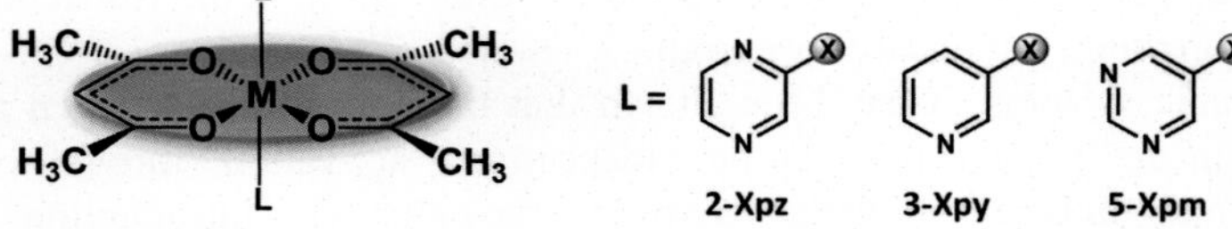

FIGURE 6.35

0-D Metal-based building unit designed to present a single potential donor atom, the halogen atom, intended as to control the assembly process (M = Co, Ni; X = Cl, Br, I).

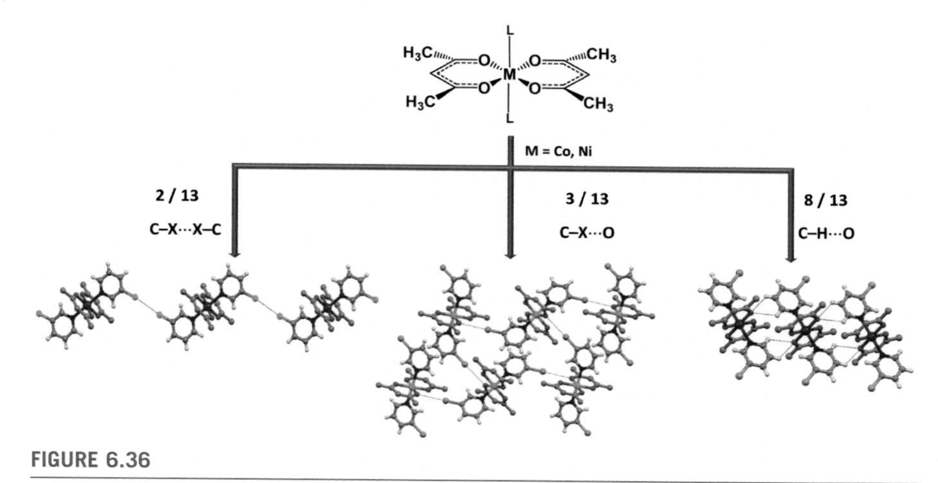

FIGURE 6.36

Structures of the assembly process of discrete 0-D metal-containing systems with competing halogen and hydrogen interactions along with frequencies of their occurrence.

Table 6.5 Calculated molecular electrostatic potential (MEP) values (kJ/mol) for two best donor sites along with the MEP differences between them and the supramolecular link materialized.

Compound	MEP(X)	MEP(H)	MEP(X) − MEP(H)	Interaction	
$[Co(acac)_2(3\text{-}Clpy)_2]$	55	68	48	**XB**	Halogen
$[Ni(acac)_2(3\text{-}Clpy)_2]$	57	65	42	**XB**	bond
$[Co(acac)_2(3\text{-}Brpy)_2]$	77	70	17	HB	*Diffuse*
$[Ni(acac)_2(3\text{-}Brpy)_2]$	80	68	12	XB	region
$[Co(acac)_2(3\text{-}Ipy)_2]$	112	70	12	HB	
$[Ni(acac)_2(3\text{-}Ipy)_2]$	115	67	7	HB	
$[Co(acac)_2(3\text{-}Clpz)_2]$	56	99	6	HB	
$[Ni(acac)_2(3\text{-}Clpz)_2]$	60	96	−12	Type I	
$[Co(acac)_2(3\text{-}Brpz)_2]$	75	101	−13	Type I	
$[Ni(acac)_2(3\text{-}Brpz)_2]$	79	99	−20	HB	
$[Co(acac)_2(3\text{-}Ipz)_2]$	107	101	−26	HB	
$[Ni(acac)_2(3\text{-}Ipz)_2]$	110	98	−36	**HB**	Hydrogen
$[Co(acac)_2(5\text{-}Brpz)_2]$	99	72	−43	**HB**	bond

7. Conclusions

Despite considerable efforts by numerous experts in experimental and theoretical chemistry and materials science, reliable bottom-up engineering strategies of new solids with tunable physical properties and responses remain largely elusive. However, in trying to bridge the gap between a molecular structure and organized 3-D solid-state assembly, calculated molecular electrostatic potentials can offer reliable and informative guidance as a practical crystal-engineering tool. This has been accomplished by using MEPs for identifying a variety of structure-directing features as well as for rationalizing supramolecular synthetic outcomes by decoding and ranking the donor/acceptor potential of prospective interacting sites. The fundamental insight that MEPs can offer in intermolecular assembly is now being transitioned into reliable supramolecular synthetic strategies for organizing molecular building blocks of both organic and metal-organic systems into extended architectures with specific dimensionalities and topologies. In particular, crystal engineering strategies that utilize hydrogen bonds or σ-hole interactions as primary synthetic vectors, are especially well supported by MEP-based hierarchies of competing donor and/or acceptor. Ultimately, materials produced via hierarchical self-assembly will play crucial roles in the development of adaptive and more sustainable technologies. As long as we can deliver structure-controlled materials, we can begin to effectively apply concepts of nanoarchitectonics for dialing-in properties that determine the performance, cost, and sustainability of new technologies and devices.

References

[1] M.D. Ward, P.R. Raithby, Functional behaviour from controlled self-assembly: challenges and prospects, Chem. Soc. Rev. 42 (2013) 1619−1636.

[2] G.R. Desiraju, Crystal Engineering: The Design of Organic Solids, Elsevier, Amsetrdam, 1989.

[3] G. Schmidt, Photodimerization in the solid state, Pure Appl. Chem. 27 (1971) 647−678.

[4] G.R. Desiraju, Supramolecular synthons in crystals engineering − a new organic synthesis, Angew. Chem. Int. Ed. 34 (1995) 2311−2327.

[5] C.R. Groom, I.J. Bruno, M.P. Lightfoot, S.C. Ward, The Cambridge Structural Database, Acta Crystallogr. B72 (2016) 171−179.

[6] E.R.T. Tiekink, J.J. Vittal, Frontiers in Crystal Engineering, John Wiley & Sons, 2006.

[7] G.R. Desiraju, Crystal engineering: a holistic view, Angew. Chem. Int. Ed. (2007) 8342−8356.

[8] E.R.T. Tiekink, J.J. Vittal, M. Zaworotko, Organic Crystal Engineering: Frontiers in Crystal Engineering, Wiley-Blackwell, 2010.

[9] G.R. Desiraju, J.J. Vittal, A. Ramanan, Crystal Engineering, Cambridge University Press India, 2011.

[10] T. Steiner, The hydrogen bond in the solid-state, Angew. Chem. Int. Ed. 41 (2002) 48−76.

[11] G. Resnati, E. Boldyreva, P. Bombicz, P. Kawano, Supramolecular interactions in the solid-state, IUCrJ 2 (2015) 675–690.

[12] C.B. Aakeröy, N.R. Champness, C. Janiak, Recent advances in crystal engineering, CrystEngComm 12 (2010) 22–43.

[13] H.-J. Schneider, Applications of Supramolecular Chemistry, CRC Press, Taylor & Francis Group, 2016.

[14] I.V. Kolesnichenko, E.V. Anslyn, Chem. Soc. Rev. 46 (2017) 2385–2390.

[15] P. Numov, S. Chizhik, M.K. Panda, N.K. Nath, E. Boldyreva, Mechanically responsive molecular crystals, Chem. Rev. 115 (2015) 12440–12490.

[16] L.R. MacGillivray, Metal-Organic Frameworks: Design and Applications, John Wiley & Sons, 2010.

[17] C.B. Aakeröy, A.S. Sinha, Co-crystals: Preparation, Characterization and Applications, RSC Publishing, pp. 1–32, 184–301, 2018.

[18] D. Braga, Crystal engineering, Where from? Where To? Chem. Commun. (2003) 2751–2754.

[19] V.R. Thalladi, B.S. Goud, V.J. Hoy, F.H. Allen, J.A.K. Howard, G.R. Desiraju, Supramolecular synythons in crystal engineering. Structure simplification, synthon robustness and supramolecular retrosynthesis, Chem. Commun. (1996) 401–402.

[20] C.B. Aakeröy, Crystal engineering: strategies and architectures, Acta Crystallogr. B53 (1997) 569–586.

[21] A. Nangia, G.R. Desiraju, Supramolecular synthons and pattern recognition, Top. Curr. Chem. 198 (1998) 57–94.

[22] C.B. Aakeröy, A.M. Beatty, Crystal engineering of hydrogen-bonded assemblies – a progress report, Aust. J. Chem. 54 (2001) 409–421.

[23] G. Desiraju, Crystal engineering: a brief overview, J. Chem. Sci. 122 (2010) 667–675.

[24] K. Marz, V. Vasylyeva, Developments and boundaries in the field of supramolecular synthons, CrystEngComm 12 (2010) 3989–4002.

[25] M.K. Corpinot, D.-K. Bučar, A practical guide to the design of molecular crystals, Cryst. Growth Des. 19 (2019) 1426–1453.

[26] D. Braga, F. Grepioni, P. Sabatino, G.R. Desiraju, Hydrogen bonding in organometallic crystals. 1. From carboxylic acids and alcohols to carbonyl complexes, Organometallics 13 (1994) 3532–3543.

[27] K. Biratha, G.R. Desiraju, D. Braga, F. Grepioni, Hydrogen bonding in organometallic crystals. 3.1 transition-metal complexes containing amido groups, Organometallics 15 (1996) 1284–1295.

[28] C.B. Aakeröy, A.M. Beatty, D.S. Leinen, A versatile route to porous solids: organic-inorganic hybrid materials assembled through hydrogen bonds, Angew. Chem. Int. Ed. 38 (1999) 1815–1819.

[29] C.B. Aakeröy, J. Desper, J. Valdés-Martínez, Controlling molecular and supramolecular structure of hydrogen-bonded coordination compounds, CrystEngComm 6 (2006) 413–418.

[30] S. Baudron, D. Salazar-Mendoza, M.W. Hosseini, Combination of primary amide and dipyrrin for the elaboration of extended architectures built upon both coordination and hydrogen bonding, CrystEngComm 11 (2009) 1245–1254.

[31] M. Đaković, D. Vila-Viçosa, M.J. Calhorda, Z. Popović, Coordination-driven self-assembly of thiocyanate complexes of Co(II), Ni(II) and Cu(II) with picolinamide: a structural and DFT study, CrystEngComm 13 (2011) 5863–5871.

[32] M. Đaković, Z. Jagličić, B. Kozlevčar, Z. Popović, Association of copper(II) isonicotinamide moieties via different anionic bridging ligands: two paths of ferromagnetic interaction in the azide coordination compound, Polyhedron 29 (2010) 1910–1917.
[33] L. Brammer, Developments in inorganic crystal engineering, Chem. Soc. Rev. 33 (2004) 476–489.
[34] L. Brammer, Metals and hydrogen bonds, Dalton Trans. (2003) 3145–3157.
[35] C.B. Aakeröy, Is there any point in making co-crystals? Acta Crystallogr. B71 (2015) 387–391.
[36] M. Garçon, C. Bakewell, G.A. Sackman, A.J.P. White, R.I. Cooper, A.J. Edwards, M.R. Crimmin, A hexagonal planar transitional-metal complex, Nature 574 (2019) 390–393.
[37] M. Đaković, M. Borovina, M. Pisačić, C.B. Aakeröy, Ž. Soldin, B.-M. Kukovec, I. Kodrin, Mechanically responsive crystalline coordination polymers with controllable elasticity, Angew. Chem. Int. Ed. 57 (2018) 14801–14805.
[38] S.M. Jane, P. Lane, T. Brinck, P. Politzer, Relationships between computed molecular properties and solute-solvent interactions in supercritical solutions, J. Phys. Chem. 97 (1993) 5144–5148.
[39] J.S. Murray, P. Lane, T. Brinck, K. Paulsen, M.E. Grice, P. Politzer, Relationships of critical constants and boiling points to computed molecular surface properties, J. Phys. Chem. 97 (1993) 9369–9373.
[40] T. Clark, M. Hennemann, J. Murray, P. Politzer, Halogen bonding: the σ-hole, J. Mol. Model. 13 (2007) 291–296.
[41] J.S. Murray, K. Paulsen, P. Politzer, Molecular surface electrostatic potentials in the analysis of non-hydrogen-bonding noncovalent interactions, Proc. Indian Acad. Sci. 106 (1994) 267–275.
[42] S.R. Gadre, S.A. Kulkarni, C.H. Suresh, I.H. Shrivastava, Basis set dependence of the molecular electrostatic potential topography. A case study of substituted benzenes, Chem. Phys. Lett. 239 (1995) 273–281.
[43] E. Scrocco, J. Tomasi, Electronic molecular structure, reactivity and intermolecular forces: an heuristic interpretation by means of electrostatic molecular potentials, Adv. Quantum Chem. 11 (1978) 115–193.
[44] P.A. Kollman, The role of the electrostatic potential in modeling hydrogen bonding and other non-covalent interactions, in: P. Politzer, D.G. Truhlar (Eds.), Chemical Applications of Atomic and Molecular Electrostatic Potentails, Springer Science + Business Media, LLC, New York, 1981.
[45] G. Naray-Szabó, G.G. Ferenczy, Molecular electrostatics, Chem. Rev. 95 (1995) 829–847.
[46] J.S. Murray, P. Politzer, Statistical analysis of the molecular surface electrostatic potential: an approach to describing noncovalent interactions in condensed phases, THEOCHEM 425 (1998) 107114.
[47] P. Politzer, J.S. Murray, M.C. Concha, The complementary roles of molecular surface electrostatic potentials and average local ionization energies with respect to electrophilic processes, Int. J. Quantum Chem. 88 (2002) 12–27.
[48] P. Kollman, J. McKelvey, A. Johansson, S. Rothenberg, Theoretical studies of hydrogen-bonded dimers. Complexes involving HF, H_2O, NH_3, HCl, H_2S, PH_3, HCN, HNC, HCP, CH_2NH, H_2CS, H_2CO, CH_4, CF_3H, C_2H_2, C_2H_4, C_6H_6, F^-, and H_3O^+, J. Am. Chem. Soc. 97 (1975) 955–965.
[49] G.G. Hall, K. Tsujinaga, The molecular electrostatic potential of some simple molecules, Theor. Chim. Acta 69 (1986) 425–436.

[50] J.S. Murray, R. Gilardi, M.E. Grice, P. Kane, P. Politzer, Structures and molecular surface potentials of high-density C, N, H systems, Struct. Chem. 7 (1996) 273–280.
[51] D. Hadži, M. Hodoščcek, D. Kocjan, T. Šolmajer, F. Avbelj, The molecular electrostatic potential as a determinant of receptor-drug recognition, Croat. Chem. Acta 57 (1984) 1065–1074.
[52] P.K. Weiner, R. Langridge, J.M. Blaney, R. Schaefer, P.A. Kollman, Electrostatic potential molecular surfaces, Proc. Natl. Acad. Sci. U.S.A. 79 (1982) 3754–3758.
[53] R.F.W. Bader, M.T. Carroll, J.R. Cheesmann, C. Chang, Properties of atoms in molecules: atomic volumes, J. Am. Chem. Soc. 109 (1987) 7968–7979.
[54] C.A. Hunter, Quantifying intermolecular interaction: guidelines for the molecular recognition toolbox, Angew. Chem. Int. Ed. 43 (2004) 5310–5324.
[55] M.H. Abraham, P.L. Grellier, D.V. Prior, R.W. Taft, J.J. Morris, P.J. Taylor, C. Laurence, M. Berthelot, R.M. Doherty, M.J. Kamlet, J.-L.M. Abboud, K. Sraidi, G. Guiheneuf, A general treatment of hydrogen bond complexation constants in tetrachloromethane, J. Am. Chem. Soc. 110 (1988) 8534–8536.
[56] M.H. Abraham, J.A. Platts, Hydrogen bond structural group constants, J. Org. Chem. 66 (2001) 3484–3491.
[57] L. Brammer, E.A. Bruton, P. Sherwood, Fluoride ligands exhibit marked departures from the hydrogen bond acceptor behaviour of their heavier halogen congeners, New J. Chem. 23 (1999) 965–968.
[58] L. Brammmer, J.K. Swearingen, E.A. Bruton, P. Sherwood, Hydrogen bonding and perhalometallate ions: a supramolecular synthetic strategy for new inorganic materials, Proc. Natl. Acad. Sci. U. S. A. 99 (2002) 4956–4961.
[59] M. Borovina, I. Kodrin, M. Đaković, Predicting supramolecular connectivity of metal-containing solid-state assemblies using calculated molecular electrostatic potential surfaces, Cryst. Growth Des. 19 (2019) 1985–1995.
[60] I. Kodrin, M. Borovina, L. Šmital, J. Valdés-Martínez, C.B. Aakeröy, M. Đaković, Exploring and predicting intermolecular binding preferences in crystalline Cu(II) coordination complexes, Dalton Trans. 48 (2019) 16222–16232.
[61] J.M.A. Robinson, B.M. Kariuki, K.D.M. Harris, D. Philp, Interchangeability of halogen and ethenyl substituents in the solid state structures od di- and tri-substituted benzenes, J. Chem. Soc., Perkin Trans. 2 (1998) 2459–2469.
[62] J.A.R.P. Sarma, F.H. Allen, V.J. Hoy, J.A.K. Howard, R. Thaimattam, K. Biradha, G.R. Desiraju, Design of an SHG-active crystal, 4-iodo-4′-nitrobiphenyl: the role of supramolecular synthons, Chem. Commun. (1997) 101–102.
[63] H.-C. Weiss, R. Boese, H.L. Smith, M.H. Haley, $\equiv$CH$\cdots\pi$ versus $\equiv$CH$\cdots$halogen interactions - the crystal structures of the 4-halogenoethynylbenzenes, Chem. Commun. (1997) 2403–2404.
[64] N. Masciocchi, M. Bergamo, A. Sironi, Comments on the elusive crystal structure of 4-iodo-4′-nitrobiphenyl, Chem. Commun. (1998) 1347–1348.
[65] B.K. Saha, A. Nangia, M. Jaskólski, Crystal engineering with hydrogen bonds and halogen bonds, CrystEngComm 7 (2005) 355–358.
[66] J.-L. Syssa-Magalé, K. Boubekeur, P. Palvadeau, A. Meerschaut, B. Schöllhorn, The tailoring of crystal structures *via* the self-assembly of organic coordination compounds by N$\cdots$I non-covalent halogen bonds: co-crystal of sterically hindered N-heterocycles and 1,4-diiodo-tetrafluorobenzene, CrystEngComm 7 (2005) 302–308.
[67] E. Bartashevich, E. Troitskaya, Á.M. Pendás, V. Tsirelson, Understanding the bifurcated halogen bonding N$\cdots$Hal$\cdots$N in bidentate diazaheterocyclic compounds, Comput. Theor. Chem. 1053 (2015) 229–237.

[68] M. Novák, C. Foroutan-Nejad, R. Marek, Asymmetric bifurcated halogen bonds, Phys. Chem. Chem. Phys. 17 (2015) 6440–6450.

[69] B. Ji, W. Wang, D. Deng, Y. Zhang, Symmetrical bifurcated halogen bond: design and synthesis, Cryst. Growth Des. 11 (2011) 3622–3628.

[70] R. Liu, Y.J. Gao, W.J. Jin, Colour-tunable phosphorescence of 1,10-phenanthrolines by 4,7-methyl/-diphenyl/-dichloro substituents in cocrystals assembled *via* bifurcated C–I···N halogen bonds using 1,4-diiodotetrafluorobenzene as a bonding donor, Acta Crystallogr. B73 (2017) 247–254.

[71] J.C. Gamekkanda, A.S. Sinha, J. Desper, M. Đaković, C.B. Aakeröy, The role of halogen bonding in controlling assembly and organization of Cu(II)-acac based coordination complexes, Crystals 7 (2017) 226–243.

[72] M. Borovina, I. Kodrin, M. Đaković, Testing the limits of halogen bonding in coordination chemistry, CrystEngComm 20 (2018) 539–549.

[73] A.V. Rozhkov, A.S. Novikov, D.M. Ivanov, D.S. Bolotin, N.A. Bokach, V.Y. Kukushkin, Structure-directing weak interactions with 1,4-diiodotetrafluorobenzene convert one-dimensional arrays of [$M^{II}(acac)_2$] species into three-dimensional networks, Cryst. Growth Des. (2018) 183626–183636.

[74] V. Stilinović, T. Grgurić, T. Piteša, V. Nemec, D. Cinčić, Bifurcated and monocentric halogen bonds in cocrystals of metal(II) acetylacetonates with *p*-dihalotetrafluorobenzenes, Cryst. Growth Des. 19 (2019) 1245–1256.

[75] S.V. Baykov, S.I. Filimonov, A.V. Rozhkov, A.S. Novikov, I.V. Ananyev, D.M. Ivanov, V.Y. Kukushkin, Reverse sandwich structures form interplay between lone pair–π-hole atom-directed C···d_{z2}[M] and halogen bond interactions, Cryst. Growth Des. 19 (2019) 995–1008.

[76] D. Musumeci, C.A. Hunter, R. Prohens, S. Scuderi, J.F. McCabe, Virtual cocrystal screening, Chem. Sci. 2 (2011) 883–890.

[77] T. Grecu, C.A. Hunter, E.J. Gardiner, J.F. McCabe, Validation of a computational cocrystal prediction tool: comparison of virtual and experimental cocrystal screening results, Cryst. Growth Des. 14 (2014) 165–171.

[78] B. Sandhu, A. McLean, S. Sinha, J. Desper, A.A. Sarjeant, S. Vyas, S.M. Reutzel-Edens, C.B. Aakeröy, Evaluating competing intermolecular interactions through molecular electrostatic potentials and hydrogen-bond propensities, Cryst. Growth Des. 18 (2018) 466–478.

[79] P.T.A. Galek, F.H. Allen, L. Fabian, N. Feeder, Knowledge-based H-bond prediction to aid experimental polymorph screening, CrystEngComm 11 (2009) 2634–2639.

[80] P.A. Wood, N. Feeder, M. Furlow, P.T.A. Galek, C.R. Groom, E. Pidcock, Knowledge-based approach to co-crystal design, CrystEngComm 16 (2014) 5839–5848.

[81] P.T.A. Galek, J.A. Chisholm, E. Pidcock, P.A. Wood, Hydrogen bond co-ordination in organic crystal structures: statistics, predictions and applications, Acta Crystallogr. B70 (2014) 91–105.

[82] D.A. Haynes, J.M. Rawson, Molecular electrostatic potential as a predictor of supramolecular synthons in non-hydrogen-bonded systems: application to heavier p-block systems, Eur. J. Inorg. Chem. 13 (2018) 3554–3564.

[83] C.B. Aakeröy, K. Epa, Controlling supramolecular assembly using electronic effects, Top. Curr. Chem. 351 (2011) 125–155.

[84] C.B. Aakeröy, A.M. Beatty, B.A. Helfrich, "Total synthesis" supramolecular style: design and hydrogen-bond-directed assembly of ternary supermolecules, Angew. Chem. Int. Ed. 40 (2001) 3240–3242.

[85] P. Vishweshwar, A. Nangia, V.M. Lynch, Supramolecular synthons in phenol–isonicotinamide adducts, CrystEngComm 5 (2003) 164–168.
[86] C.B. Aakeröy, T.K. Wijethunga, J. Desper, Molecular electrostatic potential dependent selectivity of hydrogen bonding, New J. Chem. 39 (2015) 822–828.
[87] C.B. Aakeröy, K. Epa, S. Forbes, N. Schultheiss, J. Desper, Ranking relative hydrogen-bond strengths in hydroxybenzoic acids for crystal-engineering purposes, Chem. Eur. J. 19 (2013) 14998–15003.
[88] M.C. Etter, Hydrogen bonds as design elements in organic chemistry, J. Phys. Chem. 95 (1991) 4601–4610.
[89] C.B. Aakeroy, A. Rajbanshi, J. Desper, Hydrogen-bond driven assembly of a molecular capsule facilitated by supramolecular chelation, Chem. Commun. 47 (2011) 11411–11413.
[90] M.K. Stanton, A. Bak, Physicochemical properties of pharmaceutical co-crystals: a case study of ten AMG 517 co-crystals, Cryst. Growth Des. 8 (2008) 3856–3862.
[91] P. Metrangolo, H. Neukirch, T. Pilati, G. Resnati, Halogen bonding based recognition processes: a world parallel to hydrogen bonding, Acc. Chem. Res. 38 (2005) 386–395.
[92] P. Metrangolo, F. Meyer, T. Pilati, G. Resnati, G. Terraneo, Halogen bond in supramolecular chemistry, Angew. Chem. Int. Ed. 47 (2008) 6114–6127.
[93] M. Fourmigue, Halogen bonding: recent advances, Curr. Opin. Solid State Mater. Sci. 13 (2009) 36–45.
[94] G. Cavallo, P. Metrangolo, R. Milani, T. Pilati, A. Priimagi, G. Resnati, G. Terraneo, The halogen bond, Chem. Rev. 116 (2016) 2478–2601.
[95] P. Politzer, J. Murray, T. Clark, Halogen bonding: an electrostatically-driven highly directional noncovalent interaction, Phys. Chem. Chem. Phys. 12 (2010) 7748–7757.
[96] T. Brink, J.S. Murray, P. Politzer, Surface electrostatic potentials of halogenated methanes as indicators of directional intermolecular interactions, Int. J Quantum Chem. 44 (1992) 57–64.
[97] P. Politzer, J.S. Murray, P. Lane, σ-hole bonding and hydrogen bonding: competitive interactions, Int. J. Quantum Chem. 107 (2007) 3046–3052.
[98] P. Politzer, J.S. Murray, M.C. Concha, Halogen bonding and the design of new materials: organic bromides, chlorides and perhaps even fluorides as donors, J. Mol. Model. 13 (2007) 643–650.
[99] C.B. Aakeröy, T.K. Wijethunga, M.A. Haj, J. Desper, C. Moore, The structural landscape of heteroaryl-2-imidazoles: competing halogen- and hydrogen-bond interactions, CrystEngComm 16 (2014) 7218–7225.
[100] R.B. Walsh, C.W. Padgett, P. Metrangolo, G. Resnati, T.W. Hangs, W.T. Pennington, Crystal engineering through halogen bonding: complexes of nitrogen heterocycles with organic iodides, Cryst. Growth Des. 1 (2001) 167–175.
[101] K. Raatkainen, K. Rissanen, Hierarchical halogen bonding induces polymorphism, CrystEngComm 11 (2009) 750–752.
[102] J.-Y. Le Questel, C. Laurence, J. Graton, Halogen-bond interactions: a crystallographic basicity scale towards iodoorganic compounds, CrystEngComm 15 (2013) 3212–3221.
[103] C.B. Aakeröy, M. Baldrighi, J. Desper, P. Metrangolo, G. Resnati, Supramolecular hierarchy among halogen-bond donors, Chem. Eur J. 19 (2013) 16240–16247.
[104] C.B. Aakeröy, T.K. Wijethunga, J. Desper, M. Đaković, Crystal engineering with iodoethynylnitrobenzenes: a group of highly effective halogen-bond donors, Cryst. Growth Des. 15 (2015) 3853–3861.

[105] O. Bolton, E.C. Burgess, A.J. Matzger, Unprecedented size of the sigma-holes on 1,3,5-triiodo-2,4,6-trinitrobenzene begets unprecedented intermolecular interactions, Cryst. Growth Des. 16 (2016) 1765–1771.
[106] C.B. Aakeröy, T.K. Wijethunga, J. Desper, Practical crystal engineering using halogen bonding: a hierarchy based on calculated molecular electrostatic potential surfaces, J. Mol. Struct. 1072 (2014) 20–27.
[107] C.G. Wermuth, The Practice of Medicinal Chemistry, third ed., Elsevier Ltd., China, 2008.
[108] Y. Lu, Y. Liu, Z. Xu, H. Li, H. Liu, W. Zhu, Halogen bonding for rational drug design and new drug discovery, Expet Opin. Drug Discovery 7 (2012) 375–383.
[109] (a) R. Wilcken, M.O. Zimmermann, A. Lange, F.M. Boeckler, Using halogen bonds to address the protein backbone: a systematic evaluation, J. Comput. Aided Mol. Des. 26 (2012) 935–945;
(b) H. Duan, W. Zhang, J. Zhao, D. Liang, X. Yang, S. Jin, A novel halogen bond and a better-known hydrogen bond cooperation of neonicotinoid and insect nicotinic acetylcholine receptor recognition, J. Mol. Model. 18 (2012) 3867–3875;
(c) A.K. Przybyl, M. Kubicki, The role of halogen bonding in crystal structures of 3-halogeno cytisine derivatives, J. Chem. Crystallogr. 42 (2012) 685–690;
(d) A.V. Jentzsch, D. Emery, J. Mareda, Transmembrane anion transport mediated by halogen-bond donors, Nat. Commun. 3 (2012) 905;
(e) M. Carter, A.K. Rappe, P.S. Ho, Scalable anisotroplic shape and electrostatic models for biological bromine halogen bonds, J. Chem. Theor. Comput. 8 (2012) 2461–2473.
[110] V.N.G. Lindsay, A.B. Charette, Design and synthesis of chiral heteroleptic rhodium(II) carboxylate catalysts: experimental investigation of halogen bond rigidification effects in asymmetric cyclopropanation, ACS Catal. 2 (2012) 1221–1225.
[111] D.W. Bruce, P. Metrangolo, F. Meyer, C. Präsang, G. Resnati, A.C. Whitwood, Mesogenic, trimeric, halogen-bonded complexes from alkoxystilbazoles and 1,4-diiodotetrafluorobenzene, New J. Chem. 32 (2008) 477–482.
[112] F. Zapata, A. Caballero, N.G. White, T.D.W. Claridge, P.J. Costa, V. Félix, P.D. Beer, Fluorescent charge-assisted halogen-bonding macrocyclic halo-imidazolium receptors for anion recognition and sensing in aqueous media, J. Am. Chem. Soc. 134 (2012) 11533–11541.
[113] H. Li, Y.-X. Lu, X. Zhu, CO_2 capture through halogen bonding: a theoretical perspective, Sci. China Chem. 55 (2012) 1566–1572.
[114] R. Gutzler, C. Fu, A. Dadvand, Y. Hua, J.M. MacLeod, F. Rosei, D.F. Perepichka, Halogen bonds in 2D supramolecular self-assembly of organic semiconductors, Nanoscale 4 (2012) 5965–5971.
[115] C.B. Aakeröy, T.K. Wijethunga, J. Desper, C. Moore, Halogen-bond preferences in co-crystal synthesis, J. Chem. Crystallogr. 45 (2015) 267–276.
[116] C.B. Aakeröy, J. Desper, M. Fasulo, I. Hussain, B. Levin, N. Schultheiss, Ten years of co-crystal synthesis: the good, the bad, and the ugly, CrystEngComm 10 (2008) 1816–1821.
[117] C.B. Aakeröy, T.K. Wijethunga, J. Desper, M. Đaković, Electrostatic potential differences and halogen-bond selectivity, Cryst. Growth Des. 16 (2016) 2662–2670.
[118] T. Di Paolo, C. Sandorfy, On the biological importance of the hydrogen bond breaking potency of fluorocarbons, Chem. Phys. Lett. 26 (1974) 466–469.
[119] T. Di Paolo, C. Sandorfy, On the hydrogen bond breaking ability of fluorocarbons containing higher halogens, Can. J. Chem. 52 (1974) 3612–3622.

[120] G. Ciancaleoni, Cooperativity between hydrogen- and halogen bonds: the case of selenourea, Phys. Chem. Chem. Phys. 20 (2018) 8506–8514.
[121] C.B. Aakeröy, M. Fasulo, N. Schultheiss, J. Desper, C. Moore, Structural competition between hydrogen bonds and halogen bonds, J. Am. Soc. Chem. 129 (2007) 13772–13773.
[122] B.A. DeHaven, A.L. Chen, E.A. Shimizu, S.R. Salpage, M.D. Smith, L.S. Shimizu, Synergistic effects of hydrogen and halogen bonding in co-crystals of dipyridylureas and diiodotetrafluorobenzene, Supramol. Chem. 30 (2018) 315–327.
[123] C.B. Aakeröy, J. Desper, B.A. Helfrich, P. Metrangolo, T. Pilati, G. Resnati, A. Stevenazzi, Combining halogen bonds and hydrogen bonds in the modular assembly of heteromeric infinite 1-D chains, Chem. Commun. (2007) 4236–4238.
[124] C.B. Aakeröy, N. Schultheiss, A. Rajbanshi, J. Desper, C. Moore, Supramolecular synthesis based on a combination of hydrogen and halogen bonds, Cryst. Growth Des. 9 (2009) 432–441.
[125] J. Martí-Rujas, L. Colombo, J. Lü, A. Dey, G. Terraneo, P. Metrangolo, T. Pilati, G. Resnati, Hydrogen and halogen bonding drive the orthogonal self-assembly of an organic framework possessing 2D channels, Chem. Commun. 48 (2012) 8207–8209.
[126] M.D. Perara, J. Desper, A.S. Sinha, C.B. Aakeröy, Impact and importance of electrostatic potential calculation for predicting structural patterns of hydrogen and halogen bonding, CrystEngComm 16 (2016) 8631–8636.
[127] C.A. Gunawardana, J. Desper, A.S. Sinha, M. Đaković, C.B. Aakeröy, Competition and selectivity in supramolecular synthesis: structural landscape around 1-(pyridylmethyl)-2,2'-biimidazoles, Faraday Discuss 203 (2017) 371–388.
[128] J.C. Gamekkanda, A.S. Sinha, J. Desper, M. Đaković, C.B. Aakeröy, Competition between hydrogen and halogen bonds: a structural study, New J. Chem. 42 (2018) 10539–10547.
[129] C.B. Aakeröy, C.L. Spartz, S. Dembowski, S. Dwyre, J. Desper, A systematic structural study of halogen bonding versus hydrogen bonding within competitive supramolecular systems, IUCrJ 2 (2015) 498–510.
[130] J.E. Ormond-Prout, P. Smart, L. Brammer, Cyanometallates as halogen bond acceptors, Cryst. Growth Des. 12 (2012) 205–216.
[131] M. Đaković, Ž. Soldin, B.-M. Kukovec, I. Kodrin, C.B. Aakeröy, N. Baus, T. Rinkovec, Building inorganic supramolecular architectures using principles adopted from the organic solid state, IUCrJ 5 (2018) 13–21.
[132] B.M. Kukovec, M. Malik, I. Kodrin, C.B. Aakeröy, M. Đaković, Directed assembly of acac-based complexes by deliberately fine-tuning electrostatic molecular-recognition events, Cryst. Growth Des. 16 (2016) 7308–7317.
[133] R. Wang, C.W. Lehmann, U. Englert, Weak interactions in chain polymers [M(μ-X) 2L2]∞ (M = Zn, Cd; X = Cl, Br; L = substituted pyridine) - an electron density study, Acta Crystallogr. B65 (2009) 600–611.
[134] M.C. Etter, Encoding and decoding hydrogen-bond patterns of organic compounds, Acc. Chem. Res. 23 (1990) 120–126.
[135] J. Bernstein, R.E. Davis, L. Shimoni, N.-L. Chang, Patterns in hydrogen bonding: Functionality and Graph set analysis in crystals, Angew. Chem., Int. Ed. Engl. 34 (1995) 1555–1573.
[136] C.B. Aakeröy, N. Schultheiss, J. Desper, C. Moore, Attempted assembly of discrete coordination complexes into 1-D chains using halogen bonding or halogen···halogen interactions, CrystEngComm 9 (2007) 421–426.

Index

'*Note*: Page numbers followed by "f" indicate figures.'

D

E

F

Printed in the United States
by Baker & Taylor Publisher Services